Lecture Notes in Mechanical Engineering

Lecture Notes in Mechanical Engineering (LNME) publishes the latest developments in Mechanical Engineering—quickly, informally and with high quality. Original research or contributions reported in proceedings, and post-proceedings represent the core of LNME. Volumes published in LNME embrace all aspects, subfields, and new challenges of mechanical engineering.

To submit a proposal or request further information, please contact the Springer Editor of your location:

Europe, USA, Africa: Leontina Di Cecco at Leontina.dicecco@springer.com
China: Ella Zhang at ella.zhang@cn.springernature.com
India, Rest of Asia, Australia, New Zealand: Swati Meherishi at swati.meherishi@springer.com

Topics in the series include:

- Engineering Design
- Machinery and Machine Elements
- Mechanical Structures and Stress Analysis
- Automotive Engineering
- Engine Technology
- Aerospace Technology and Astronautics
- Nanotechnology and Microengineering
- Control, Robotics, Mechatronics
- MEMS
- Theoretical and Applied Mechanics
- Dynamical Systems, Control
- Fluid Mechanics
- Engineering Thermodynamics, Heat and Mass Transfer
- Manufacturing Engineering and Smart Manufacturing
- Precision Engineering, Instrumentation, Measurement
- Materials Engineering
- Tribology and Surface Technology

Indexed by SCOPUS, EI Compendex, and INSPEC.

All books published in the series are evaluated by Web of Science for the Conference Proceedings Citation Index (CPCI).

To submit a proposal for a monograph, please check our Springer Tracts in Mechanical Engineering at https://link.springer.com/bookseries/11693.

Mohamed Soula · Naoufel Ben Moussa ·
Ali Trabelsi · Noureddine Ben Yahia ·
Farhat Ghanem · Issam Dridi
Editors

Advances in Additive Manufacturing: Materials, Processes, and Applications II

Selected Contributions to the 3rd Advances in Additive Manufacturing Conference (AIAM' 2024), May 17–19, 2024, Hammamet, Tunisia

Springer

Editors
Mohamed Soula
National Engineering School of Tunis
Tunis, Tunisia

Naoufel Ben Moussa
Preparatory Institute of Engineering
University of Carthage
Tunis, Tunisia

Ali Trabelsi
High National School of Engineers of Tunis
Tunis, Tunisia

Noureddine Ben Yahia
High National School of Engineers of Tunis
Tunis, Tunisia

Farhat Ghanem
High National School of Engineers of Tunis
Tunis, Tunisia

Issam Dridi
High National School of Engineers of Tunis
Tunis, Tunisia

ISSN 2195-4356 ISSN 2195-4364 (electronic)
Lecture Notes in Mechanical Engineering
ISBN 978-3-031-94888-6 ISBN 978-3-031-94889-3 (eBook)
https://doi.org/10.1007/978-3-031-94889-3

This Springer imprint is published by the registered company Springer Nature Switzerland AG
The registered company address is: Gewerbestrasse 11, 6330 Cham, Switzerland

Preface of the AIAM'2024 Conference

Additive manufacturing (AM) is being used to create an increasing number of custom parts and sub-assemblies with complex geometry, low waste, short lead-time, and optimal strength-to-weight ratios. The AM technology has then brought valuable solutions to the problems that traditional machining has. Although AM processing techniques are now widely used in consumer construction, automotive, healthcare, energy, aerospace, and other industrial sectors, sustained research and development in the foreseeable future are still required. This covers issues such as tolerance transfer and stack-up, geometric distortion, residual stresses, porosity and cracking defects, heat treatment, mechanical and microstructural characterization, and post and preprocessing, to name a few. The prediction and optimization techniques for the AM processes are primarily rooted in statistical, metaheuristic, data-driven, and hybrid approaches. However, these methods face challenges when dealing with nonlinearity from different sources, including material, friction, geometry, and heat distribution.

The third International Conference on Advances in Additive Manufacturing (AIAM'2024) aimed to answer problems encountered in AM processes. It spans a wide range of fields, including material, process, and technology. By gathering researchers and professionals in the field, we hope to promote interdisciplinary cooperation and stimulate creative answers to the multifaceted problems in AM. The AIAM'2024 conference has presented the latest achievements in the field of AM and provided an opportunity to discuss and exchange views and opinions in the areas of material characterization, microstructure, surface integrity, software and hardware, serviceability, and tolerancing problems.

The AIAM'2024 was held in Hammamet, Tunisia, on May 17–19, 2024. This book, the result of the three-day conference and months of preparation, compiles outstanding articles from the AIAM'2024 oral sessions. Sixty papers have been selected for oral presentation, and only 21 articles were chosen to appear in the AIAM'20224 Springer proceeding. The selection score for each paper consists of 30% from the oral presentation jury and 70% from a peer-review process (double-blind review). Three reviewers process each paper; two nationals and one external expert. The external experts originate from six countries: Kuwait, Morocco, France, the Kingdom of Saudi Arabia, Switzerland, and Belgium. Moreover, all papers are checked using the Turnitin program and achieved a similarity index of less than 20%. The AIAM'2024 conference covered five major topics:

Topic 1: Characterization of AM parts: Microstructures & Material behavior.
Topic 2: Material grades for AM applications.
Topic 3: Mechanical behavior for 3D printing processes.
Topic 4: Automation, Artificial Intelligence & post-processing for AM processes.
Topic 5: Computer-Aided Design for AM processes.

During the plenary sessions, internationally renowned keynote speakers have animated the debate about the recent developments in additive manufacturing technology, processes, and materials. These include:

1. Pr. Moncef HAMMADI, Professor at the Institut Supérieur de Mécanique de Paris (Isae-Supméca), Quartz Laboratory. "Advancing Additive Manufacturing through emerging computational techniques for optimized simulation and control".
2. Pr. Driss AMEGOUZ, Professor at the École Supérieure de Technologie - Université Sidi Mohamed Ben Abdellah, TSI Laboratory. "Additive Manufacturing and 3D Printing: The Case of Concrete Formulations".
3. Pr. Issam DOGHRI, Professor at the Université catholique de Louvain (UCLouvain), Belgium. "Multiscale modeling of polymer solids and structures additively manufactured with the Selective Laser Sintering process".
4. Mr. Maher SKHIRI, Head of the Research Valorization File Processing Service at the Ministry of Higher Education and Scientific Research, Tunisia. "Research Results Valorization Process and Knowledge Transfer: Tools for Scientific Research Valorization & Value Creation".
5. Pr. Mounir SAYADI, Professor at the École Nationale Supérieure d'Ingénieurs de Tunis, University of Tunis, Director of the Signal Image and Energy Management Laboratory, Tunisia. "Texture Images Analysis with Image Processing and Artificial Intelligence".

In addition to the plenary and oral presentations, a **competition event** has been organized to increase awareness and share experiences among professionals and students from different engineering schools and training centers.

We want to express our gratitude to each and every member of the scientific and organization committees for their tireless efforts in making the AIAM'2024 conference a successful event and this book a reality. We acknowledge all participants who contributed to the production of the AIAM'2024, especially keynote speakers, authors, reviewers, and sponsors.

March 2025

Mohamed Soula
Naoufel Ben Moussa
Ali Trabelsi
Noureddine Ben Yahia
Farhat Ghanem
Issam Dridi

Organization

Steering Committee

General Chair

Noureddine Ben Yahya	High National School of Engineers of Tunis, Tunisia

Chairs

Mohamed Soula	National School of Engineers of Tunis, Tunisia
Naoufel Ben Moussa	Preparatory Institute of Engineering University of Carthage, Tunisia
Ali Trabelsi	High National School of Engineers of Tunis, Tunisia
Issam Dridi	High National School of Engineers of Tunis, Tunisia

Program Committee

Chairman

Mohamed Toumi Nasri	LMPE High National School of Engineers of Tunis, Tunisia

Members

Hafedh Dhifelaoui	LMMP-High National School of Engineers of Tunis, Tunisia
Khalil Mansouri	LMAI - National School of Engineers of Tunis, Tunisia
Slah Chayoukhi	LMPE - High National School of Engineers of Tunis, Tunisia
Issam Dridi	LMPE - High National School of Engineers of Tunis, Tunisia
Chokri Ben Aissa	LMMP-High National School of Engineers of Tunis, Tunisia

Ons Marzougui	LMMP-High National School of Engineers of Tunis, Tunisia
Khaled Boulehem	LMAI - National School of Engineers of Tunis, Tunisia
Jihed Zghal	LEME - Paris Nanterre University, France
Jmal Hamdi	ICube - University of Strasbourg, France
Driss Amegouz	LTSI Laboratory Fès, Morocco

Competition Committee

Members

Atef Boulila	LMPE - High National School of Engineers of Tunis, Tunisia
Ali Bejaoui	ISET.Bizerte, Tunisia
Radhouen Ncibi	ISET.Sidi.Bouzid, Tunisia
Jihed Zghal	LEME - Paris Nanterre University France
Samuel Berthe	ICube - University of Strasbourg France
Faouzi Slimani	ISET.Sidi.Bouzid, Tunisia
Chokri Ben Aissa	ISET.Nabeul, Tunisia

Scientific Committee

Abbassi Fethi	American University of the Middle East, Kuwait
Aifaoui Nizar	National Engineering School of Monastir, Tunisia
Amegouz Driss	Université Euromed of Fès, Morocco
Ammar Amine	High National School of Arts et Métiers, France
Anas Bouguecha	National Engineering School of Sfax, Tunisia
Arfaoui Makrem	National School of Engineers of Tunis, Tunisia
Asad Muhammad	Prince Mohammad Bin Fahd University, KSA
Ayed Yessine	High National School of Arts et Métiers, France
Ayedi Mahfoudh	National School of Engineers of Bizerte, Tunisia
Baccar Mounir	National School of Engineers of Sfax, Tunisia
Bahloul Riadh	National School of Engineers of Monastir, Tunisia
Baili Maher	National School of Engineers of Tarbes, France
Barkaoui Abdelwahed	Université Internationale de Rabat, Morocco
Belhadi Salim	University of Guelma, Algeria
Ben Ameur Tarek	National School of Engineers of Monastir, Tunisia
Ben Fathallah Brahim	National School of Engineers of Tunis, Tunisia
Ben Fredj Nabil	High National School of Engineers of Tunis, Tunisia

Ben Moussa Naoufel	High National School of Engineers of Tunis, Tunisia
Ben Rhouma Amir	High National School of Engineers of Tunis, Tunisia
Ben Salem Sahbi	Preparatory Institute of Engineering studies of Nabeul, Tunisia
Ben Salem Wacef	High Institute of Applied Sciences and Technology of Mahdia, Tunisia
Ben Yahia Nourdine	High National School of Engineers of Tunis, Tunisia
Bennour Sami	National School of Engineers of Sousse, Tunisia
Ben Souf Mohamed Amine	National School of Engineers of Sfax, Tunisia
Bergheau Jean-Michel	Centrale School of Lyon and ENISE-St-Etienne, France
Bouaziz Slim	National School of Engineers of Sfax, Tunisia
Bouaziz Zoubeir	National School of Engineers of Sfax, Tunisia
Boulila Atef	High Institute of Applied Sciences and Technology of Tunis, Tunisia
Bouraoui Tarak	National School of Engineers of Monastir, Tunisia
Carreño-Morelli Efraín	High School of Engineering - HES-SO Valais Wallis, Switzerland
Chatti Sami	National School of Engineers of Monastir, Tunisia
Chatti Sami	High Institute of Applied Sciences and Technology of Sousse, Tunisia
Charfeddine Mrad	National School of Engineers of Tunis, Tunisia
Chinesta Francisco	High National School of Arts et Métiers, ParisTech, France
Courbon Cédric	National School of Engineers of Saint-Étienne, France
Dammak Fakhreddine	National School of Engineers of Sfax, Tunisia
Dhiflaoui Hafeth	High National School of Engineers of Tunis, Tunisia
Doghri Issam	Catholique University of Louvain, Belgium
Elleuch Khaled	National School of Engineers of Sfax, Tunisia
Elleuch Riadh	Preparatory Institute of Engineering Studies of Sfax, Tunisia
Ghanem Farhat	High National School of Engineers of Tunis, Tunisia
Furet Benoit	Nantes University, France
Germain Guenael	High National School of Arts et Métiers, Angers France
Gharmezi Noamen	National School of Engineers of Sfax, Tunisia
Hajlaoui Abdessalem	National School of Engineers of Tunis, Tunisia

Hamdi Adel	National School of Engineers of Tunis, Tunisia
Hamdi Hedi	National School of Engineers of Saint-Étienne, France
Hammadi Moncef	SUPMECA –Paris, France
Hor Anis	High Institute of space and Aeronautics Toulouse, France
Jarraya Abdessalem	National School of Engineers of Sfax, Tunisia
Rebiere Jean Luc	High Institute of Mechanics and Aeronautics Maine-University, France
Khlifi Kawther	Preparatory Institute of Engineering Studies of Manar, Tunisia
Kouba Sana	National School of Engineers of Sfax, Tunisia
Kouki Rahim	Preparatory Institute of Engineering Studies of Manar, Tunisia
Lotfi Sai	Preparatory Institute of Engineering Studies of Nabeul, Tunisia
Makhlouf Kamel	Preparatory Institute of Engineering Studies of Nabeul, Tunisia
Marouani Haykel	National School of Engineers of Monastir, Tunisia
Mehdi Kamel	Preparatory Institute of Engineering Studies of Manar, Tunisia
Mezlini Salah	National School of Engineers of Monastir, Tunisia
Mezni Taoufik	High National School of Engineers of Tunis, Tunisia
Moakher Maher	National School of Engineers of Tunis, Tunisia
Najjar Walid	High National School of Engineers of Tunis, Tunisia
Soula Mohamed	National School of Engineers of Tunis, Tunisia
Terres Mohamed Ali	High Institute of Applied Sciences and Technology of Sousse, Tunisia
Trabelsi Ali	High National School of Engineers of Tunis, Tunisia
Trabelsi Mohamed	High National School of Engineers of Tunis, Tunisia
Wali Monder	National School of Engineers of Sfax, Tunisia
Zemzemi Farhat	National School of Engineers of Sousse, Tunisia

Special Sessions and Special Session Organizing Committee

Characterization of Am PartsMicrostructures and Material Behavior

Farhat Ghanem	High National School of Engineers of Tunis, Tunisia
Moncef Hammadi	SUPMECA –Paris, France

Material Grades for Am and Civil Applications

Mohamed Soula	National School of Engineers of Tunis, Tunisia
Naoufel Ben Moussa	High National School of Engineers of Tunis, Tunisia

Mechanical Behavior for 3D-Printed

Mohamed Ali Terres	High Institute of Applied Sciences and Technology of Sousse, Tunisia
Adel Hamdi	National School of Engineers of Tunis, Tunisia

Automation, Artificial Intelligence and Post Processing for Am

Ali Trabelsi	High National School of Engineers of Tunis, Tunisia
Mounir Sayedi	High National School of Engineers of Tunis, Tunisia

Computer Aided Design for Am

Farhat Zemzemi	National School of Engineers of Sousse, Tunisia
Sami Chatti	High Institute of Applied Sciences and Technology of Sousse, Tunisia

Biomechanics and Other Applications in Am

Karem Dhouib	High National School of Engineers of Tunis, Tunisia
Med Ali Rezgui	High National School of Engineers of Tunis, Tunisia

Contents

Characterization of AM Parts: Microstructures and Additive Manufacturing Processes; Material Behavior

Study of Additive Manufactured Macroscopic Graphene Structures Behavior Under a Compressive Load ... 3
Hamza Jmii, Mohamed Ali Rezgui, Ali Bejaoui, Ali Trabelsi, and Jihed Zghal

Experimental Investigation of FDM-Printed SCFR-PLA Composites Used to Fabricate Blades for Medium-Sized Wind Turbines ... 10
Wissem Zghal and Mondher Wali

3D Printing of Bio-inspired Fluidic Conduct Based on Tracheal Geometry ... 23
Marwa Jaziri, Badreddine Larbi, Khaled Boulahem, Mohamed Toumi Nasri, Olfa Trabelsi, and Mahfoudh Ayadi

Experimental and Numerical Analysis of the Mechanical Characteristics of 3D Printed ABS Specimens Produced by the FDM Process ... 44
Shiri Seddik, Boulahem Khaled, and Habouba Mohamed Hichem

Material Grades for AM Applications

Exploration of 3D Printed Anti-trichiral Meta-materials Based on PLA/Wood Flour Bio-composites Using Dynamic Mechanical Analysis and Artificial Neural Networks (ANN) Technique ... 57
Mondher Nasri, Mohamed Toumi Nasri, Badreddine Larbi, Houda Khaterchi, Salah Hamzi, Ali Zghal, Olivier Dalverny, and Amir Ben Rhouma

3D Printed Auxetic Structures Based on Re-entrant Hexagon Design for Biomedical Application ... 69
Nada Ben Hariz, Atef Boulila, and Mahfoudh Ayadi

Dynamic Behavior of Honeycomb Composite Structures: Effects of Filling Density Using Split Hopkinson Compression Bars ... 83
Manel Chihi, Mostapha Tarfaoui, Sami Chatti, and Yumna Qureshi

Influence of Repetitive Recycling of 3D Printing Filament on Mechanical Properties of PLA and PLA/Spent Coffee Grounds Composite ... 92
Oumaima Boughanmi, Lamis Allegue, and Haykel Marouani

Mechanical Behavior for 3D Printing Processes

Use of the RDPP-SF Method to Analyze Rheology Variation in an Am-Cement-Based Process 105
Marwen Amdouni, Helmi Nasraoui, Mohamed Ali Rezgui, and Ali Trabelsi

MSLA 3D Printing of Wax Pattern for Rapid Investment Casting Inconel 718-Superalloy Aeronautical Part 115
Chams Eddine Dakhli, Brahim Ben Fathallah, Mohamed Soula, and Badreddine Aouadi

Taguchi Optimization of FDM Process Parameters on Ultimate Tensile Strength of 3D-Printed PLA Parts 133
Boulahem Khaled, Souli Imen, Badreddine Larbi, Mohamed Toumi Nasri, and Ayadi Mahfoudh

Overall Equipment Effectiveness Modelling and Assessment of Additive Manufacturing Printing Machines 149
Amel Hajji, Karem Dhouib, and Ali Gharbi

Automation, Artificial Intelligence and Additive Manufacturing Processes; Post-processing for AM Processes

Neural Network Optimization for 3D Printing Parameters in Biomedical Applications: An Anti-trichiral Honeycomb Structure Case Study 165
Houda Khaterchi, Mondher Nasri, Nasri Mouhmed Toumi, Baderddine Larbi, Mondher Zidi, Ated Ben Khlifa, and Hédi Belhadjsalah

New Energy Consumption Prediction Aid Model Applied to Additive Manufacturing Process 181
Kamel Bousnina, Anis Hamza, and Noureddine Ben Yahia

PLA Damage Properties in FDM Printing: A Hybrid Experimental and Machine Learning Study 186
Manel Dhouioui, Boutheina Ben Fraj, Hamdi Hentati, Mounir Ben Amar, and Mohamed Haddar

Effect of Post-treatment on Additively Manufactured Medical Implant: Numerical Study 195
Oumayma Bougadouha, Mounir Frija, Raouf Fathallah, and Malika Khodja

Comparative Analysis of Simulation Models for LENS and LWAM Processes in Ti-6Al-4V Additive Manufacturing 203
Larbi Yousfi, Amir Guizani, Moncef Hammadi, Slim Bouaziz, and Mohamed Haddar

Computer-Aided Design for AM Processes

Explanation of the Slowing Infill in 3D Printing and the Influence of CAM Patterns 219
El Bechir Msaddek, Zeinab Hmila, Zoubeir Bouaziz, and Maher Baili

Design for Additive Manufacturing (DfAM) in the Automobile Industry 231
Anis Hamza, Kamel Bousnina, Issam Dridi, and Noureddine Ben Yahia

Investigating Compensation Techniques for Extruded Microstructured Optical Fibers Preforms Using 3D Printed Die 244
Manel Abda, Mohamed Trabelssi, and Noureddine Ben Yahia

Addressing Geometrical and Dimensional Accuracy Challenges in 3D Concrete Printing 260
El Mesoudy Mouad, Foulki Rida, and Amegouz Driss

Author Index 271

Characterization of AM parts: Microstructures and Material Behavior

Study of Additive Manufactured Macroscopic Graphene Structures Behavior Under a Compressive Load

Hamza Jmii[1], Mohamed Ali Rezgui[1(✉)], Ali Bejaoui[1], Ali Trabelsi[1], and Jihed Zghal[2]

[1] Laboratory of Mechanics, Production and Energetics (LR18ES01), University of Tunis, Higher National Engineering School of Tunis, 5 Av. Taha Hussein, BP. 56. Bâb Manara, 1008 Tunis, Tunisia
mohamedali.rezgui@esstt.rnu.tn

[2] Energy Laboratory Mechanical Electromagnetism (LEME-EA4416), Paris Nanterre University, Campus, 50, rue de Sèvres, 92410 Ville d'Avray, France
jzghal@parisnanterre.fr

Abstract. This paper explores the mechanical behavior of additive-manufactured macroscopic graphene structures, focusing on their response to compressive loads. Leveraging the exceptional mechanical properties of graphene and the precision of additive manufacturing, we investigate the structural integrity and deformation mechanisms of these structures. Our study aims to bridge theoretical simulations with experimental results, providing validation and insights for researchers in the field. Additionally, we delve into the mechanical properties of carbon nanotubes, highlighting their remarkable strength and stiffness. By simulating structures inspired by the unique architecture of Single-Walled Carbon Nanotubes, we analyze the effects of varying parameters on mechanical performance. Through computational modeling, we explore how structural variations influence properties such as stiffness and strength. A visual representation of our simulation demonstrates the deformation of a modeled structure under compressive load, offering valuable insights into mechanical responses.

Keywords: Additive manufacturing · Single-Walled Carbon Nanotubes · Graphene structure · Macroscopic scale · Structures behavior · Compressive load

1 Introduction

Additive manufacturing (AM) plays a crucial role in fabricating complex structures, enabling precise control over geometry and material selection. We enhance the structural integrity of the fabricated structures, mimicking the exceptional properties of natural Single-Walled Carbon Nanotubes (SWCNTs). Overall, this study contributes to advancing our understanding of additive-manufactured graphene structures and their potential applications in various fields. Graphene, with its exceptional mechanical properties, has gained significant attention for various applications, including structural materials. In theory, the formation of a SWCNT involves cutting out an infinite graphene sheet and then

M. Soula et al. (Eds.): AIAM 2024, LNME, pp. 3–9, 2025.
https://doi.org/10.1007/978-3-031-94889-3_1

rolling it up, or "wrapping" it into a tube. The way the graphene sheet is cut determines the resulting structure or chirality of the SWCNT. This process can be envisioned as a mapping from the two-dimensional graphene sheet onto the three-dimensional SWCNT (Fig. 1) [1, 2].

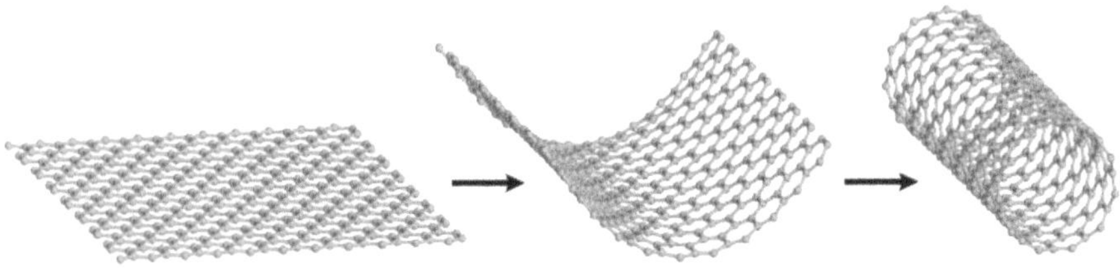

Fig. 1. Creating a SWCNT by rolling a graphene sheet into a tube figure.

Additive manufacturing offers a promising approach to fabricating complex graphene structures with precise control over geometry and microstructure. In this study, we explore the response of such structures to compressive loads, aiming to enhance our understanding of their mechanical behavior. We intend to establish robust connections between our computational findings and real-world experimental results. This bridging of theory and practice will not only validate the accuracy of our models but also provide valuable insights for experimentalists working in the field.

2 Mechanical Properties

Carbon nanotubes (CNTs) have long been regarded as exceptional building blocks for high-performance engineering materials, thanks to their outstanding mechanical properties and low density. Both theoretical and experimental investigations into the mechanical properties of individual CNTs or bundles of CNTs have been conducted extensively since the late twentieth century. It has been found that the modulus (stiffness) and strength of a typical SWCNT can reach remarkable values of around 640 and 37 GPa, respectively, surpassing the properties of all existing materials. However, a significant challenge arises when attempting to maintain such exceptional mechanical properties when CNTs are assembled into macro-architectures, such as fibers and films [2].

A SWCNT can be uniquely identified by its chiral indices (n, m) or equivalently by its chiral angle θ, denoted as:

$$\theta = \arctan\sqrt{\frac{3m}{2n+m}}$$

It's worth noting that due to symmetry, we can interchange the chiral indices, i.e., (n, m) can be swapped with (m, n). Furthermore, symmetry constraints limit the chiral angle to fall within the range of 0° to 30°. Based on chirality, SWCNTs can be categorized into three distinct types or classes (Fig. 2) [3, 4]:

$$Armchair : m = n, \theta = 30^\circ \quad (1)$$

$$Zigzag : m = 0, \theta = 0^\circ \tag{2}$$

$$Chiral : 0 < m < n, 0^\circ < \theta < 30^\circ \tag{3}$$

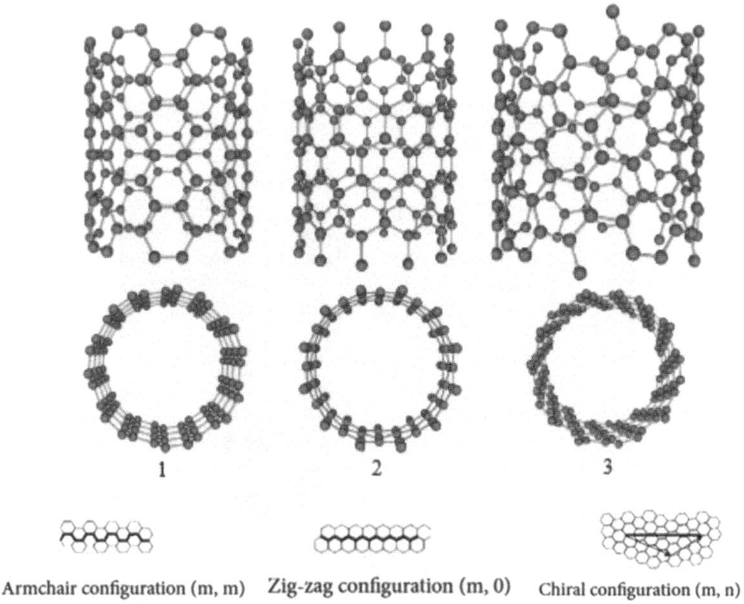

Fig. 2. Various geometric configurations of SWCNTs.

3 Simulating SWCNT Mechanical Behavior

In our study, we delve into the simulation of structures inspired by the unique architecture of SWCNTs to investigate their mechanical behavior. These simulated structures mimic the cylindrical geometry and atomic arrangement of SWCNTs while varying parameters such as size, chirality, and defect distribution. By analyzing the mechanical response of these SWCNT-inspired structures, we aim to uncover design principles that optimize their performance in specific applications. Through computational modeling, we examine how variations in structural parameters influence mechanical properties such as stiffness, strength, and flexibility.

Simulations were performed by Abaqus, employing a static analysis approach to capture the behavior of the structures under a compressive load, focusing on the equilibrium state of the structure. The loading conditions applied included a pressure-type load with a uniform distribution, emphasizing compressive pressure. Additionally, the boundary conditions were specified as fixed (recessed connection) in one extremity and free on the other, indicating a mixed boundary condition setup. These boundary conditions help simulate real-world scenarios by constraining the movement of the structures in specified directions. The solver settings utilized an implicit solver, which is suitable for gradual changes in displacement and provides accurate solutions for static problems. Figure 3 illustrates the Simulation of S-Mises Stress Distribution, with stress values labeled in megapascals (MPa), and Fig. 4 represents the Simulation of Reaction Force Distribution in the structure under compressive loading conditions, with force values labeled in Newton (N) (Figs. 3 and 4).

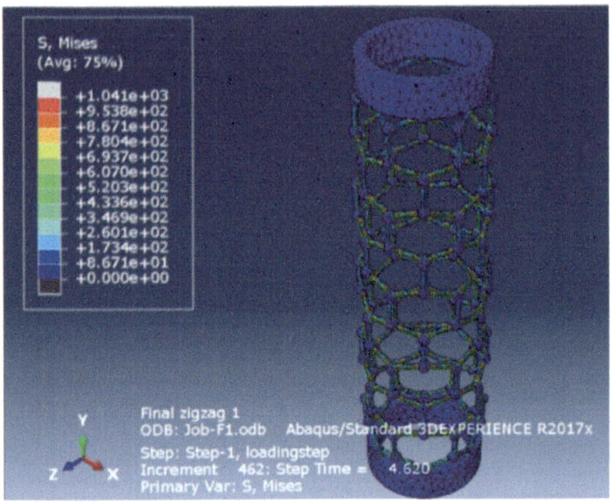

Fig. 3. Simulation of S-Mises Stress Distribution in Structure under Compressive Load.

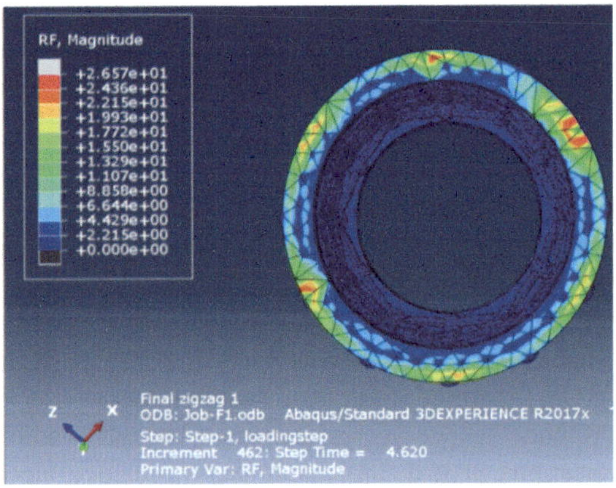

Fig. 4. Simulation of Reaction Force Distribution in Structure under Compressive Load.

4 3D Printing

Additive manufacturing plays a pivotal role in the realization of the simulated structures and enhances their unique characteristics. This capability allows us to replicate the intricate architecture of Single-Walled Carbon Nanotubes (SWCNTs) and their derivatives, thereby providing a platform for studying their mechanical behavior in a controlled environment. Moreover, additive manufacturing offers versatility in material selection, enabling the incorporation of advanced materials into the fabricated structures. This integration enhances the structural integrity and mechanical properties of the simulated structures, mimicking the enhanced strength and resilience observed in natural SWCNTs [5].

The detailed visualization of the 3D-printed structure presented offers valuable insights into the additive manufacturing process and the resulting architecture. Figure 5 highlights the overall geometry and complexity achieved with high precision. Figure 6 offers a focused partial view, allowing for a closer examination of the internal structure and layering pattern, thereby providing insights into the layer-by-layer buildup of the printed object. Figure 7 emphasizes the interaction between the 3D printing nozzle and the material deposition process, providing a glimpse into the additive manufacturing technique. Collectively, these images offer a comprehensive insight into this process and its role in fabricating complex structures with intricate geometry (Figs. 5, 6 and 7) [6, 7].

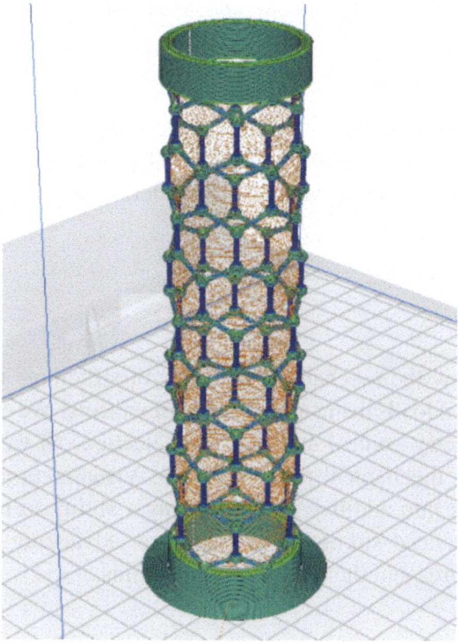

Fig. 5. Overall Geometry View.

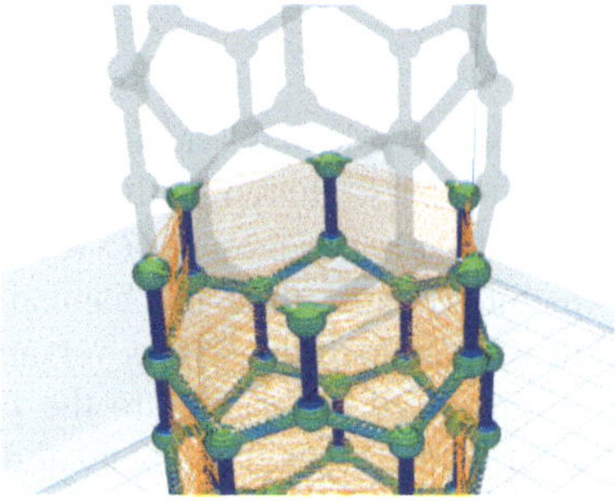

Fig. 6. Internal Structure Focus.

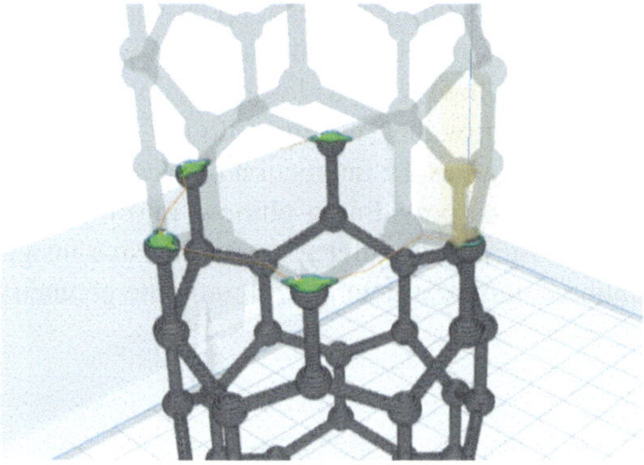

Fig. 7. Nozzle-Material Interaction.

By replicating the unique features of SWCNT-based structures through additive manufacturing, researchers can advance their understanding of mechanical behavior and explore potential applications.

5 Conclusion

The study explored how the mechanical behavior of structures inspired by SWCNTs responds to compressive loads using additive manufacturing. We emphasized the importance of precise geometry control and material selection, as well as the influence of varying parameters on mechanical properties. The Visualization of 3D-printed structures provided insights into additive manufacturing's ability to replicate SWCNT-like architectures with enhanced strength and resilience, advancing our understanding of graphene-based structures and their diverse applications.

References

1. Dimitrakakis, G.K., Tylianakis, E., Froudakis, G.E.: Pillared graphene: a new 3-D network nanostructure for enhanced hydrogen storage. Nano Lett. **8**(10), 3166–3170 (2008). https://doi.org/10.1021/nl801417w
2. Hedman, D.A.: Theoretical Study: The Connection Between Stability of Single-Walled Carbon Nanotubes and Observed Products. Thesis for: Licentiate Advisor: Andreas Larsson & Sven Öberg. Hedman's Lab (2017)
3. Saito, R., Fujita, M., Dresselhaus, G., Dresselhaus, M.S.: Electronic structureof chiral graphene tubules. Appl. Phys. Lett. **60**(18), 2204–2206 (1992). https://doi.org/10.1063/1.107080
4. Sidi, S.L., Ouslimani, N., Bousba, D., Huynen, I., Danlée, Y., Aksas, H.: Carbon nanotubes (CNTs) from synthesis to functionalized (CNTs) using conventional and new chemical approaches. J. Nanomater. **14**, 1–31 (2021)
5. Königshofer, M., Stoiber, M., Unger, E., Grasl, C., Moscato, F.: Mechanical and dimension-al investigation of additive manufactured multi-material parts. Front. Phys. **9**, 635736 (2021). https://doi.org/10.3389/fphy.2021.635736
6. Farbman, D., McCoy, C.: Materials testing of 3D printed ABS and PLA samples to guide mechanical design. In: Proceedings of the ASME2016 11th International Manufacturing Science and Engineering Conference, 27 June–1 July 2016, vol. 2. MSEC, Blacksburg, VA (2016)
7. Kumar, S.A., Narayan, S.Y.: Tensile testing and evaluation Of 3D printed PLA specimens as per ASTM D638 Type-IV standard. In: 3rd International Conference on Innovative Design and Development Practices in Aerospace and Automotive Engineering, 22–24 February, IDAD, Chennai, India (2018)

Experimental Investigation of FDM-Printed SCFR-PLA Composites Used to Fabricate Blades for Medium-Sized Wind Turbines

Wissem Zghal[1](✉) and Mondher Wali[2]

[1] Laboratory of Electromechanical Systems (LASEM), National Engineering School of Sfax—Higher Institute of Industrial Management of Sfax (ISGIS), University of Sfax, Sfax, Tunisia
wissem.zghal@isgis.usf.tn

[2] LEE Laboratory, National Engineering School of Sfax, Higher School of Sciences and Technologie of Hammam Sousse, University of Sousse, Sousse, Tunisia

Abstract. New blade designs are created using ABS material. Compared to ABS, PLA composites are easier to print. Therefore, the main objective of this work is to optimize the 3D printing parameters for fabricating small and medium-sized horizontal-axis wind turbine blades using SCFR-PLA material. The fused deposition modeling (FDM) technology is gaining significant attention for its capability to rapidly prototype high-performance parts across various fields, particularly in the manufacture of blades for small and medium-sized wind turbines. In this paper, FDM is utilized to print SCFR-PLA composites. The study primarily focuses on experimental tensile tests to explore the impact of FDM parameters on the mechanical properties of the 3D-printed samples, specifically targeting Young's modulus (E) and ultimate tensile strength (UTS). Mechanical properties are analyzed in relation to process variables, which are the carbon reinforcement percentage, the layer thickness, the raster angle, and the infill percentage. This study investigates the effects of different FDM parameters on the printing of SCFR-PLA composite parts and optimizes the FDM parameters to enhance the mechanical performance of SCFR-PLA composites, which will be used to fabricate blades for small and medium-sized wind turbines.

Keywords: Numerical Study · New configuration of Blades · SCRF-PLA composites · FDM process · Mechanical properties

1 Introduction

Understanding the impact of manufacturing parameters on various processes is crucial, as it significantly affects both the geometric precision and mechanical properties of the produced components [1–3]. Various studies have explored the influence of manufacturing parameters on traditional processes [4–6]. These parameters determine the final dimensions, and structural integrity, ultimately influencing the performance of the components in their applications. In contrast to the traditional manufacturing processes, 3D printing technology, called additive manufacturing has also been recently

M. Soula et al. (Eds.): AIAM 2024, LNME, pp. 10–22, 2025.
https://doi.org/10.1007/978-3-031-94889-3_2

developed. This technology offers the capability to create complex structural designs, quickly prototype, and manufacture a diverse range of products at a relatively low cost. When it comes to composites with polymers, additive manufacturing opens up new possibilities. Composites are made from two or more constituent materials with significantly different physical and chemical properties, which when combined, produce a material with characteristics different from the individual components. Research in polymer-based composites delves into optimizing materials by selecting suitable polymer matrices and reinforcement materials and using advanced characterization methods to increase mechanical properties. Through experimentation and numerical modeling [7, 8], scientists aim to understand the complex interplay of factors affecting composite behavior at different scales, from molecular interactions to macroscopic performance. The utilization of 3D printing with thermoplastic polymers has become prevalent in various prototyping and modeling endeavors. In particular, polylactic acid (PLA), owing to its useful properties, stands out as one of the notable choices among printing thermoplastics [9, 10]. Its popularity stems from factors such as biodegradability, ease of printing, and suitability for a wide range of applications, making it a favored material in the additive manufacturing processes. PLA stands out for its high tensile strength, and relatively low flexural strength compared to other polymers [11, 12]. Despite these advantages, its utilization is constrained, particularly when we require novel materials with specific mechanical performance criteria. To address this limitation, one proposed solution involves incorporating carbon fibers into these polymers.

Recent research has focused on additive manufacturing processes in the aerospace, automotive, and energy sectors [13–15], particularly Fused Deposition Modeling (FDM) [16, 17]. The main goal is to investigate the intricate relationship between composite mechanical behavior and FDM process factors. In recent investigations, Ammar et al. [18] studied the influence of selected FDM parameters on the mechanical properties of the Short Carbon Fiber reinforced PLA composites (SCFR-PLA). Ramírez-Elías et al. [19] seek to determine the properties of the continuous Carbon Fiber reinforced PLA composites (CCFR-PLA) obtained through FDM. By incorporating continuous carbon fibers, the study seeks to further increase the mechanical strength of the composite material. Maqsood and Rimasauskas [20] conducted experiments on 3D-printed pure PLA, SCFR-PLA, and CCFR-PLA. They found that continuous CFR-PLA exhibited superior tensile and flexural properties compared to the other materials.

From an energy point of view and essentially the exploitation of wind energy through small wind turbines, several authors have used additive manufacturing techniques as a preferred approach to easily produce wind turbine components. Indeed, Seralathan Sivamani et al. [21] analyzed the dimensionless performance parameters of a wind turbine produced using fused deposition modeling using Acrylonitrile Butadiene Styrene (ABS) material. In their research study, Marwane Rouway et al. [22] examined the design considerations of small turbines, including material properties and aerodynamic parameters. The turbine blade was printed using the selective laser sintering (SLS) method with polyamide 12 (PA12) and polyether ether ketone (PEEK) polymers reinforced with carbon beads (CB) and glass beads (GB). Suresh Arivalagan et al. [23] presented a numerical and experimental study whose objective is the optimization of a horizontal axis wind turbine blade 3D printed in PLA material through the identification of the different parameters relating to 3D printing.

This paper delves into the key challenges posed by additive manufacturing in production processes, particularly focusing on the utilization of the FDM process. Our study explores the novel advancements in polymer matrix composites, revealing enhanced performances that render these materials increasingly suitable for targeted applications. In particular, we seek to provide valuable insights into optimizing FDM parameters for enhancing the mechanical properties of SCFR-PLA used for printing small and medium-sized blades for horizontal axis wind turbines.

This study is part of a research project that aims to improve the efficiency of a horizontal axis wind turbine: by increasing the power coefficient and reducing the start-up speed of a wind turbine through:

- The proposal for a new design of wind turbine blades
- The determination of new materials and manufacturing procedure of the blades

This project has four stages presented in Fig. 1. The first stage has been completed and this paper presents the results of the second stage.

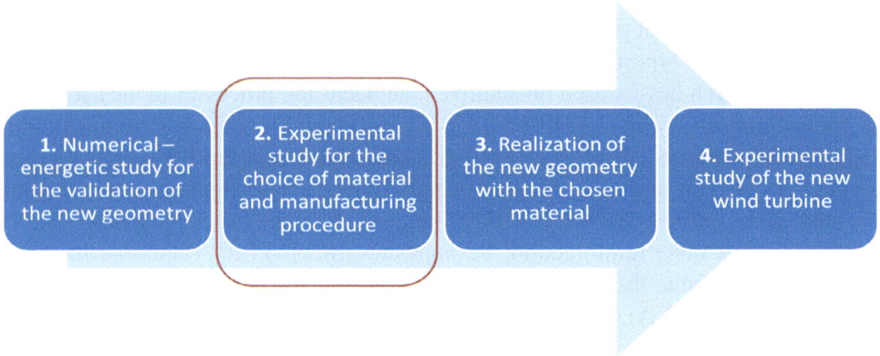

Fig. 1. Stages of the wind turbine efficiency improvement project.

2 Numerical Study of the New Geometry

In this paragraph, we will try to present an overview of the numerical study of energy aspects for the validation of the new geometry. The digital validation of the advantages of the new geometry of wind turbine blades was carried out in two phases:

- The first phase consists of validating the proposed numerical code through a comparison of the results with those of previous work by Balasem et al. [24];
- The second phase led to the presentation of the improvement in the efficiency of the new proposed wind turbine by exploiting the validated code.

Before showing the results of our numerical simulations, we present the main equations that characterize the operation of a wind turbine. Indeed, Wind energy is extracted by the turbine rotor which converts it into mechanical power. Thus, the rotor power Pr is the product of the rotor torque (T) and the angular speed (ω) of the rotor.

$$P_r = T \cdot \omega \tag{1}$$

The wind turbine performance is represented by the power coefficient (Cp) which is defined as:

$$C_p = \frac{T \cdot \omega}{0.5 \cdot \rho \cdot A \cdot V_w^3} \tag{2}$$

where:

- A is the area swept by the rotor
- ρ is the air density
- Vw is the wind speed

For wind turbines, the torque and power coefficients depend on the specific speed λ determined by the following equation where R is the rotor radius.

$$\lambda = \frac{\omega \cdot R}{V_w} \tag{3}$$

Phase 1: the numerical study was carried out on the same profile operated by Balasem, the characteristics of which are presented in Table 1. Indeed, the objective of this part is to validate our developed numerical simulation model to exploit it in a second phase with our new wind turbine configuration.

Table 1. Technical characteristics of the wind turbine (Balasem et al.).

Designation	Characteristic
Wind turbine diameter	154 mm
Number of blades	3
Blade profile	NACA SD8000
Material	ABS

The results of this study are illustrated in Fig. 2.

Figure 2 shows good agreement between the results of our numerical simulation and the results of the numerical and experimental studies of Balasem et al.

Phase 2: a new geometry has been developed. Figure 3 presents the characteristics of this geometry.

The simulation was conducted at a wind speed of 8 m/s at different specific speed ranges. The values of the torque coefficient C_T and power C_p of the new wind turbine determined through the numerical simulation code are shown in Table 2.

The simulation results obtained are shown in Fig. 4.

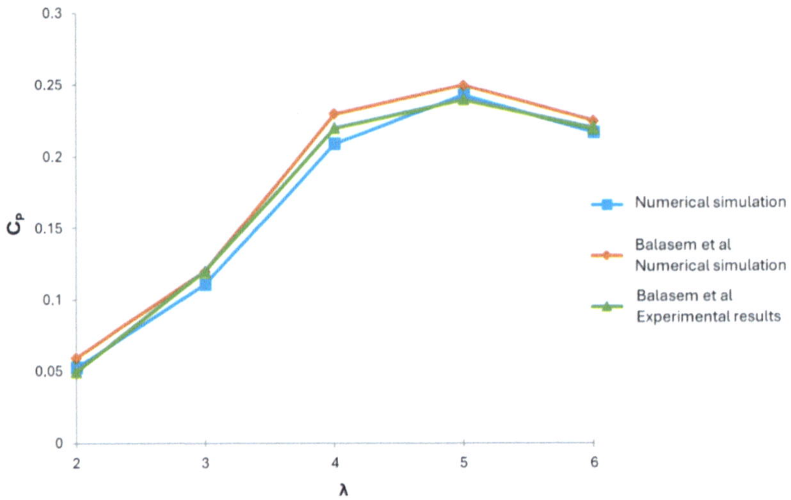

Fig. 2. Validation of the numerical analysis.

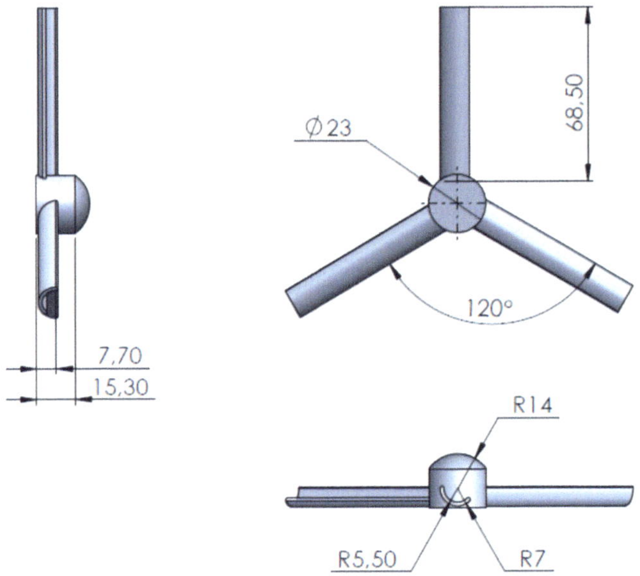

Fig. 3. Geometric characteristics of the new configuration.

- The simulations carried out for the configuration of the new wind turbine predict a $C_p = 0.462$ (i.e. 46%) for $\lambda = 6$.
- This result shows an achievement in terms of the efficiency of this new wind turbine configuration.

Table 2. Numerical simulation of the new configuration at $V_v = 8$ m/s.

λ	2	3	4	5	6
ω (rad/s)	207	312	415	519	623
C_T	0.00130	0.0041	0.00433	0.0047	0.0044
C_P	0.0463	0.192	0.305	0.40	0.462

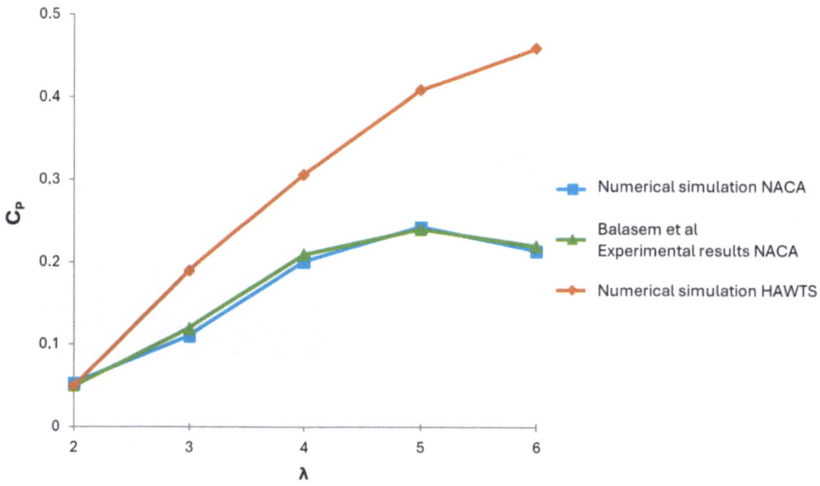

Fig. 4. Results of the new wind turbine configuration.

After verifying and evaluating the new configuration of the blades of the horizontal axis wind turbine through numerical simulations, we approached the following experimental study to choose the best material and the best manufacturing procedure for these blades.

3 Materials and Methods

In the manufacturing process described, the SCFR-PLA (Short Carbon Fiber Reinforced Polylactic Acid) filament is fed into a nozzle, which is outfitted with an induction heater that maintains a temperature of 210 °C. This elevated temperature is necessary to ensure proper melting and flow of the filament. Once heated, the filament is extruded through the nozzle and deposited layer by layer onto the XY build platform, gradually building up the desired printed part. The schematic of the SCFR-PLA with FDM process is illustrate in Fig. 5. The composite filament used has a diameter of 1.75 mm.

A crucial aspect of the printing process is selecting a well-defined raster angle, which determines the orientation of the printed layers relative to each other. Raster angles include 0°, 45°, and 90° are printed. The raster angle influences the mechanical properties of the printed part, principally its strength and stiffness. Furthermore, different infill percentages are utilized during printing, which raises the density of the internal

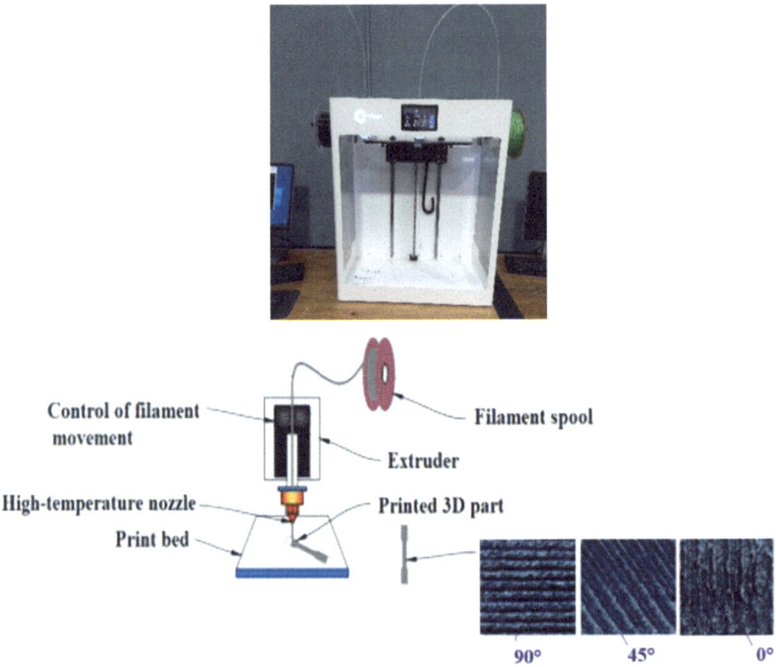

Fig. 5. Schematic of the SCFR-PLA with FDM process.

structure of the printed part. Higher infill percentages result in a denser internal structure increasing the strength and the stiffness of the part. By carefully selecting the raster angle and infill percentage, manufacturers can modify the mechanical properties and overall performance of the printed parts to meet specific requirements for different applications. This optimization process is crucial for achieving desired outcomes in terms of the strength of the final printed parts. According to ASTM 638-02a, the standard specimen for tensile test is presented in Fig. 6.

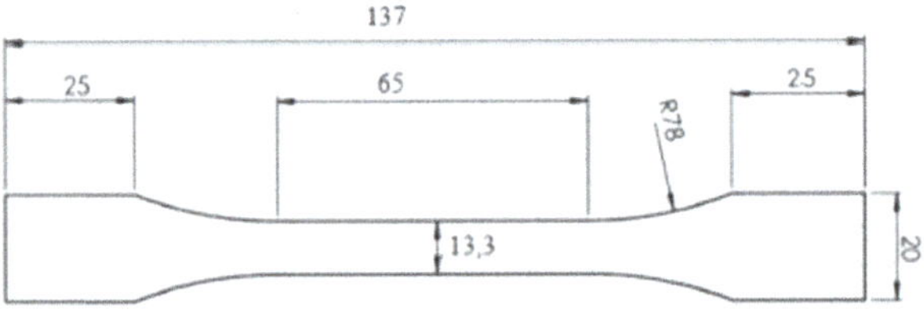

Fig. 6. Dimensions (mm) of standard specimen for 3D printing.

Uniaxial tensile tests are conducted on all SCFR-PLA samples, employing various printing orientations, different carbon fiber proportions, and different infill percentages, until failure occurs. These tests are conducted at room temperature (25 °C). Young's modulus (E) and ultimate tensile strength (UTS) were determined from conventional stress-strain curves. In our study, we employed a full factorial experimental design with a total of 36 tests (3 levels for carbon content, 2 levels for infill density, 2 levels for layer thickness, and 3 levels for raster angle). The results from these repeated tests were averaged to minimize experimental error. Each tensile test (Fig. 7) was conducted twice for every set of printing parameters to minimize the contrast of the measurements, and the average values were used to plot the stress-strain curve.

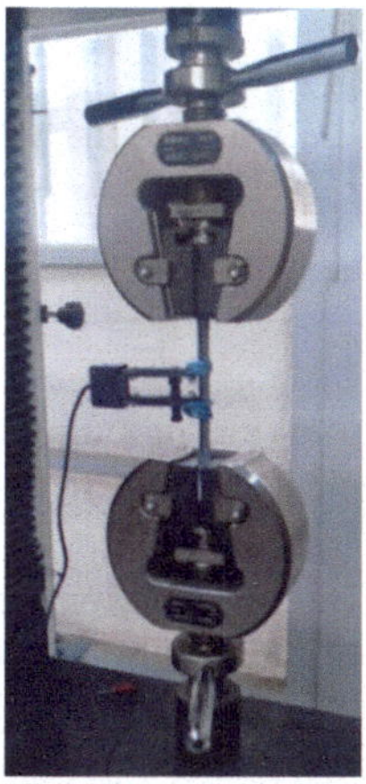

Fig. 7. Experimental uniaxial tensile test.

We present in Table 3 some experimental results by varying the values of carbon content, layer thickness, raster angle, and infill percentage. To minimize the number of 3D tests while still studying the influence of the most influential parameters on the mechanical properties of the composite, we conducted 72 tests. A constant printing speed of 50 mm/s was selected.

Table 3. Parameters of FDM process for tensile test.

Specimen	C%	Layer thickness (mm)	Infill (%)	Raster angle (°)
S1	0	0.3	100	0
S2	0	0.3	100	90
S3	15	0.4	100	0
S4	15	0.3	100	0
S5	30	0.4	100	0
S6	30	0.4	100	45

(*continued*)

Table 3. (*continued*)

Specimen	C%	Layer thickness (mm)	Infill (%)	Raster angle (°)
S7	30	0.4	100	90
S8	0	0.3	50	0
S9	30	0.4	50	0
S10	15	0.4	100	45
S11	30	0.3	100	0
S12	0	0.4	100	0

4 Experimental Results

Young's Modulus (E), known also as the modulus of elasticity and ultimate tensile strength (UTS) are crucial mechanical properties for composite materials, indicating their ability to withstand external forces and stresses. Young's modulus measures the stiffness or rigidity of a material. The ultimate tensile strength refers to the maximum stress a material can withstand before it fails or breaks. It indicates the composite's ability to resist stretching or pulling forces. An experimental study is conducted to characterize the mechanical behavior of the printed SCFR-PLA parts based on tensile tests. Figure 8 presents an example of a tensile curve for Specimen 9.

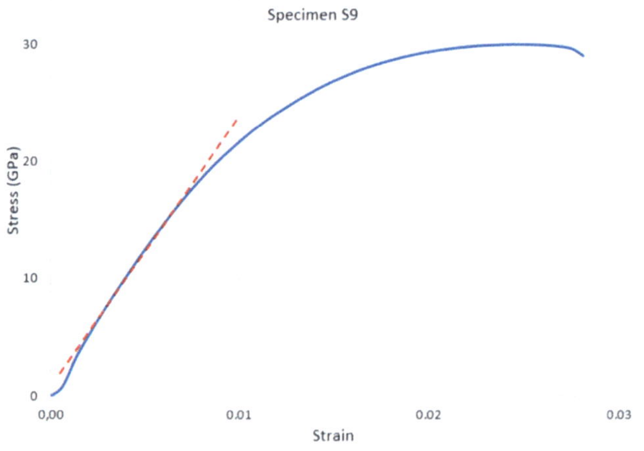

Fig. 8. Tensile curve for specimen 9.

From this curve (Fig. 8), we observe that the Young's modulus (E) is equal to 1.25GPa and ultimate tensile strength (UTS) is around 18.2 MPa. We illustrate in Table 4 the obtained results.

Table 4. Experimental results of tensile test.

Specimen	E(GPa)	UTS (MPa)
S1	3.6	57
S2	3.3	43
S3	2.17	31
S4	7.66	55
S5	2.27	30
S6	1.83	22.2
S7	1.13	16.5
S8	1.02	15.2
S9	1.25	18.2
S10	1.8	23
S11	7.6	48.5
S12	1.02	34

Based on Figs. 9 and 10, we show that Young's modulus (E) and ultimate tensile strength (UTS) are influenced by infill percentage, layer height, raster angle, and principally the carbon percentage.

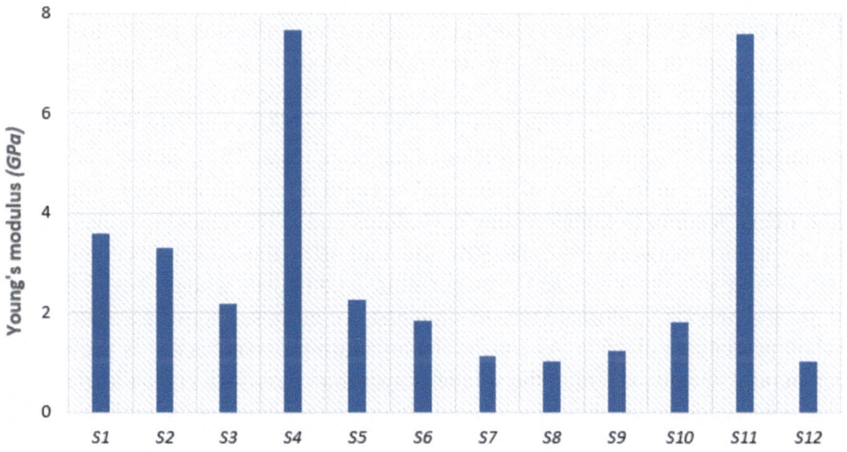

Fig. 9. Experimental results: Young's modulus for different specimens.

Infill percentage refers to the density of material within the interior of a printed composite part. The influence of the infill percentage on the properties of the composite. Therefore, higher infill percentages typically result in greater material density and improved mechanical properties, including higher Young's Modulus and ultimate tensile strength. However, excessively high infill percentages may lead to increased material

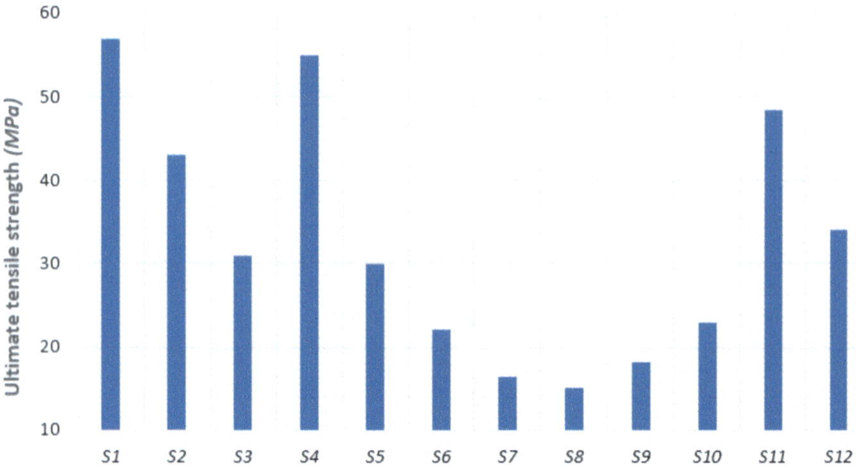

Fig. 10. Experimental results: Ultimate tensile strength for different specimens.

usage and longer print times without significant improvements in mechanical properties. In addition, this study emphasizes how the printing orientation angle affects the mechanical characteristics of SCFR-PLA. With an increase in the raster angle, there is a notable weakening in both Young's modulus (E) and ultimate tensile strength (UTS). This decline indicates a reduction in stiffness, suggesting that the structural integrity of the material weakens as the printing orientation deviates from the optimal configuration. In the same context, layer height affects the printed layer in the SCFR-PLA composite part. Finer layer heights generally result in better interlayer adhesion, leading to enhanced mechanical properties, principally higher Young's Modulus and UTS. Finally, the carbon percentage in SCFR-PLA composite plays a crucial role in determining mechanical properties. By comparing (S1 – S4) and (S3 – S5), we can determine the influence of the raster angle on the mechanical properties of the composite. Thus, adding carbon fiber to the PLA results in improved stiffness and strength due to the inherent properties of carbon fibers, leading to higher Young's Modulus and UTS values. However, there is a slight variation between 15% and 30%. Beyond 30%, the material becomes brittle. Based on these interpretations and to manufacture a composite with the best E (modulus of elasticity) and UTS (ultimate tensile strength), we propose choosing a material with a carbon percentage of 15%. As for the manufacturing parameters, it is preferable to select an infill density of 100%, the smallest layer thickness, and a raster angle of 0°.

5 Conclusion

In this study, FDM was used to print PLA and SCFR-PLA composites with different printing parameters. Four parameters were chosen: the infill percentage, the layer height, the raster angle, and the carbon percentage. The tensile test results showed that the mechanical properties were influenced by these FDM parameters. Increasing the infill percentage improved the tensile strength of the printed composite. Changing the raster angle from 0° to 90°, notably decreased stiffness in SCFR-PLA samples, with significant reductions in strength and strain at failure observed as well. The carbon percentage has a significant influence on E (modulus of elasticity) and UTS (ultimate tensile strength) because carbon fibers are renowned for their exceptional mechanical properties. As the carbon content increases, mechanical performance is enhanced. However, achieving the optimal balance of carbon content is essential, as excessively high percentages may also introduce challenges such as increased brittleness.

The results of this study can contribute to emphasizing the importance of using PLA and SCFR-PLA composites for additive manufacturing in various fields, such as the production of blades for small and medium-sized wind turbines.

Conflict of Interest. The authors have no relevant financial or non-financial interests to disclose. They have no competing interests to declare that are relevant to the content of this article.

References

1. Guerra Silva, R., Salinas Estay, C., Morales Pavez, G., Zahr Viñuela, J., Torres, M.J.: Influence of geometric and manufacturing parameters on the compressive behavior of 3D printed polymer lattice structures. Dent. Mater. **14**, 1462 (2021)
2. Moakhar, S., Hentati, H., Barkallah, M., et al.: Modeling of the ductile damage—application for bar shearing. Mater. Werkst. **50**, 1353–1363 (2019)
3. Hentati, H., Naceur, I.B., Bouzid, W., Maalej, A.: Numerical analysis of damage thermo-mechanical models. Adv. Appl. Math. Mech. **7**(5), 625–643 (2015)
4. Ilyas, M., Hussain, G., Rashid, H., Alkahtani, M.: Influence of forming parameters on the mechanical behavior of a thin aluminum sheet processed through single point incremental forming. Metals. **10**, 1461 (2020)
5. Ben Fraj, B., Kamoun, T., Hentati, H., Trabelsi, M., Ghazouani, N., Ahmed, M.: Optimization of forming force and Erichsen index using Taguchi design of experiments: Mathematical models and experimental validation. Proc. Inst. Mech. Eng. B J. Eng. Manuf. **238**(9), 1316–1326 (2023)
6. Moakhar, S., Hentati, H., Barkallah, M., Louati, J., Haddar, M.: Parametric study of aluminum bar shearing using Johnson-Cook material modeling. Proc. Inst. Mech. Eng. B J. Eng. Manuf. **235**(9), 1399–1411 (2021)
7. Yamato, H., Keiichi, S., Ryo, H., Yoshiaki, K., et al.: Experimental and numerical investigation of the relationship between material defects and elastoplasticity behavior of 3D-printed carbon-fiber-reinforced thermoplastics under compressive loading. Compos. Sci. Technol. **241**, 110–116 (2023)
8. Georgantzinos, S.K., Giannopoulos, G.I., Stamoulis, K., Markolefas, S.: Composites in aerospace and mechanical engineering. Dent. Mater. **16**, 7230 (2023)
9. Sunil, B., Roberto, A.L.A., Douglas, J.G.: Enhancing the interlayer tensile strength of 3D printed short carbon fiber reinforced PETG and PLA composites via annealing. Addit. Manuf. **30**, 100922 (2019)
10. Hsueh, M.H., Lai, C.J., Wang, S.H.: Effect of printing parameters on the thermal and mechanical properties of 3D-printed PLA and PETG, using fused deposition modeling. Polymers. **13**(11), 1758 (2021)
11. Shady, F., Daniel, G.A., Robert, L.: Physical and mechanical properties of PLA, and their functions in widespread applications—a comprehensive review. Adv. Drug Deliv. Rev. **107**, 367–392 (2016)
12. Raja, S.A., Muthukumaran, E., Jayakrishna, K.: A case study of 3D printed PLA and its mechanical properties. Mater. Today. Proc. **5**, 11219–11226 (2018)

13. Salmi, M.: Additive manufacturing processes in medical applications. Dent. Mater. **14**, 191 (2021)
14. Sun, C., Wang, Y., McMurtrey, M.D., Jerred, N.D., Liou, F., Li, J.: Additive manufacturing for energy: a review. Appl. Energy. **282**, 116041 (2021)
15. Wang, H., Lamichhane, T.N., Paranthaman, M.P.: Review of additive manufacturing of permanent magnets for electrical machines: a prospective on wind turbine. Mater. Today Phys. **24**, 100675 (2022)
16. Allouch, M., Ben Fraj, B., Dhouioui, M., Kesssentini, A., Hentati, H., Wali, M., Ferhi, M.: Mechanical, microstructural and numerical investigations of 3D printed carbon fiber reinforced PEEK. Proc. Inst. Mech. Eng. C J. Mech. Eng. Sci. **238**(6), 2131–2139 (2023)
17. Rajan, K., Samykano, M., Kadirgama, K., et al.: Fused deposition modeling: process, materials, parameters, properties, and applications. Int. J. Adv. Manuf. Technol. **120**, 1531–1570 (2022)
18. Ammar, S., Ben Fraj, B., Hentati, H., Saouab, A., Ben Amar, M., Haddar, M.: Mechanical performances of printed carbon fiber-reinforced PLA and PETG composites. Proc. Inst. Mech. Eng. L J. Mater. Des. Appl. **238**(8), 1488–1499 (2024)
19. Ramírez-Elías, V.A., Damian-Escoto, N., Choo, K., Gómez-Martínez, M.A., Balvantín-García, A., Diosdado-De la Peña, J.A.: Structural analysis of carbon fiber 3D-printed ribs for small wind turbine blades. Polymers. **14**, 4925 (2022)
20. Maqsood, N., Marius, R.: Characterization of carbon fiber reinforced PLA composites manufactured by fused deposition modeling. Compos. Part C. **4**, 100112 (2021)
21. Seralathan, S., Mukesh, N., Kameshwaran, R., et al.: Analysis of cross-axis wind turbine blades designed and manufactured by FDM-based additive manufacturing. Mater. Today Proc. **33**(7), 3504–3509 (2020)
22. Marwane, R., Mourad, N., Tarfaoui, M., et al.: 3D printing: rapid manufacturing of a new small-scale tidal turbine blade. Int. J. Adv. Manuf. Technol. **115**, 61–76 (2021)
23. Suresh, A., Rajakumar, S., Robert, C., Mahalingam, S.: Optimization and experimental investigation of 3D printed micro wind turbine blade made of PLA material. Materials. **16**, 2508 (2023)
24. Al-Quraishi, B.A., Nor Zelawati, B., Mohammed Najeh, N., Salih Meri, A.: Experimental and simulation investigation for performance of a small-scale model of bare and shrouded HAWT. Int. J. Mech. Eng. Technol. **10**(1), 434–449 (2019)

3D Printing of Bio-inspired Fluidic Conduct Based on Tracheal Geometry

Marwa Jaziri[1(✉)], Badreddine Larbi[1,3], Khaled Boulahem[2,3], Mohamed Toumi Nasri[1,3], Olfa Trabelsi[4], and Mahfoudh Ayadi[1,3]

[1] LR18ES01, LMPE, University of Tunis, ENSIT, 5 Avenue Hussein, BP 56, 1008 Bâb Manara, Tunisia
jazirimarwa90@gmail.com

[2] National School of Engineers of Tunis, ENIT, Laboratory of Applied Mechanics and Engineering, University of Tunis El Manar, 1002 Tunis the Belvedere, Tunisia

[3] University of Carthage, ENIB, Menzel Abderrahman University Campus, 7035 Bizerte, Tunisia

[4] CNRS, Biomechanics and Bioengineering, Centre de Recherche Royallieu, University de Technologie de Compiègne, CS 60319, 60203 Compiègne Cedex, France
olfa.trabelsi@utc.fr

Abstract. The mechanical behavior of the trachea in living organisms is critical to respiratory function. Utilizing these biological characteristics, this study investigates the design of conduct fabricated through 3D printing to replicate the stiffness and elasticity of the trachea for applications in medicine and industry. This project marks a significant advancement in developing efficient biologically inspired medical and industrial devices. Conducts are produced with 3D printing technology tailored to the specific requirements of each field. To examine the mechanical properties of these conducts numerical simulations were conducted using ABAQUS to analyze conduct movement under maximum simulated loads during a standard respiratory cycle. This analysis provides valuable insights into the conduct's behavior under realistic conditions facilitating its practical application. Additionally, Design of Experiments (DOE) modeling was employed to evaluate stiffness and elasticity exploring the impact of various parameter combinations. This approach demonstrates a strong correlation between experimental results and DOE predictions supporting the performance and reliability of the manufactured conducts. A numerical compression modeling study was also conducted where forces determined by a DMA machine corroborated the experimental test results.

Keywords: Trachea · conduct · breathing · 3D Printing

1 Introduction

The trachea is a conduct-shaped organ whose role is to efficiently transport air to the lungs due to its adapted structure. Indeed, the specific geometric shape of the trachea in biology is tailored to the physiological needs of air transport and supports the respiratory

M. Soula et al. (Eds.): AIAM 2024, LNME, pp. 23–43, 2025.
https://doi.org/10.1007/978-3-031-94889-3_3

functions of the living organism. Fiorentino's study in 2016 [1] demonstrated that the curved shape of the trachea resists displacement and minimizes movement. In reality, the trachea is not cylindrical; it is "D" shaped. This configuration is not arbitrary but aims to reduce displacement as shown by the studies of Evila et al. [2] and Matthieu et al. [3]. Upon examination of the trachea it is observed that it does not have uniform thickness; the cartilage is thicker than the muscles. This observation is supported by Evila et al. [2] and Freitag et al. [4] studies which emphasized the composition of the trachea consisting of muscles and cartilage.

The geometry of the trachea was considered as a model for designing industrial conducts. The trachea's unique shape tailored for efficient and turbulence-free airflow provides valuable insights applicable to industrial conduct design. By optimizing conduct geometry fluid transportation in industrial environments can be streamlined reducing resistance and minimizing energy losses [5]. Furthermore by mirroring the trachea's flexible yet robust structure industrial conducts can be engineered to withstand external pressures such as internal fluid pressure ensuring mechanical resilience [6]. The turbulence-reducing design of the trachea also serves as a model for industrial applications where reducing turbulence is essential to avoid premature wear and maintain transport efficiency [7]. Moreover, akin to the trachea's adaptability to physiological demands industrial conducts can be tailored to respond to variations in environmental conditions such as pressure and temperature. Thereby enhancing their versatility and performance. This proposal investigates the feasibility of replicating these adaptive characteristics through advanced manufacturing techniques like 3D printing.

The trachea is far from being static; it undergoes movements essential to every phase of respiration whether it's inspiration or expiration. Research conducted by Holden et al. [8] has revealed that during deep inspiration the trachea extends to about one-fifth of its length and widens by about one-tenth of its diameter while during expiration it shortens, narrows, and compresses oppositely. During coughing episodes, Holden also observed that the diameter of the trachea decreases by a quarter during moderate coughs but can reach half or even a third of its initial size during paroxysms. This remarkable flexibility and elasticity prompt us to consider the fabrication of conduct with similar characteristics to those of the trachea to facilitate fluid flow without structural alteration and to minimize the risks of failure or premature wear. This study aims to develop conduct inspired by the structure of the human trachea to achieve comparable performances. This conduct made of PLA holds potential applications in various fields such as the food industry, the environment, and agriculture particularly irrigation fluid transport in the chemical industry and especially in medicine where PLA is recognized as a biomedical material as indicated by a prior study by Rifky et al. [9] regarding the production of bone screws. This conduct could be utilized for manufacturing intubation tubes or tracheal catheters thus providing a lightweight and biocompatible alternative to conventional medical devices. Additionally, in the industrial domain it could find its place in ventilation systems for the conveyance of gases or liquids resisting chemicals as well as durability. The approach centers on replicating the structure and properties of the trachea using polylactic acid (PLA). By adjusting variables such as printing temperature infill percentage, layer thickness, and geometric thickness, the design can be optimized to achieve efficiency and functionality comparable to that of the biological trachea. Two main characteristics of the living organism's trachea are its ability to expand and contract as well as its simultaneous elasticity and rigidity.

In this study, a numerical simulation using ABAQUS was conducted to evaluate the movements of the conduct under the effect of maximum simulated loads during a normal breathing cycle to design conduct that could be used in industrial applications. Furthermore, 13 test samples with varied parameters underwent printing and compression tests on a DMA machine. The DMA equipment model METRAVIB DMA 50 operated at a frequency range of 1 Hz to 200 Hz with a heating rate of 3 °C/min and a temperature range from −150 °C to 500 °C. This was accompanied by a parametric study aimed at identifying the most influential parameter on the rigidity of the conduct.

At the same time a Design of Experiments (DOE) model was implemented to evaluate rigidity, and elasticity and understand the impact of different parameter combinations.

Finally, additional ABAQUS models simulated the compression already performed on the DMA machine to adjust the mechanical properties of the conduct and determine its position as well as its deformation values.

Polylactic acid (PLA) was selected for manufacturing the test specimen via 3D printing due to its numerous key advantages in the industry. It has several advantages for sustainability-focused industries. Firstly, its biodegradability makes it an environmentally friendly solution addressing growing concerns about sustainability [10]. Secondly, its affordability helps to reduce conduct manufacturing costs compared to other materials providing a cost-effective alternative. Additionally, its lightweight nature makes conducting installation and handling easier while also reducing the load on support structures. Finally, its compatibility with the 3D printing environment makes it a natural choice for modern printing technologies offering a versatile and practical solution for various applications [11].

Furthermore, PLA is a biomedical material. as indicated in the Olga et al. study [12]. This study demonstrated that the use of PLA films with small integrated chambers can allow for the creation of coatings for tracheal endoprostheses. These coatings slowly release medications effectively preventing the formation of granulation tissue in the trachea.

2 Numerical Modeling

2.1 The Geometry of the Proposed Model

The anatomical complexity of the tracheal structure along with variations in its dimensions among individuals complicates the design process. Consequently, certain simplifying assumptions were necessary when formulating the process model. To achieve conduct with characteristics similar to those of the human trachea, the design incorporated geometry inspired by the true tracheal structure. The cross-sectional shape of the cartilaginous rings was modeled with an elliptical arrangement in the anterior region while the posterior closed by the insertion of smooth muscle. Dimensions were based on the findings of Bagnoli et al. [13]. The geometric parameters of the tracheal rings are illustrated in Fig. 1.

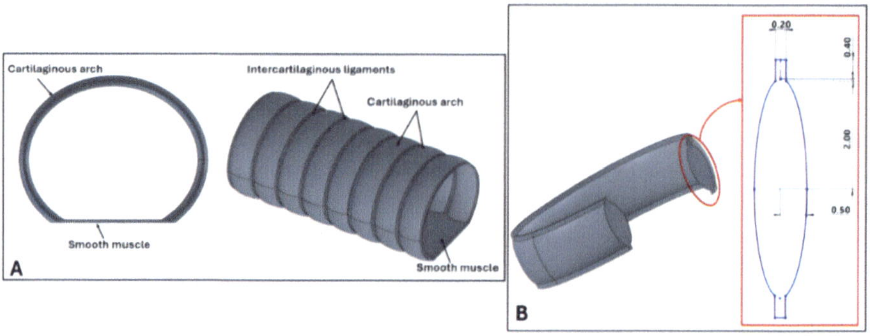

Fig. 1. **A**: The geometric model of conduct inspired by the trachea. **B**: The geometric dimensions of the model [mm].

2.2 Material Behavior Law

The objective of this study is to model the behavior of the PLA fluidic duct whose geometry is inspired by the trachea when subjected to the same conditions as the human trachea. To achieve this, it is necessary to first understand the behavior of the trachea. In a study conducted by Trabelsi et al. [14] experimental cyclic tensile tests conducted on cartilage samples revealed nonlinear behavior. During the loading phase the cartilage initially exhibited a quasi-linear response up to 5% strain indicating reversible elastic behavior. During the unloading phase the cartilage demonstrated viscous behavior with energy dissipated throughout the loading-unloading cycle [14].

To model the viscoelastic behavior of tracheal cartilage, the experimental study conducted by Trabelsi et al. [14] employed the Neo-Hookean model. Model parameters were fitted to the experimental data with excellent correlation suggesting that the Neo-Hookean model accurately captures the stiffness and nonlinear mechanical properties of tracheal cartilage. Following this, the cartilage's behavior was assumed to be hyperelastic and isotropic [14].

The results of cyclic tensile tests conducted on tracheal smooth muscle samples in the study by Trabelsi et al. [14] revealed nonlinear behavior with greater stiffness in the longitudinal direction than in the transverse direction. The mechanical constants and collagen fiber parameters ($k1 = 0.154$ kPa, $k2 = 34.157$ for longitudinal direction and $k3 = 0.347$ kPa, $k4 = 13.889$ for the transverse direction) were used to model this behavior applying the Holzapfel strain energy density function [14]. The annular ligament of the trachea is composed of type I and type III collagen. This composition with a higher proportion of type III collagen compared to tendons provides the ligament with a structure that promotes both flexibility and strength. Longitudinal extensibility is necessary to accommodate the changes in tracheal diameter during inhalation and exhalation. The experimental study conducted by Malvè et al. [15] shows that the annular ligament exhibits a nonlinear and isotropic mechanical behavior with hyperelastic properties notably prominent in the longitudinal direction. To model these properties the isotropic Holzapfel model was applied to the tissue.

In the subsequent step, air pressures applied to the model were based on a complete respiratory cycle including inspiration and expiration as indicated in Fig. 2. The pressure values were referenced from the literature study by Del Palomar et al. [16]. The maximum values for both phases are 5 kPa in tension (inspiration) and −5 kPa in compression (expiration).

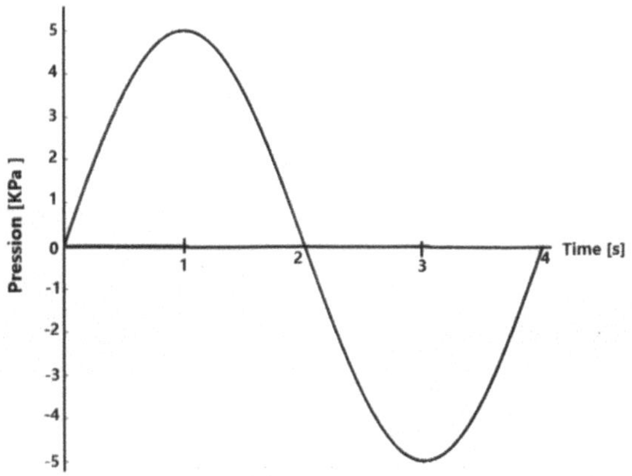

Fig. 2. The normal respiratory.

The duct must exhibit three mechanical behaviors analogous of those of the trachea: one portion that is hyperelastic and isotropic inspired by the cartilage and the second portion that is hyperelastic and anisotropic inspired by the tracheal muscle and a third portion is hyperelastic and isotropic inspired by the ligaments as shown in Fig. 3. The mesh used for this calculation consists of linear hexahedral elements of type C3D8 (15164 elements). To model the isotropic hyperelastic part inspired by the cartilage the Neo-Hookean mathematical formulation has been used. It is widely used in finite element simulations to represent materials such as rubber, polymers, and other elastic materials. This model is based on a strain energy function:

$$\Psi = C_1\left(\bar{I}_1 - 3\right) + \frac{K}{2}(J-1)^2 \tag{1}$$

C_1: The material constant related to the ground substance. According to the experimental study [16], the constant C_1 is taken equal to 577.7 kPa.
K: The compressibility modulus is typically expressed in pascals (Pa).
J: The volume change factor. It represents the ratio of the final volume to the initial volume of a material during deformation. In this case, the trachea is considered an incompressible material. so J = 1.
$\bar{I}_1$: The first invariant which is defined as the trace of the deviatoric Cauchy strain tensor ($\overline{C}$), where $\overline{C}$ is the matrix of the Cauchy stress tensor.

To model the anisotropic hyperelastic part inspired by the tracheal muscle, the ABAQUS software was used employing the Holzapfel mathematical model. This model is based on a strain energy function:

$$\Psi(\bar{I}_1.\bar{I}_2.\bar{I}_{41}.\bar{I}_{42}) = C_{10}(\bar{I}_1 - 3) + \frac{K_1}{2K_2}\left\{\exp\left[K_2\left(\bar{I}_{41} - 1\right)^2\right] - 1\right\} + \frac{K_3}{2K_4}\left\{\exp\left[K_4\left(\bar{I}_{42} - 1\right)^2\right] - 1\right\} \quad (2)$$

All parameters of this model were determined from the experimental study conducted by Trabelsi et al. [14]. The isotropic behavior of the smooth muscle matrix is described by a stiffness-constant $C_{10} = 0.877$ kPa. The parameters of the Holzapfel strain energy density function are as follows: for the longitudinal direction $K_1 = 0.154$ kPa and $K_4 = 34.157$ while for the transverse direction the values of these parameters were: $K_2 = 34.157$ kPa and $K_3 = 0.347$.

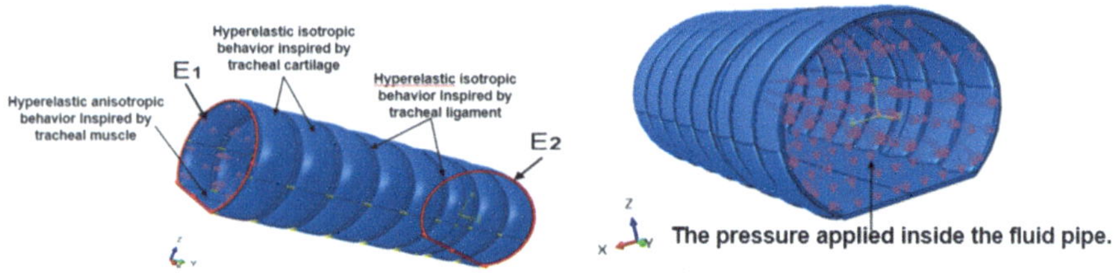

Fig. 3. The boundary conditions the mechanical behaviors of the pipe inspired by the human trachea and the radial pressures applied to the fluid pipe.

$\bar{I}_1$ and $\bar{I}_2$ are the standard invariants of the Cauchy-Green tensor associated with the isotropic material behavior. $\bar{I}_{41}$ and $\bar{I}_{42}$ are pseudo-invariants that characterize the constitutive response of the fibers in the longitudinal and transversal directions.

The following Table 1 summarizes the mechanical constants obtained experimentally for the main constituents of the human trachea which will be used in modeling the fluid conduct behavior.

Table 1. Summary of the mechanical constants inserted in logic for the main constituents of the human trachea specifying the constituent models used [14].

	Constitute model	C_{10} (KPa)	K_1 (KPa)	K_2	K_3 (KPa)	K_4
Cartilage ring	Neo-Hookean	577.7	–	–	–	–
Smooth muscle	Holzapfel	0.877	0.154	34.157	0.347	13.889

2.3 Boundaries Conditions

Different constraints were applied to the model to eliminate displacements that in reality are restricted by the presence of surrounding tissues. Fixtures were applied to the ends (E1, E2) (Fig. 3). It is essential to note that during breathing the only movement allowed is radial.

3 Results of Numerical Modeling

The fluidic conduct was subjected to conditions nearly identical to those of the human trachea during a normal respiratory cycle (inhalation and exhalation) lasting 4 s. The modeling yielded significant results. The deformation of the conduct is illustrated during normal breathing in the phases of inspiration and expiration. To validate the model, the results will be compared with in vivo data from Malvè et al. [16]. In that study, a single cartilage ring was identified with a red dashed line projected around the tracheal wall to highlight the ring's deformation. This deformation is illustrated in Fig. 4 for normal breathing and compared to the calculated deformation of the numerical model of the conduct.

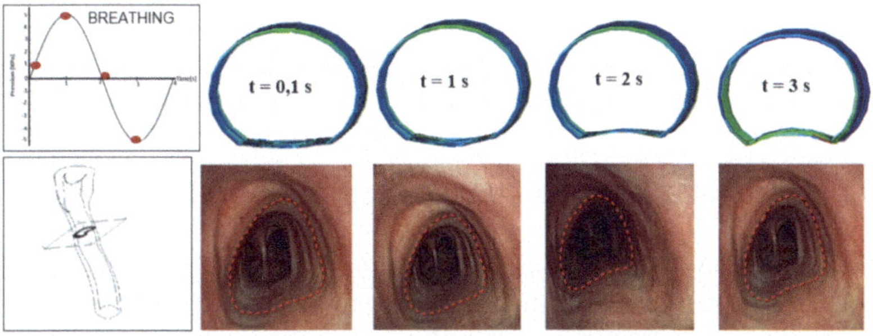

Fig. 4. Deformed shape of a section of the duct during normal breathing. The correlation between our numerical prediction and the actual deformation can be established.

Figure 4 presents the qualitative validation of the deformed shape of the conduct during the phases of maximum inspiration and expiration in comparison to the endoscopic image from the study [16]. The results of the numerical model demonstrate similar trends for both the inspiration and the expiration phases. This numerical model has been validated through comparison with the existing results [16].

This study simulated the mechanical behavior of a fluid conduct subjected to the typical pressure conditions of a human respiratory cycle. Although this conduct is not intended for direct insertion into the body it could find valuable applications in various fields.

For instance, the conduct could be utilized in medical ventilation devices to deliver air or respiratory gases more efficiently by mimicking the natural geometry of the airways. Additionally, it could have biomedical applications such as in perfusion systems, dialysis circuits, or other scenarios requiring precise and efficient fluid transfer.

The results of this simulation allow for the evaluation of the feasibility and robustness of such fluid conduct under dynamic stresses. The cyclic deformations of the conduct diameter in response to variations in internal pressure were observed. During inhalation, a drop in pressure leads to a slight dilation of the diameter with a positive displacement of 1.86 mm. Conversely, during exhalation, the increase in pressure caused a slight compression with a negative displacement of −1.86 mm (Fig. 5).

These cyclic diameter variations are normal and expected for this type of conduct subjected to pressure fluctuations demonstrating that the conducting structure adapts

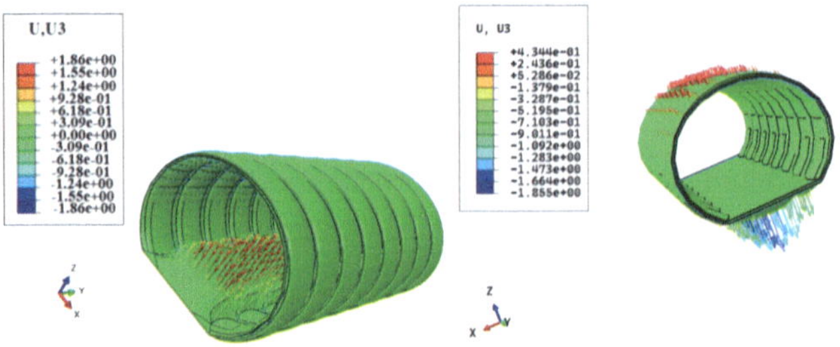

Fig. 5. The variation of the conduct during the respiratory cycle. A: inspiration phase B: exhalation phase

appropriately to stresses without undergoing excessive deformations. This result indicates that the design is suitable for withstanding the pressure variations it will encounter in service whether in medical ventilation applications or other biomedical fields.

4 3D Printing of the Proposed Geometry with an Experiment Plan

3D printing also known as additive manufacturing has revolutionized the industrial field by offering a range of significant advantages. Its importance lies in its ability to enable rapid and cost-effective prototyping thereby facilitating the design and iteration process. Moreover, it allows for the fabrication of complex and customized parts that would otherwise be difficult to produce. The final design of the test specimen was manufactured using the TRONXY-XY-02 3D printer as indicated in Fig. 6.

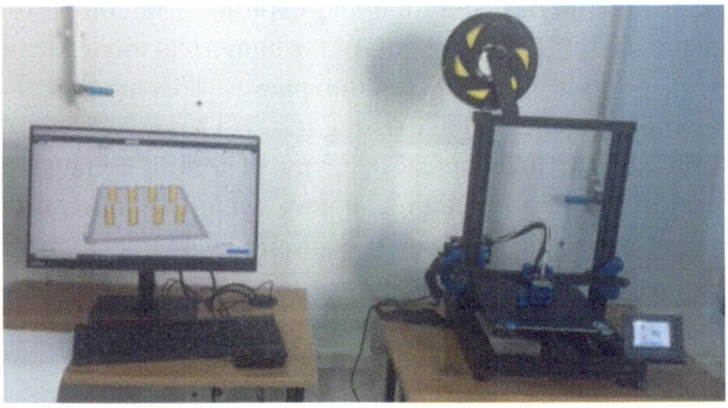

Fig. 6. The TRONXY-XY-02 3D PRINTER.

Initially, the 3D model created in SolidWorks 2023 was saved as an STL file and subsequently exported to the printing machine. The STL file was imported into the Ultimaker Cura software resulting in the shape presented in Fig. 7.

This software allows for adjustments to the 3D printing parameters which are crucial for achieving optimal results. Among these parameters, layer thickness plays an essential role in determining the detail fidelity of the printed object. Additionally, the choice of

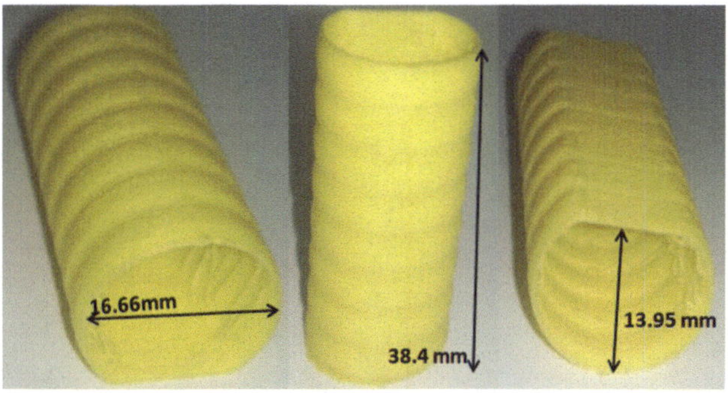

Fig. 7. The conduct obtained through 3D printing.

material and its printing temperature significantly affect layer adhesion and print quality. The infill percentage influences the strength of the final model by adjusting the amount of material within the part.

All these parameters have a substantial impact on the mechanical properties of the material used thereby affecting the quality and functionality of the 3D-printed object. Four specimen geometries were designed by modifying the thickness either of the thickest or the thinnest areas. These four models will be printed by adjusting the parameters of the 3D printer each time including layer thickness, infill percentage, and PLA temperature as detailed in Table 2. Consequently, a total of 13 models will be produced as illustrated in Figs. 7 and 8 which will be characterized through the application of the Design of Experiments.

Table 2. The different parameters were modified during the printing experience.

	E_l	T (°C)	%	E_{Lig}	E_c
The minimum value	0.15	195	20	0.1	0.8
The maximum value	0.25	205	100	0.3	1.2

E_l: The thickness of the layer (mm). T°: The temperature of PLA (°C). %: The percentage of filling. E_c: Thickest part thickness (mm). E_{Lig}: Thinnest part thickness (mm)

4.1 Mechanical Characterization

4.1.1 Characterization of Printed Samples

This section delves into the mechanical properties of printed materials using the Dynamic Mechanical Analysis (DMA) machine to characterize their viscoelastic properties. The polymer industry widely uses this method to measure the mechanical response of materials under various conditions including compression, bending, and tension. This analysis facilitates a better understanding of how the additive manufacturing parameters influence the properties of the samples and how these properties affect their mechanical behavior.

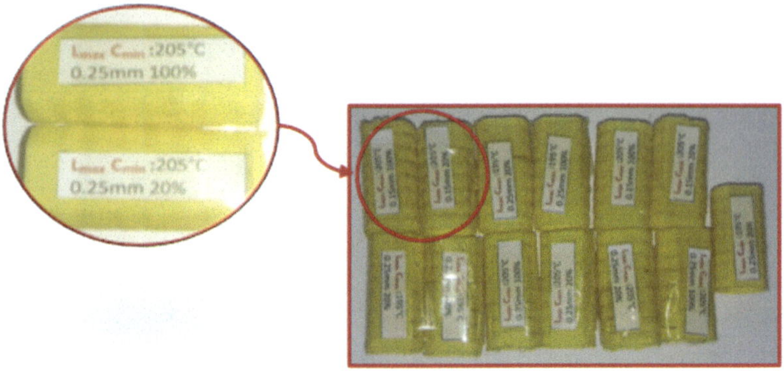

Fig. 8. The 13 specimens were designated for mechanical characterization.

The analysis characterizes the sample properties through the elastic modulus (E′) which indicates its capacity to store and release energy the loss modulus (E″) which reflects the material's tendency toward viscous energy and the Damping Factor (Tan δ) obtained across a specified temperature range. The DMA equipment used for this analysis is the Metravib DMA50 operated at a frequency of 1 Hz (Fig. 9).

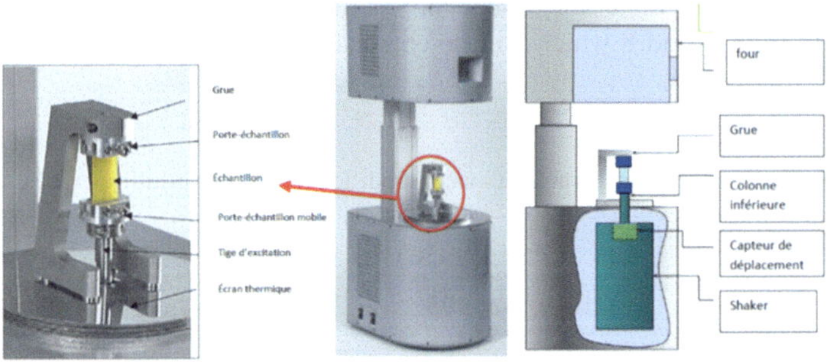

Fig. 9. The DMA polymer characterization machine.

5 Results of Experimental and Numerical Compression Test

5.1 Results of Dynamic Mechanical Behavior DMA

Following the compression tests, various measurements of mechanical properties were collected including Young's modulus, loss modulus, storage modulus, and the tangent delta (Tan δ) of the samples. The analysis focused on the scenario where both the ligament-inspired and cartilage-inspired sections were very thin. By adjusting the printing temperature, the Tan δ curves (blue) were analyzed leading to the determination of the glass transition temperature Tg. Tg is defined as the point at which the material transitions from a rubbery or viscoelastic state to a glassy state. Below Tg, the material exhibits rigid and brittle behavior while above this temperature, it becomes more flexible and deformable.

It was observed that when the sample was printed at a temperature of 205 °C. Tg was reached at 78 °C indicating a transition from a rigid to a rubbery state. Conversely, When the sample was printed at 195 °C. Tg was reached at 122 °C as illustrated in Fig. 10.

In another scenario, the examination focused on the case where the sections inspired by ligaments and those inspired by cartilage both had maximum thickness. By maintaining the printing temperature at 205 °C and varying the filling percentage (R%), the Tan δ curves (in blue) were analyzed leading to the determination of the glass transition temperature (Tg). It was observed that when the sample was printed at a temperature of 205 °C with a filling percentage of R = 20%. Tg was reached at 118 °C indicating a transition from a rigid to a rubbery state. Conversely, when the sample was printed with a filling percentage of R = 100%. Tg was reached at 71 °C as illustrated in Fig. 11.

The compression tests conducted with the DMA machine allowed for the determination of Young's modulus E for 13 samples as a function of temperature as illustrated in Fig. 12. The results are categorized into three temperature ranges: the first range (1) [25 °C, 50 °C], the second range (2) [50 °C, 60 °C] and the third range (3) [75 °C, 120 °C].

In the first range, where the temperature is below 50 °C. Young's modulus varies from 20 MPa to 90 MPa. In the second range between 50 °C and 60 °C. E decreases to about 10 MPa. Beyond 75 °C, E becomes significantly low except for specimens 5 and 13 which have specific geometric and printing parameters indicated in Table 3. Consequently, the scale was adjusted to obtain the curves illustrated in part A.

This behavior of Young's modulus as a function of temperature reflects the mechanical properties of the tested materials and facilitates the identification of the critical temperature ranges for their application.

In the temperature range (1), specimen 2 which has the lowest printing parameters and optimal geometric parameters exhibits a higher Young's modulus than specimens 3 and 7. Specimens 2 and 3 share the same printing parameters but their geometric parameters differ in terms of the intermediate thickness (the thickness of the inspired part of the ligament). Specifically, specimen 2 has a greater intermediate thickness than specimen 3 indicating that a lower intermediate thickness reduces the rigidity of the structure compared to a thicker intermediate layer. For specimen 7, the thickness of the inspired part of the cartilage was reduced and the thickness of the intermediate part was increased to the maximum by adjusting the infill percentage resulting in intermediate rigidity with an average value of 44 MPa.

In the temperature range (2), where the DMA temperature lies between 50 °C and 60 °C a comparison of the Young's moduli of specimens 3 and 2 reveals that the Young's modulus of specimen 2 remains consistently higher. This suggests that reducing the rigidity of the structure requires adjusting the thickness of the intermediate layer. The primary difference between specimens 2 and 3 lies in the thickness of the intermediate part and the infill percentage. When comparing specimen 1 with specimen 3 both are subjected to the same conditions except for a minimum infill percentage (10%) a progressive increase in Young's modulus is observed as the infill percentage decreases. These

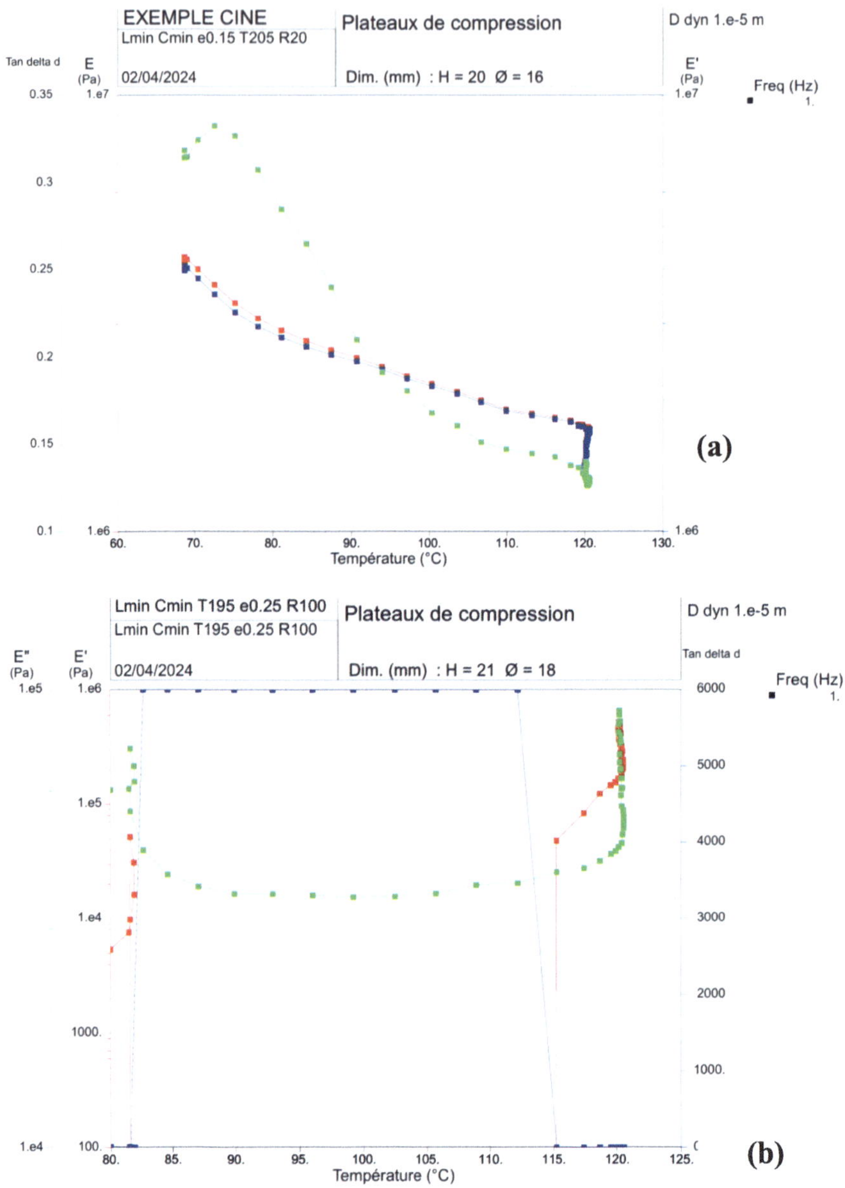

Fig. 10. Variation of Tan δ. E. E′ and E″ as a function of temperature: (a): in the case where the specimen is printed at T = 205 °C. (b): in the case where the specimen is printed at T = 195 °C.

results indicate that while the infill percentage plays a significant role in the variation of Young's modulus. The thickness of the intermediate part remains the most influential factor in reducing it.

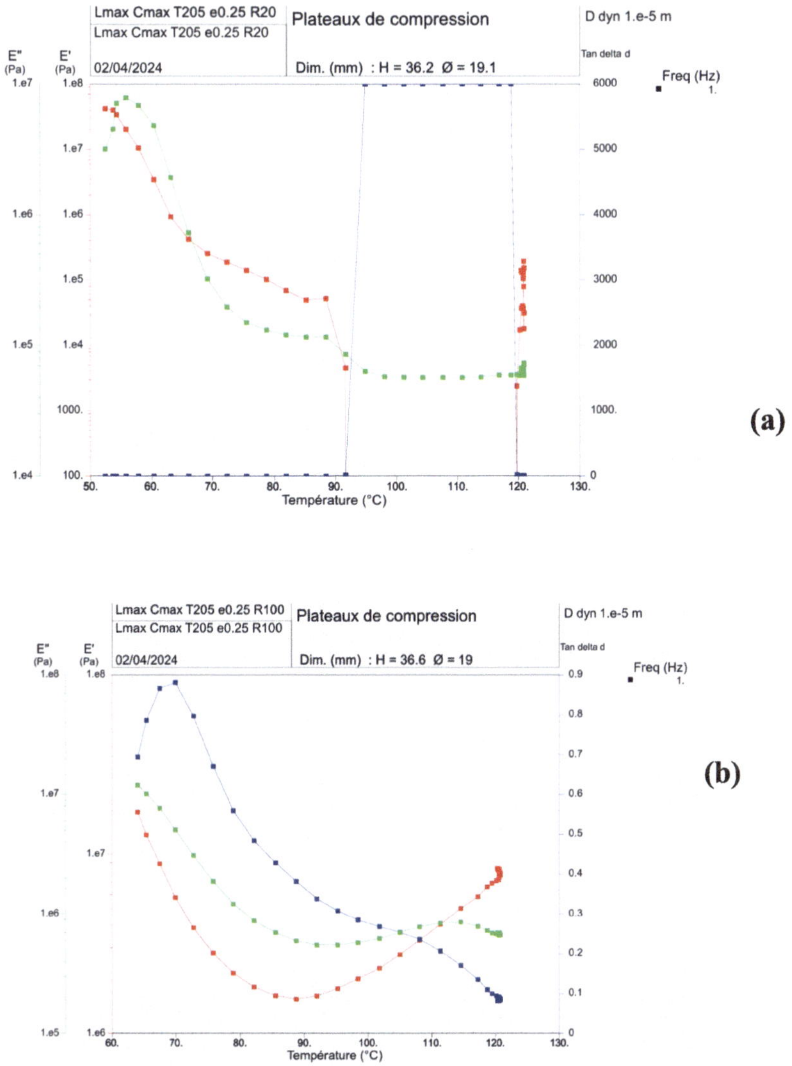

Fig. 11. Variation of Tan δ E. E′ and E″ as a function of temperature: (a): in the case where the specimen is printed at R = 20%. (b): in the case where the specimen is printed at R = 100%.

A comparison of the Young's modulus of specimens 9 and 12 shows that according to Table 3 these specimens have the same geometric parameters but differ in printing parameters. They have the same layer thickness but different printing temperatures: minimal for specimen 9 and maximal for specimen 12. The curves demonstrate that increasing the printing temperature results in a corresponding increase in Young's modulus.

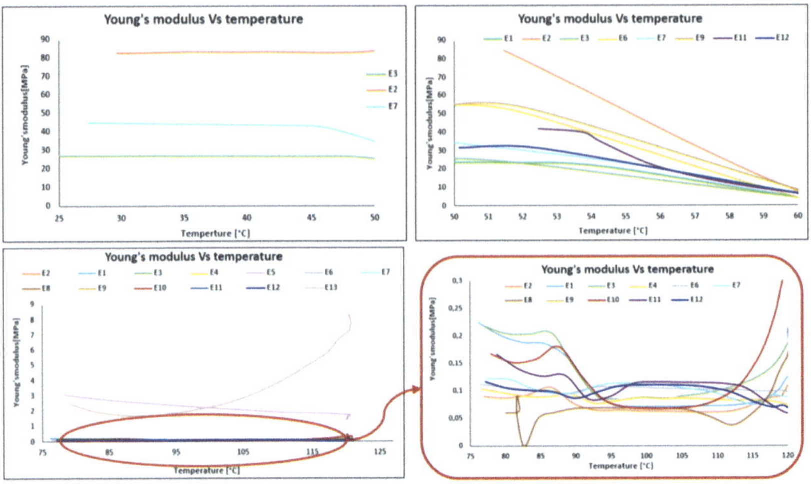

Fig. 12. The Young's modulus variation (E) in [MPa] of 13 test specimens. 1: temperature range (1). 2: temperature range (2). 3: temperature range (3). A: Zoom on the evolution of the Young's modulus E as a function of temperature for the range (3).

Table 3. The different parameters of 13 printed specimens.

Specimen N°	E_l	T (°C)	%	E_{Lig}	E_c
1	0.15	195	20	0.1	0.8
2				0.3	1.2
3			100	0.1	
4				0.3	
5		205	20	0.1	0.8
6				0.3	
7			100	0.3	
8	0.25	195		0.1	
9				0.3	
10		205	20	0.3	
11				0.3	1.2
12			100	0.1	0.8
13				0.3	1.2

A comparison between Young's modulus of specimens 7 and 12 reveals that according to Table 3 both specimens have the same geometric parameters but differ in printing parameters. While their printing temperature and infill percentage are identical the layer thickness varies with specimen 12 having the maximal thickness and specimen 7 the minimal. The curves indicate that despite the increase in layer thickness, the Young's modulus of both specimens follows nearly the same trend. However, with an increase in DMA temperature in the temperature range (3). Young's modulus of specimen 7 becomes slightly higher. This suggests that layer thickness does not significantly influence the variation of Young's modulus in this scenario.

In conclusion, the analysis shows that the thickness of the intermediate part is the most influential parameter on the rigidity of the structure and the variation of Young's modulus surpassing the impact of the infill percentage and layer thickness under the examined conditions.

5.2 Interactions of Operating Factors

A study was conducted to examine the interaction between operating factors: layer thickness, printing temperature, and fill percentage and their influence on transition temperature. Table 4 represents the design of the experiments.

Table 4. Design of experiments table.

Test	Layer thickness (mm)	Print temperature (T°)	Fill percentage (%)	Glass transition temperature (°C)
1	0.15	195	20	122
2	0.15	195	100	122
3	0.15	205	20	78
4	0.15	205	100	78
5	0.25	195	20	122
6	0.25	195	100	122
7	0.25	205	20	78
8	0.25	205	100	71

The study in Fig. 13 reveals that both the layer thickness and the printing temperature significantly influence the material's transition temperature. For a thickness of 0.15 mm, the transition temperature increases from 80 °C to 90 °C when the printing temperature rises from 195 °C to 205 °C representing a 12.5% increase. In contrast, for a thickness of 0.25 mm, the transition temperature rises from 75 °C to 82 °C under the same conditions. These results highlight that higher printing temperatures enhance thermal stability particularly for thinner layers where thickness also affects the transition temperature.

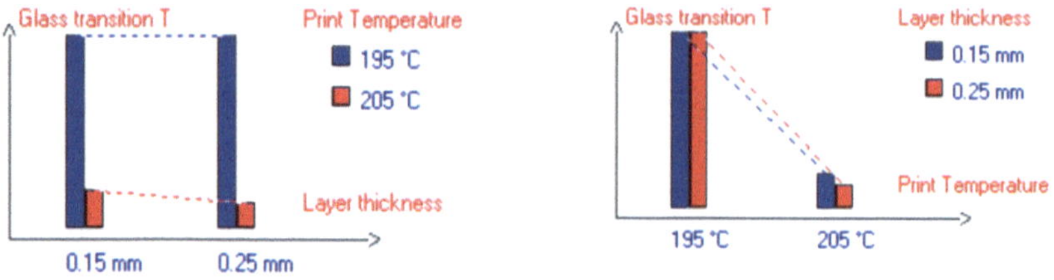

Fig. 13. Graphical study of the layer thickness/print temperature interaction.

Figure 14 shows the impact of print temperature (195 °C vs. 205 °C) and fill percentage (20% and 100%) on the glass transition temperature (Tg) of a material. At 195 °C, both fill percentages achieve a high Tg with minimal difference between them. However, at 205 °C, Tg drops significantly for both fill levels with the 100% fill maintaining a slightly higher Tg than the 20% fill. This suggests that higher print temperatures reduce thermal stability and a higher fill percentage slightly mitigates this effect.

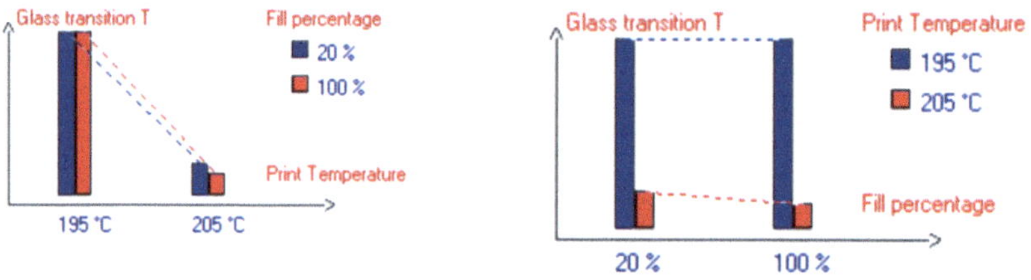

Fig. 14. Graphical study of the Fill percentage/print temperature interaction.

5.3 Results of the Numerical Compression Test

5.3.1 Model

According to the tests conducted on dynamic mechanical behavior with the DMA machine specimen 2 exhibits the highest Young's modulus compared to the other 12 specimens. This compression test was validated using an ABAQUS model incorporating the variations in Young's modulus from the DMA tests based on temperature, the applied forces as well as the Poisson's ratios determined from Eq. (3).

Since the material is isotropic, the elastic moduli are related by the following equations:

$$E = 3K(1 - 2\nu) \quad \nu = \frac{E}{2k} - 1 \tag{3}$$

k: The stiffness modulus [N/mm]; E: The Young's modulus [MPa]; ν : The Poisson's ratio.

5.3.2 Boundaries Conditions

At this stage, a compressive load was applied to the section of the conduct (E2) while a fixed support was implemented at the other end (E1) as illustrated in Fig. 15. The software utilized for modeling is ABAQUS. The mesh for this calculation comprises 25540 quadratic brick elements of type C3D20.

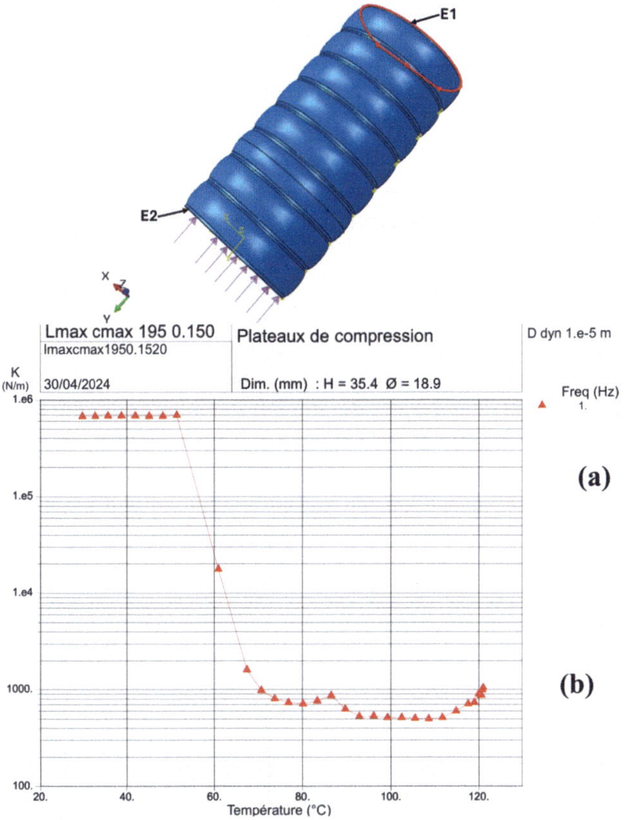

(a)

(b)

Fig. 15. The boundary conditions and the test forces determined by the DMA applied to the fluid pipe during compression.

5.3.3 Results

The compression experiment of specimen 2 lasted 800 s as indicated by the DMA results. Figure 16 presents two curves: curve 1 represents the simulated displacement of the specimen while curve 2 represents the experimental displacement. It is interesting to note that the simulated curve (1) has a similar profile to the experimental curve (2) over 800 s. The specimen reaches its maximum displacement in less than 300 s. and then this displacement decreases and stabilizes until the end of the test. Figure 17 shows the displacement of the specimen after compression modeling lasting 800 s. It is observed that the most significant displacement occurs radially along the U3 axis. The analysis of the results reveals several distinct phases. Initially, at 20 s the U_3 displacement is 9.4×10^{-4} mm. indicating a rapid response of the specimen to the load. This displacement gradually increases to 1.0×10^{-3} mm at 110 s. suggesting continuous deformation. At 225 s, the displacement slightly decreases to 9.9×10^{-4} mm. which may indicate a readjustment of the specimen. Between 110 and 225 s, a phase of structural readjustment seems to occur. At the halfway point, at 350 s, the displacement significantly decreases to 7.5×10^{-5} mm indicating stabilization of the deformation. This stabilization is confirmed at 700 s where the displacement is 7.7×10^{-5} mm close to the value at 350 s. The simulations show a concentration of deformation in the thinner part of the specimen consistent with theoretical expectations. The deformations are very small in value which is confirmed by the actual image thus validating the simulation model. In conclusion, the specimen deforms significantly at the beginning of the compression before reaching a stable state confirming the effectiveness of the compression model used in ABAQUS to predict the behavior of the specimen. These observations suggest that the compression modeling performed corresponds well to the experimental results.

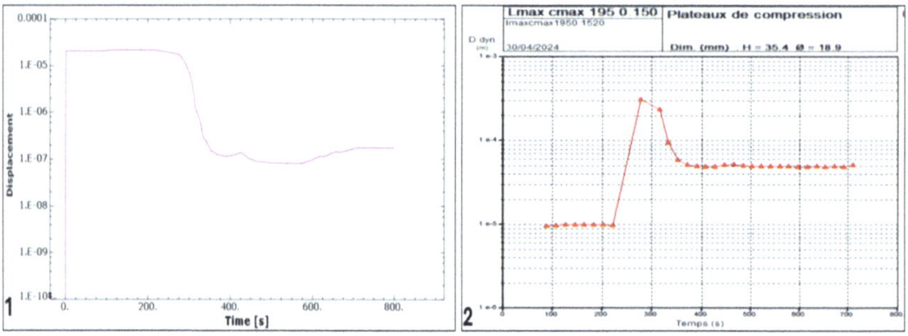

Fig. 16. The shape of the displacement curve during the compression test: Curve N°1 represents the simulated displacement of the specimen. Curve N° 2: represents the experimental displacement.

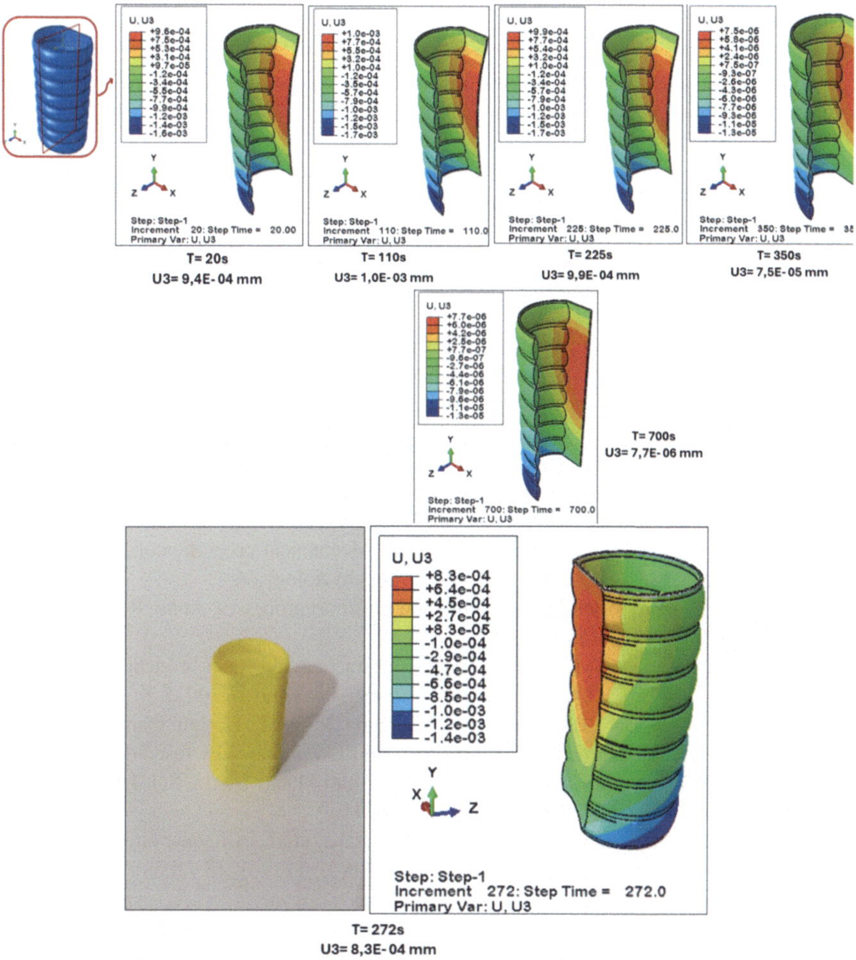

Fig. 17. Specimen Displacement in ABAQUS Compression Modeling over 800 s.

6 Conclusion

In conclusion, this study has successfully combined numerical simulations and experimental approaches to evaluate the behavior of conduct inspired by the human trachea. The simulation results provided valuable insights into the conduct's displacement under simulated respiratory conditions. Additionally, the mechanical characterizations of the 13 samples manufactured via 3D printing allowed exploration of the influence of machine parameter combinations on conduct mechanical properties. Furthermore, the numerical compression modeling enabled obtaining an accurate description of the conduct's behavior by modifying its geometric dimensions or the applied compression efforts. This project opens exciting prospects in the field of bioengineering and could potentially lead to significant advancements in both medical and industrial domains.

References

1. Antonio, F., Cesare, P., Ceretti, E.: Anti-migration enhanced tracheal stent design rapid manufacturing and experimental tests. Rapid Prototyp. J. **22**(1), 178–188 (2016). https://doi.org/10.1108/rpj-06-2014-0072
2. Melgoza, E.L., Sereno, L., Rosell, A., Ciurana, J.: An integrated parameterized tool for designing a customized tracheal stent. Comput. Aided Des. **44**(12), 1173–1181 (2012). https://doi.org/10.1016/j.cad.2012.07.004
3. IEEE Engineering in Medicine and Biology: Annual International Conference of the IEEE Engineering in Medicine and Biology Society, Osaka, Japan, 3-7 July 2013. https://doi.org/10.1109/EMBC.2013.6610978
4. Freitag, L., Gördes, M., Zarogoulidis, P., Darwiche, K., Franzen, D., Funke, F., Hohenforst-schmidt, W., Dutau, H.: Towards individualized tracheobronchial stents: technical, practical and legal considerations. Respiration. **94**(5), 442–456 (2017). https://doi.org/10.1159/000479164
5. Atzeni, C., Lesma, G., Dubini, G., Masi, M., Rossi, F., Bianchi, E.: Computational fluid dynamic models as tools to predict aerosol distribution in tracheobronchial airways. Sci. Rep. **11**(1), 1109 (2021). https://doi.org/10.1038/s41598-020-80241-0
6. Rains, J.K., Bert, J.L., Roberts, C.R., Pare, P.D.: Mechanical properties of human tracheal cartilage. J. Appl. Physiol. **72**, 19 (1992) http://jap.physiology.org
7. Grotberg, J.B.: Respiratory fluid mechanics and transport processes. Annu. Rev. Fluid Mech. **3**, 421 (2001) www.annualreviews.org
8. Holden, W.S., Ardran, G.M.: Movements of the trachea and main bronchi. Br. J. Radiol. **30**(350), 267–272 (1957). https://doi.org/10.1016/S0368-2242(57)80027-X
9. Ismail, R., Fitriyana, D.F., Bayuseno, A.P., Munanda, R., et al.: Design, manufacturing, and characterization of biodegradable bone screw from PLA prepared by fused deposition modelling (FDM) 3D printing technique. J. Adv. Res. Fluid Mech. Therm. Sci. **103**(2), 205–215 (2023). https://doi.org/10.37934/arfmts.103.2.205215
10. Taib, N.A.A.B., Rahman, M.R., Huda, D., Kuok, K.K., et al.: A review on poly lactic acid (PLA) as a biodegradable polymer. Polym. Bull. **80**(2), 1179–1213 (2023). https://doi.org/10.1007/s00289-022-04160-y
11. Morales, M.A., Maranon, A., Hernandez, C., Michaud, V., Porras, A.: Colombian sustainability perspective on fused deposition modeling technology: opportunity to develop recycled and biobased 3D printing filaments. Polymers (Basel). **15**(3), 528 (2023). https://doi.org/10.3390/polym15030528
12. Sindeeva, O.A., Prikhozhdenko, E.S., Schurov, I., et al.: Patterned drug-eluting coatings for tracheal stents based on PLA, PLGA, and PCL for the granulation formation reduction: in vivo studies. Pharmaceutics. **13**(9), 1437 (2021). https://doi.org/10.3390/pharmaceutics13091437
13. Bagnoli, P., Acocella, F., Di Giancamillo, M., Fumero, R., Costantino, M.L.: Finite element analysis of the mechanical behavior of preterm lamb tracheal bifurcation during total liquid ventilation. J. Biomech. **46**(3), 462–469 (2013). https://doi.org/10.1016/j.jbiomech.2012.10.036
14. Trabelsi, O., Del Palomar, A.P., López-Villalobos, J.L., Ginel, A., Doblare, M.: Experimental characterization and constitutive modeling of the mechanical behavior of the human trachea. Med. Eng. Phys. **32**(1), 76–82 (2010). https://doi.org/10.1016/j.medengphy.2009.10.010

15. Malve, M., Del Palomar, A.P., López-Villalobos, J.L., Ginel, A., Doblare, M.: FSI analysis of the coughing mechanism in a human trachea. Ann. Biomed. Eng. **38**(4), 1556–1565 (2010). https://doi.org/10.1007/s10439-010-9951-3
16. Del Palomar, A.P., Trabelsi, O., Men, A., López-Villalobos, J.L., Doblare, M., Ginel, A.: Patient-specific models of human trachea to predict mechanical consequences of endoprosthesis implantation. Philos. Trans. R. Soc. A Math. Phys. Eng. Sci. **368**(1921), 2881–2896 (2010). https://doi.org/10.1098/rsta.2010.0092

Experimental and Numerical Analysis of the Mechanical Characteristics of 3D Printed ABS Specimens Produced by the FDM Process

Shiri Seddik[1,2](✉), Boulahem Khaled[1,3], and Habouba Mohamed Hichem[2]

[1] National Engineering School of Bizerte (ENIB), Carthage University, 7035 Bizerte, Tunisia
seddik.shiri@enib.ucar.tn, khaled.boulahem@enit.utm.tn

[2] Mechanics, Production and Energetics Laboratory (LMPE), ENSIT, 1008 Tunis, Tunisia
hichem.habouba@akwel-automotive.com

[3] Laboratory of Applied Mechanics and Engineering (LAMI), ENIT, 1002 Tunis, Tunisia

Abstract. This paper presents an experimental and numerical analysis of the mechanical properties of ABS samples produced through 3D printing using the FDM (Fused Deposition Modeling) process. Three types of mechanical tests — tensile, flexural, and notched Charpy impact tests — were performed on samples manufactured with identical production parameters to evaluate their behavior under various loading conditions. In parallel, numerical simulations of these tests were conducted using the LS-DYNA computational code to accurately replicate the experimental conditions. The results reveal a significant agreement between the experimental data and the numerical simulations.

Keywords: ABS · FDM · mechanical properties · experimental tests · numerical simulations

1 Introduction

Additive manufacturing, also known as 3D printing, has revolutionized many industrial sectors by offering an innovative and flexible method for producing complex and customized parts. Among the various additive manufacturing technologies, the Fused Deposition Modeling (FDM) process is widely used for its ease of use and versatility [7, 8]. The application of FDM in producing parts from Acrylonitrile Butadiene Styrene (ABS), a polymer widely used in various industrial fields, has garnered particular interest due to its numerous potential applications.

A thorough understanding of the mechanical properties of ABS parts produced by the FDM process is essential to ensure their quality and performance. This study combines an experimental and numerical approach to analyze the mechanical characteristics of these parts. The paper is structured into three main parts: a first part for the analysis of experimental tests, a second part for the numerical study, and finally, the conclusions and perspectives.

M. Soula et al. (Eds.): AIAM 2024, LNME, pp. 44–53, 2025.
https://doi.org/10.1007/978-3-031-94889-3_4

Existing literature on this topic highlights several experimental and numerical studies that have contributed to improving our understanding of the performance of ABS parts manufactured by FDM [1, 2].

Kolekar et al. [3]'s study investigates the impact of process parameters on the impact strength and hardness of FDM-printed ABS parts. Using a 3D printer, test specimens were produced considering various printing factors such as orientation, pattern, and infill density. The results highlight correlations between infill density, structure, printing orientation, and the impact strength and Shore D hardness of FDM-printed ABS items. Sahoo et al. [4]'s study investigates the impact of mesostructural characteristics on the tensile strength of ABS specimens produced via FDM. Their experimental analysis reveals that specific processing parameters, such as layer thickness, raster angle, and build orientation, significantly influence the mesostructure and subsequently affect the mechanical properties.

Numerical studies employ advanced simulation tools to anticipate the intricate dynamics of FDM processes, enabling the modeling of mechanical responses of parts across diverse loading and stress scenarios. Al Rashid and Koç [5] conducted a comprehensive review of state-of-the-art scientific achievements in numerical simulations for the FDM process of polymers and their composites. Their study delves into the physical phenomena involved in this process and highlights its multi-level complexity. They also discuss various research efforts and numerical simulation techniques reported in the literature for FDM.

The literature review [1–5] also highlights the challenges and opportunities in the field of additive manufacturing, emphasizing the growing importance of an integrated approach combining experimental and numerical studies for a comprehensive characterization of the mechanical properties of ABS parts produced by FDM. This study aims to contribute to this knowledge base by offering an in-depth analysis of the mechanical characteristics of 3D-printed ABS samples produced by the FDM process.

2 Experimental Characterization

A filament made of Acrylonitrile Butadiene Styrene (ABS), with mechanical properties specified in Table 1, was utilized for fabricating specimens using the FDM process. The samples for different tests were 3D printed according to ISO 527, ISO 178, and ISO 179 standards for tensile, flexural, and Charpy impact tests, respectively (refer to Fig. 1 for specimen dimensions (mm)).

All samples were printed using a VOLUMIC STREAM30 PRO MK2 printer under the following conditions: 100% infill, printing speed of 60 mm/s, nozzle diameter of 0.4 mm, extrusion temperature of 235 °C, fill orientation at 0°, and layer height of 0.2 mm. Five specimens were manufactured for each test type, and Table 2 outlines the fixed printing parameters for all specimens.

2.1 Tensile Test

Tensile tests on the five specimens obtained by FDM were conducted using an "Ibertest TESTCOM" tensile testing machine (see Fig. 2).

Table 1. Mechanical properties of ABS filament.

ABS Properties	Values	Units
Filament diameter	1,75	Mm
Density	1,04	g/cm^3
Tensile strength	43	MPa
Impact strength	58	KJ/m^2
Printing temperature	220–250	°C
Bed temperature	95–110	°C

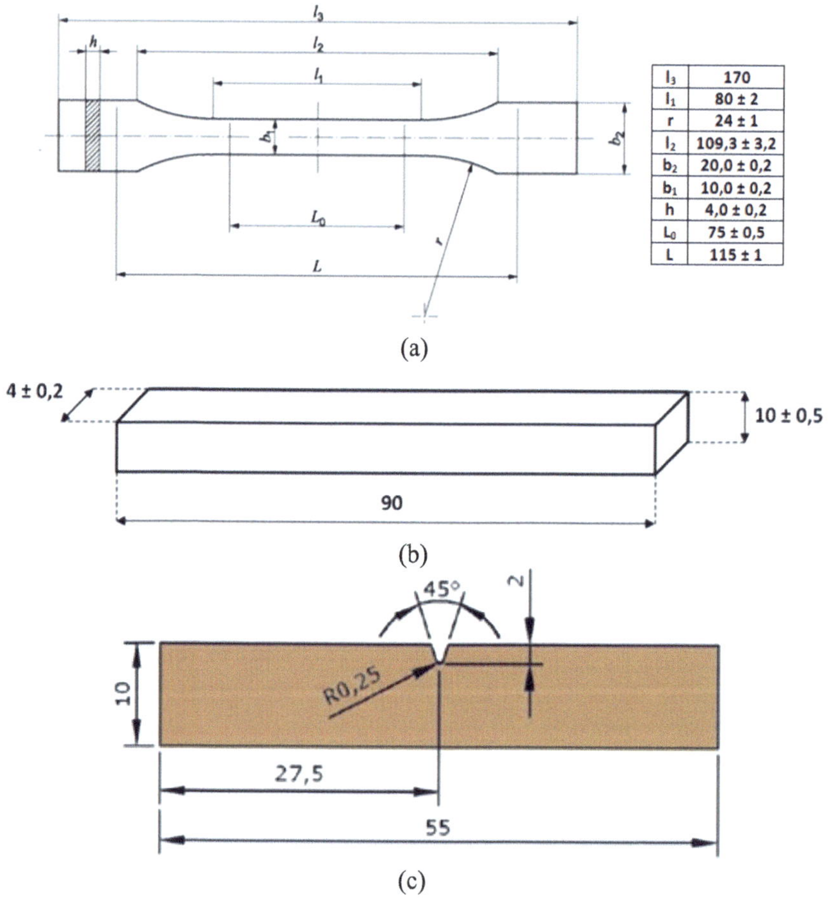

Fig. 1. Dimensions of specimens for (a) Tensile, (b) Flexural, and (c) Charpy impact tests.

Table 2. Printing parameters for specimen fabrication.

Fixed Factors	Values	Units
Filament diameter	1.75	Mm
Nozzle diameter	0,4	Mm
Number of outline layers	3	–
Outline overlap	0,6	Mm
Fill density	100	%
Overlap	15	%
Fill speed	60	Mm/s
Fill pattern	Rectilinear	–
Build plate temperature	60	°C
Top solid layers	5	–
Bottom solid layers	5	–
Build orientation	Flat in X direction	–

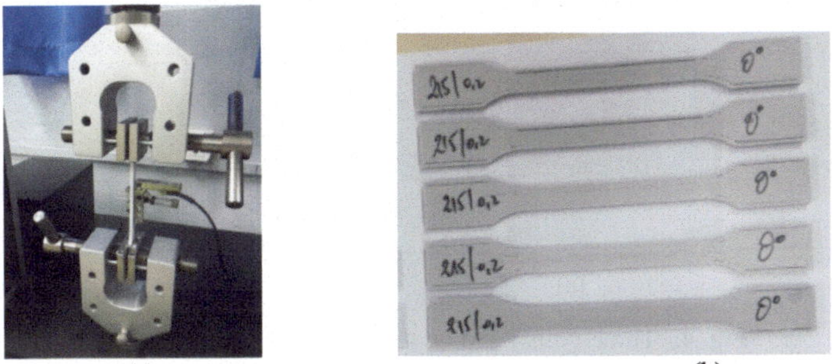

(a) (b)

Fig. 2. Tensile test (a) setup (b) specimens.

The loading was applied by controlling the crosshead displacement at a speed of 5 mm/min and an ambient temperature of 25 °C. Figure 3 shows the force-displacement curve of the different tests conducted, indicating good repeatability of the tests and allowing us to determine the necessary mechanical properties to subsequently feed into numerical test models.

2.2 Three-Point Bending Test

Three-point bending test is performed to assess the flexibility and strength of ABS specimens obtained by FDM. During this test, a specimen is placed on two fixed supports spaced 35 mm apart, while a force is applied to the center of the specimen until it bends, with the assistance of mobile support advancing at a speed of 1 mm/min at room temperature of 25 °C (see Fig. 4).

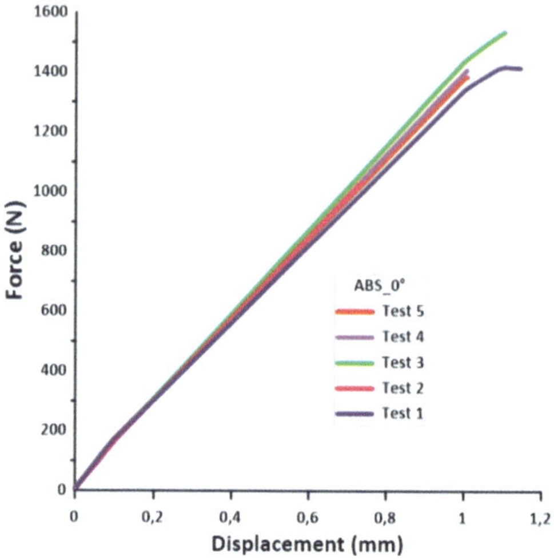

Fig. 3. Force-displacement curves of tensile Tests.

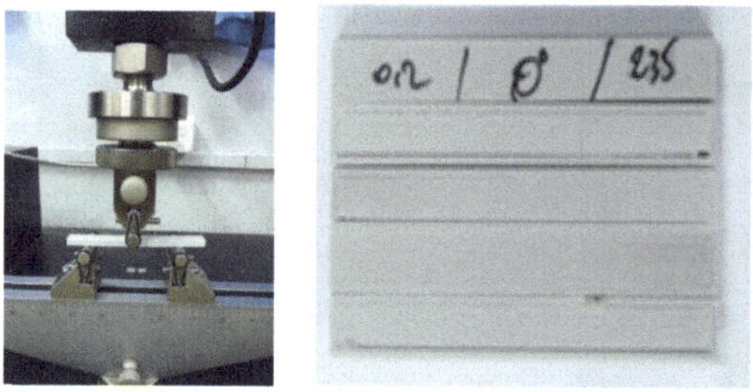

(a) (b)

Fig. 4. Three-point bending test (a) setup (b) specimens.

Figure 5 illustrates the force-displacement curve of the various tests conducted, demonstrating good reproducibility of results.

2.3 Charpy Impact Test

The Charpy notched impact test was used to evaluate the resilience of ABS specimens obtained by FDM (see Fig. 6). During this test, the notched specimen of a specific shape is subjected to a rapid and punctual impact at room temperature. The amount of energy absorbed by the specimen upon rupture is measured to assess its resistance to impact fracture. An average impact energy of 118.34 KJ/m^2 was recorded for the five tested specimens.

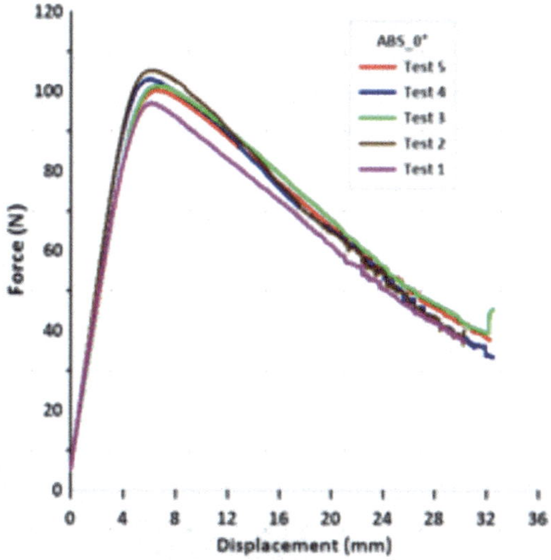

Fig. 5. Force-displacement curves.

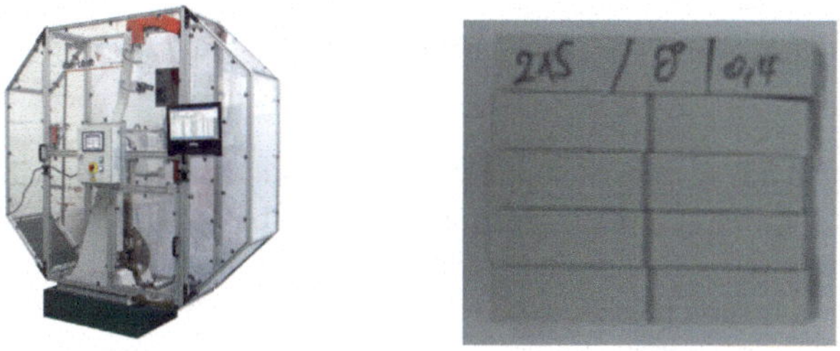

(a) (b)

Fig. 6. Charpy impact test test (a) setup (b) specimens.

3 Numerical Study

To simulate the experimental tests on ABS specimens obtained by FDM as described in the previous section, FE models were developed using the LS-DYNA computational code. Initially, a finite element model taking into account the symmetry of the tensile test was established to validate the elastoplastic behavior of ABS using the *MAT_003 (PLASTIC_KINEMATIC) material law [6]. The parameters of material model 03 used in LS-Dyna are presented in Table 3 [9]. A plastic deformation rupture criterion, utilizing the erosion technique, was employed for modeling rupture during numerical simulations. This approach involves removing mesh elements that exceed a critical value of equivalent plastic deformation (FS = 5%). Subsequently, two FE models were developed to simulate three-point bending tests and Charpy impact tests. The dies were defined as rigid bodies, and the contact between specimens and rigid dies was treated using the penalty method.

The numerical models were set up using the LS-Prepost tool (see Fig. 7). The specimens were discretized using solid hexahedral elements with full integration and a small element size to accurately represent flexural behavior. For the Charpy impact test, the area near the impact point was meshed with finer elements, with an aspect ratio close to unity.

Table 3. Plastic Kinematic Material Model used in Numerical Simulations.

Parameters	Density (kg/m^3)	Elastic Modulus (GPa)	Poisson's Ratio	Yield Strength (GPa)	FS	Tangent modulus (GPa)
Value	1040	1.967	0.3	0.03385	0.05	0

Fig. 7. Finite element models for simulating conducted tests.

One of the main advantages of simulation is to provide data that are difficult to obtain through experimentation. Figure 8 illustrates the distribution of Von Mises stress in the specimens for the three conducted test models after deformations.

The simulation results provide a better understanding of the kinematics of the phenomenon and allow estimating the forces involved in the structure. The simulated displacements and forces are compared to the experimental results for the tensile and three-point bending tests and are presented in Fig. 9. The simulations of both tests show a strong agreement with the real results obtained from the experiments. For the Charpy impact test, the numerical impact energy is 120 KJ/m^2, which is in good agreement with the experimental value (118.34 KJ/m^2).

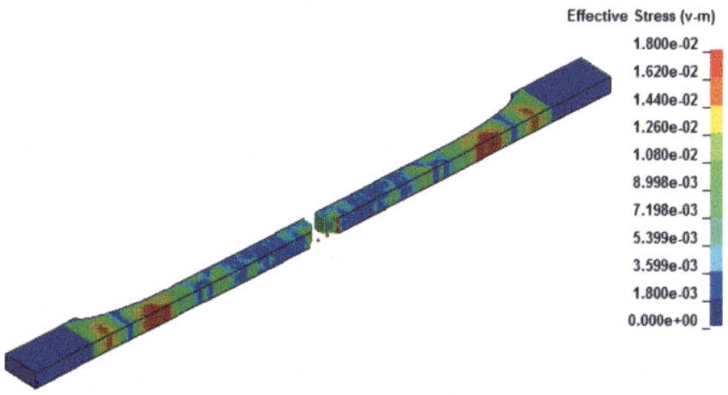

(a)

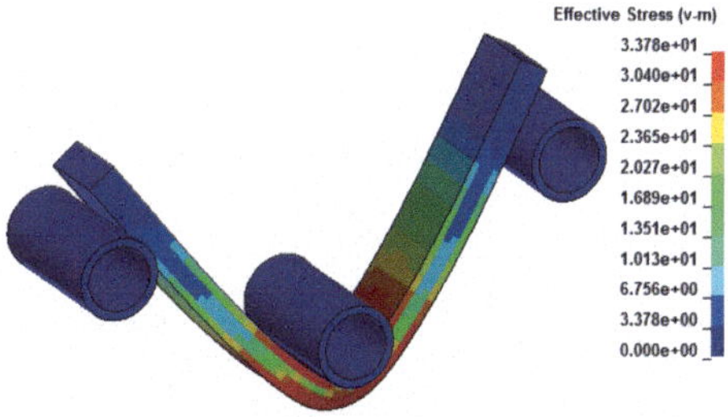

(b)

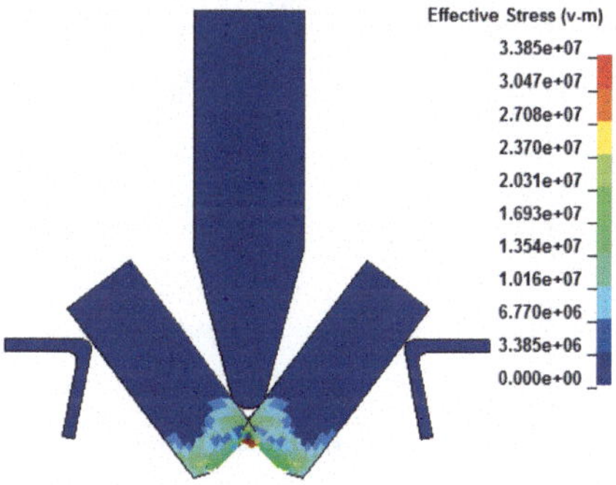

(c)

Fig. 8. Distribution of Von Mises stress (MPa) in the specimens for the (a) tensile test, (b) three-point bending test, and (c) Charpy impact test.

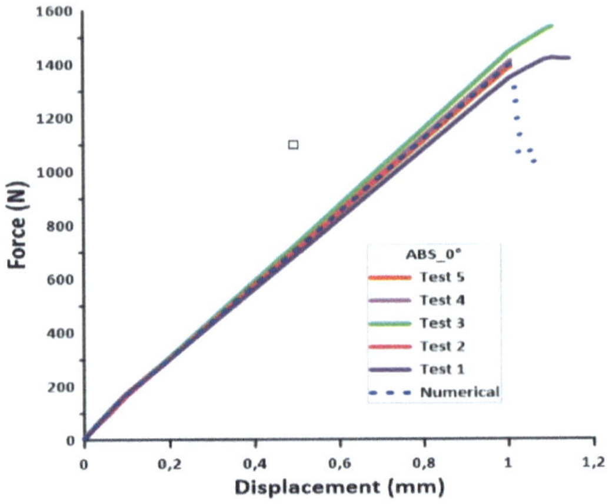

(a)

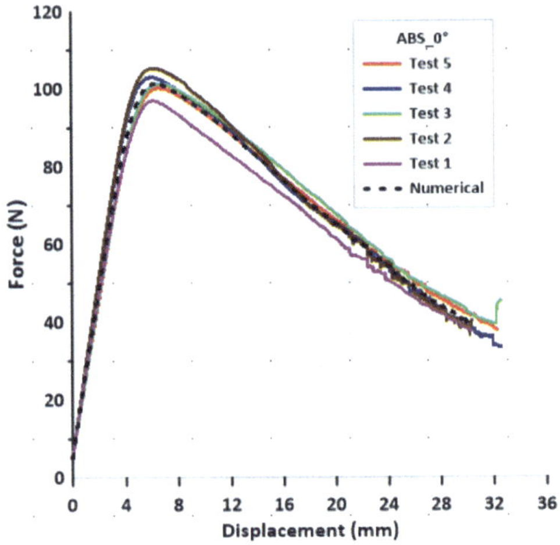

(b)

Fig. 9. Comparison of simulated and experimental stress-strain curves for the (a) tensile test, and (b) three-point bending test.

4 Conclusion

Additive manufacturing raises concerns regarding the quality and reliability of produced parts, highlighting the importance of understanding the mechanical properties of the materials used. Experimental tests including tensile, flexural, and Charpy impact tests were conducted on ABS samples manufactured with identical fabrication parameters. Concurrently, numerical simulations of these tests were performed using the LS-DYNA code.

The results demonstrated a satisfactory correspondence between experimental and simulated numerical results. With this in mind, we plan to redo the tests by varying the fabrication process parameters, which will allow us to better understand the influence of these parameters on the mechanical properties of the produced parts. Additionally, we intend to calibrate the corresponding numerical models to accurately reflect the various fabrication conditions.

These steps will contribute to a better understanding of the FDM process and enable us to optimize the topology of the produced structures, aiming for continuous improvement in the quality, reliability, and performance of parts manufactured through additive manufacturing.

References

1. Sandhu, H.S., Alam, A.S., Jeevan, K., Yadav, A.K., Sai, G., Korapram, V.M.: A review of the mechanical properties of 3D printed and electroplated ABS parts. In: AIP Conference Proceedings, vol. 2962. AIP Publishing (2024)
2. Sandhu, H.S., Raj, T.R., Venkatasubramaniam, L., Tiwari, P.: Optimization of process parameters for a better surface finish of ABS parts prepared by fused deposition modeling (FDM): a comprehensive review. Recent Trend Eng. Design Select Proc. ICAST. **2020**, 311–320 (2021)
3. Kolekar, A.B., Bhosale, S.M., Salunkhe, M.S., Raut, N.P.: Effect of process parameters on impact strength and hardness of FDM printed ABS parts. Int. J. Mater. Prod. Technol. **68**(1–2), 107–121 (2024)
4. Sahoo, S., Panja, S.C., Sarkar, D., Saha, R., Mandal, B.B.: Analysis of mesostructure characteristics and their influence on tensile strength of ABS specimens manufactured through fused deposition modeling. Int. J. Adv. Manuf. Technol. **132**(1), 349–363 (2024)
5. Rashid, A.A., Koç, M.: Fused filament fabrication process: a review of numerical simulation techniques. Polymers. **13**(20), 3534 (2021)
6. Hallquist, J.O.: LS-DYNA theory manual. Livermore Softw. Techn. Corp. **3**, 25–31 (2006)
7. Islam, M.A., Mobarak, M.H., Rimon, M.I.H., Al Mahmud, M.Z., Ghosh, J., Ahmed, M.M.S., Hossain, N.: Additive manufacturing in polymer research: Advances, synthesis, and applications. Polym. Test. **108364** (2024)
8. Radhika, C., Shanmugam, R., Ramoni, M., Gnanavel, B.K.: A review on additive manufacturing for aerospace application. Mater. Res. Exp. (2024)
9. Sabah, F., Wahid, A., Kartouni, A., Chakir, H., Elghorba, M.: Failure analysis of acrylonitrile butadiene styrene (ABS) materials and damage modeling by fracture. Int. J. Performabil. Eng. **15**(9), 2285 (2019)

Material Grades for AM Applications

Exploration of 3D Printed Anti-trichiral Meta-materials Based on PLA/Wood Flour Bio-composites Using Dynamic Mechanical Analysis and Artificial Neural Networks (ANN) Technique

Mondher Nasri[1(✉)], Mohamed Toumi Nasri[1,2], Badreddine Larbi[1,2], Houda Khaterchi[3], Salah Hamzi[4], Ali Zghal[1], Olivier Dalverny[5], and Amir Ben Rhouma[6]

[1] University of Tunis, ENSIT, LR18ES01, LMPE, 5 Avenue Hussein, BP, 56, 1008 Bâb Manara, Tunisia
mondher.nasri@yahoo.fr
[2] University of Carthage, ENIB, Menzel Abderrahman University Campus, 7035 Bizerte, Tunisia
[3] University of Monastir, ENIM LGM, Avenue Ibn El Jazzar, 5019 Monastir, Tunisia
[4] Higher Institute of Technology Studies ISET, University of Monastir, BP 68 Av. Hadj Ali Soua, 5070 Ksar Hellal, Tunisia
[5] University of Toulouse, UTTOP, ENIT, LGP, 47 Avenue d'Azereix, 65016 Tarbes, France
[6] University of Tunis, ENSIT, LR 99ES05, LMMP, 5 Avenue Hussein, BP, 56, 1008 Bâb Manara, Tunisia

Abstract. This paper presents the dynamic mechanical characterization of a honeycomb-shaped meta-material . This meta-material is developed using 3D printing techniques, using a composite based on bio-sourced and biodegradable materials, in this case, polylactic acid reinforced with wood flour fibers (PLA/wood flour). The study focused on analyzing the effects of 3D printing parameters such as printing velocity, nozzle temperature, and density, highlighting its anti-trichiral performance. The influence of the three factors is studied by using the Dynamic Mechanical Analyzer (DMA). The storage modulus, loss modulus, and tangent delta tests were carried out on this architecture as a function of these printing parameters. A Dynamic variation of Young's modulus as a function of temperature is presented. Artificial neural network (ANN) is used for training and optimization purposes.

Keywords: Meta-Materials · 3D Printed · Sensitivity study · storage modulus · loss modulus · dynamic mechanical analysis

1 Introduction

The honeycomb motifs present in nature have inspired the design of lightweight yet extremely rigid structures, suitable for a variety of industrial applications such as architecture, transportation, mechanical engineering, aerospace, chemical engineering,

M. Soula et al. (Eds.): AIAM 2024, LNME, pp. 57–68, 2025.
https://doi.org/10.1007/978-3-031-94889-3_5

biomedicine, and many others [1]. This structure exhibits high flexural strength despite being low-density and also demonstrates a significant capacity to absorb energy, particularly in the context of sandwich composite structures [2, 3]. Several studies have been carried out to improve the mechanical properties of these honeycomb structures by exploring optimal designs for the cell structure. The geometric parameters of the unit cell can be adjusted to obtain specifically targeted mechanical properties [4]. The technology of additive manufacturing (AM) by fused filament deposition, also known by the acronym FFF (Fused Filament Fabrication), has enjoyed increasing diffusion and popularity since 2008 [5]. A variety of polymers, such as PCL (poly-capro-lactone), ABS (acrylonitrile butadiene styrene), and PLA (poly lactic acid), have been extensively tested, the latter being particularly widespread today. The widespread adoption of this technology, especially within user communities in collaborative spaces such as FabLabs [6], has been a driving force behind continuous improvement. The open-source software industry has also been a key accelerator of the spread of this technology through the sharing of experience [7]. According to work by Yedige Tlegenov and al [8]., lower nozzle temperature tends to cause more clogging than higher nozzle temperature, while K. G. Jaya Christian and al [9]. note an increase in tensile strength and a decrease in print speed. In addition, research by Y. Wang and his team has shown that a higher screen angle reduces tensile strength, with all three printing parameters having a significant impact on the material's mechanical performance. In our study, we opted to use IdeaMaker software, which generates a file in STL format for specimen design. We used a RAISE3D Pro2 Plus printer to produce specimens with a pure PLA and PLA/wood flour architectural structure. These specimens were designed from a unit cell with the geometric dimensions defined in Table 1. Experimental tests were carried out on these specimens to assess their mechanical behavior as a function of machine parameters. Mechanical properties were determined in dynamic mode by dynamic mechanical analyzer DMA50. They therefore deduced that the selection of these parametric criteria in the additive manufacturing process could lead to satisfactory performance in terms of quality and reliability, by minimizing profile error. This research aims to use a parametric experiment) to analyze the effect of FFF parameters on the performance of printed specimens produced, and then to evaluate it using the Artificial Neural Network (ANN) approach. Indeed, dynamic mechanical analysis (DMA) is an extensively employed methodology for assessing the viscoelastic attributes of materials. Evaluating these properties across varying strain rates and temperatures necessitates a comprehensive experimental undertaking comprising numerous tensile tests. In DMA assessments, the in-phase and out-of-phase correlations between stress and strain, under sinusoidal loading, serve to establish the storage (E′) and loss (E″) moduli, respectively. DMA has found application in delineating the characteristics of PLA/Wood composites as well as Pure PLA.

Table 1. Parameters of the composite architectural materials PLA/Wood

l (mm)	H(mm)	t(mm)	e(mm)	r(mm)
4.17	25	5	0.6	1.2

2 Materials and Methods

2.1 Manufacturing a Specimen Using 3D Printing with PLA/Wood Flour

The material used in this study is polylactic acid (PLA) reinforced with natural wood flour fibers, with a density of 1.19 g/cm^3 and a Young's modulus of 35.5 MPa. Wood flour filament is an entirely organic product. It also requires a nozzle temperature of between 190 and 220 °C. Since wood filament is a composite material, it may be necessary to adjust certain parameters of the slicing machine to ensure reliable printing. It is essential to note that nozzles smaller than 0.4 mm are not recommended. Otherwise, the wood fibers in the filament may get caught in the nozzle and burn, leading to clogging. Also, remove the wood filament from the print head as soon as possible after printing is complete. If you neglect this step and the remaining filament remains in the nozzle, it can become excessively hot, generating a burnt odor and clogging the nozzle. Figure 1a shows the geometric configuration of the anti-tri chiral honeycomb structure, illustrating its various geometric parameters, namely L (length of the specimen), r (radius of the cylindrical components), and e (thickness of the cell wall). The cell plane is defined by the axes (X, Y), while thickness is represented along the axis (Z). The dimensions H (specimen width in Y direction) and t (thickness) complete the description. The geometric parameters are summarized in Table 1. Figure 1b shows the RAISE3D Pro2 Plus printer used in this study. This printer belongs to a new generation of robust, professional 3D printers, with capabilities that enable the production of precise, high-quality components.

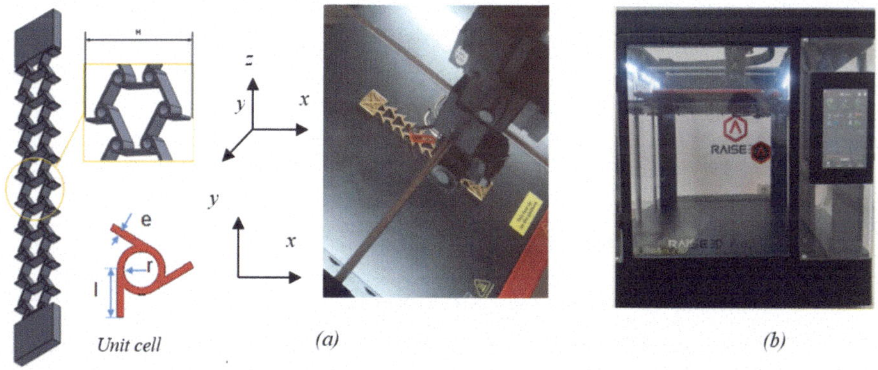

Fig. 1. Specimen preparation: (a) anti-tri chiral honeycomb structure, (b) 3D printing process

The 3D model was developed using SolidWorks software, in compliance with ASTM standards for tensile dimensions and test specifications [10, 11]. Next, the computer-aided design (CAD) was converted into an (.STL) file for processing by the 3D cutting software. Some variations in manufacturing time were noted during the preparation of the test samples, depending on the different construction parameters. The manufacture of PLA/wood flour test specimens using 3D printing involves several important steps, including variable parameters such as filling speed, density, and printing temperature. The steps involved are described below:

- Preparing the 3D model: First, we need to design the specimen model using 3D modeling software. This model must be precisely designed to include all the necessary features and dimensions.
- Choice of material: In our case, the main material is PLA (polylactic acid) combined with wood flour. This material is a natural biodegradable polymer filament commonly used in 3D printing.

Setting print parameters: Before we start printing, we need to adjust several key parameters:

- Fill speed: This refers to the speed at which the material is deposited to fill the inside of the test specimen. A higher filling speed can speed up the printing process, but can also affect the quality and strength of the part.
- Density: Specimen density can be controlled by adjusting the fill percentage of the internal structure. A higher fill percentage will increase the density of the part, but may also require more material and more printing time.
- Printing temperature: The extrusion temperature of the PLA/wood flour material must be optimized to ensure proper fusion of the layers and good adhesion between them.
- Preparing the 3D printer: Make sure your 3D printer is properly calibrated and ready for operation. Also, check that the printing plate is clean and properly leveled to ensure optimum adhesion of the first layer. There are two configurations of the test specimen, either vertical or horizontal. For the preparation of our test piece, we use the horizontal configuration, which requires a printing time of between 56 minutes and 1 hour, whereas the first vertical configuration requires an excessively long printing time of around 6 hours for each test piece.
- Start printing: Once everything is ready, start the printing process by sending the 3D model file to the printer. During printing, keep a close eye on the process to detect any possible problems, such as layer mismatch or poor adhesion.

The tensile specimens were printed following Table 2 (see Fig. 2).

Table 2. Factors used for Experience.

Operating Parameters	Value		
Filling speed (mm/s)	40	60	80
Infill density (%)	5	15	25
Issue temperature DMA (°C)	45	50	60
Temperature of printer (°C)	200		

2.2 Experimental Protocol

The Experimental protocol used to conduct this study is defined by Four parameters. Table 3 shows the factors and the values used.

Fig. 2. Preparation of specimens with variations in the three parameters

Table 3. Printing and issues parameters

	Type of sample	Infill Density	Filling velocity mm/s	Issue Temperature (°C)	E(MPa)	E′(MPa)	E″(MPa)
1	1	5	40	45	6.5	6	0.55
2		5	40	50	5	4	0.85
3		5	40	55	1.5	2	0.9
4		5	40	60	0.4	1	0.35
5	1	25	80	45	5.6	5	0.3
6		25	80	50	5	4.9	0.5
7		25	80	55	1.2	2	0.45
8		25	80	60	0.3	0.078	0.3
9	2	15	40	45	7.9	7.9	0.25
10		15	40	50	7.5	7	0.27
11		15	40	55	6.1	5.1	0.55
12		15	40	60	1	0.3	0.29
13	2	5	60	45	8.9	9	0.24
14		5	60	50	8.5	8	0.29
15		5	60	55	6	7	0.54
16		5	60	60	0.7	0.45	0.58

Young modulus E; Storage modulus E'; Loss modulus E"

2.3 Dynamic Mechanical Analysis DMA

A series of dynamic tests, in the honeycomb pattern, were carried out to determine the desired elastic components of the Meta-Material, which are then fed into the numerical simulation. The analysis aims to measure the viscoelastic properties of a material, the data can be acquired in terms of strain and stress, as these quantities are directly related to the storage and loss moduli of the material according to ASTM D638 [10, 11], taking into account various specimen manufacturing parameters such as speed, temperature, and density. The DMA equipment model Metravib DMA50 was operating at the frequency of 1 Hz, with a heating range of 3 °C/min, and temperature ranging from 20–120 °C. Figure 3 shows details of the set-up used for tensile testing of PLA/Wood specimens.

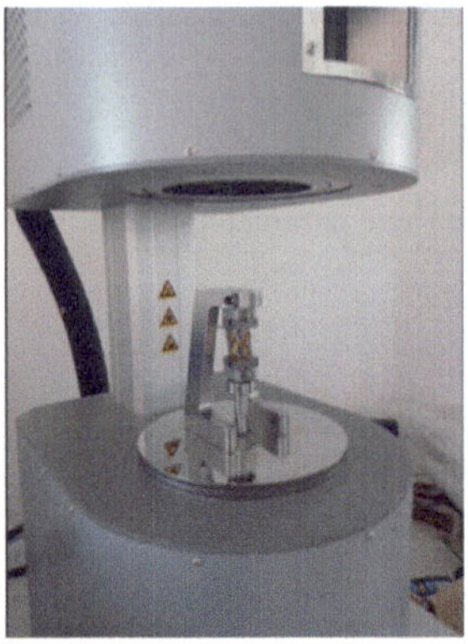

(b)

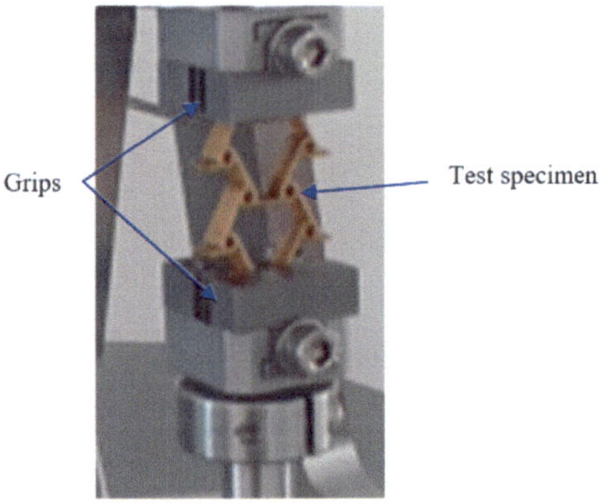

(a)

Fig. 3. (a) Experimental tensile test equipment (b) Tensile test specimen

3 Parametric Sensitivity Study: Results and Discussion

In this section, we present the definitions of the parameters to be examined during DMA characterization:

a- Storage Modulus E':

The storage modulus E' is a component of the complex modulus $E^* = E' + iE''$ measured during DMA. It represents the elastic portion of a material's viscoelastic response, indicating its stiffness and ability to resist deformation while retaining elastic energy. During DMA testing, the storage modulus E' is determined based on the frequency and amplitude of the stress applied to the material. It reflects the material's response to periodic stresses by measuring its ability to store and release elastic energy. A high storage modulus E' indicates a rigid material that resists deformation, while a low storage modulus suggests a more flexible and deformable material. Thus, E' is crucial for evaluating a material's mechanical properties and stability under different loading conditions.

b- Loss Modulus E":

The loss modulus is directly related to a material's damping capacity, which is its ability to absorb mechanical energy and dissipate it as heat. Materials with higher loss moduli have greater damping capacities, making them useful in applications requiring vibration damping or noise reduction.

c- Tangente δ:

In polymer science, tangent delta is a key parameter for characterizing polymer blends, copolymers, and composites. It provides insights into the molecular structure, chain mobility, and overall viscoelastic behavior of polymers, which are crucial for material design and processing. The peak value of the tangent delta often corresponds to the glass transition temperature (Tg) in polymers and other amorphous materials. The Tg represents a change in the material's viscoelastic properties, and the corresponding peak in the tangent delta signifies a transition from a glassy to a rubbery state.

The variation in Young's modulus as a function of transition temperature for the average of the five specimens is shown in Fig. 4, for each result of the biocomposite (PLA reinforced with wood flour) and pure PLA specimens. This first analysis shows the significant effect of the proportion of wood flour for the same parameters of the specimens in terms of density, filling speed, and cooling temperature (E1 and E4).

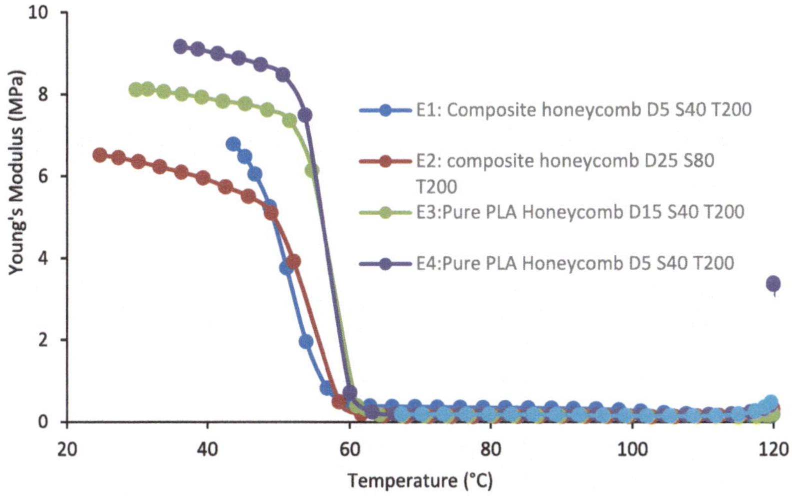

Fig. 4. Dynamic variation of Young's modulus as a function of temperature

The detailed experimental results for determining the storage modulus E' are presented in Fig. 5. The storage modulus for our samples is limited to 9 MPa. Figures 6 and 7 illustrate the effect of density and filling velocity on the transition temperature of the biocomposite. Indeed, for the PLA/Wood honeycomb D5 S40 T200, the transition temperature is 57 °C, whereas for PLA/Wood honeycomb D25 S80 T200, the transition temperature Tg is around 62 °C. Additionally, controlling the same pure PLA material reveals a significant effect of filling density and filling velocity on Tg. For Pure PLA honeycomb D15 S40 T200, the transition temperature is 61 °C, whereas for Pure PLA honeycomb D5 S60 T200, Tg is 60 °C (see Figs. 8 and 9).

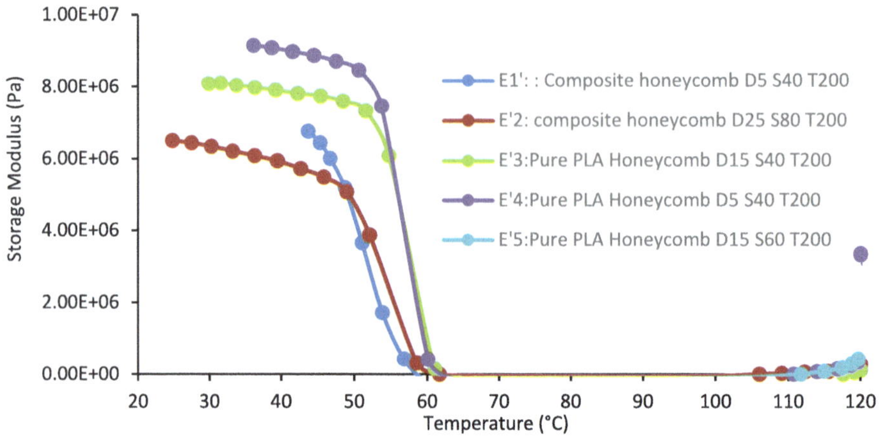

Fig. 5. Dynamic variation of Storage's modulus as a function of temperature.

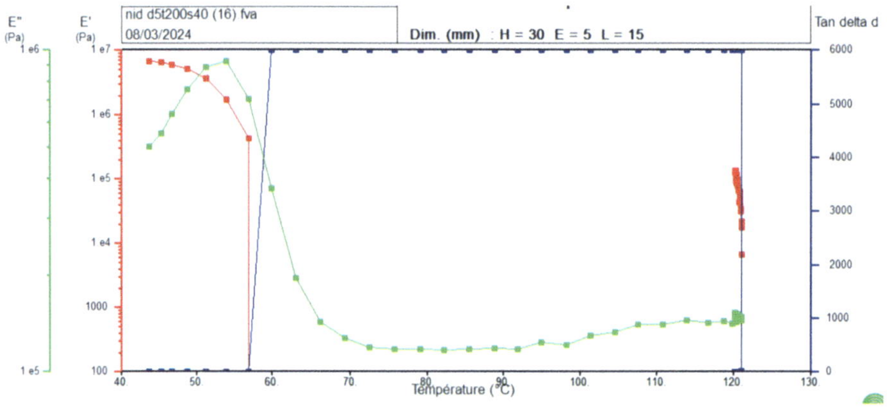

Fig. 6. PLA/Wood honeycomb D5 S40 T200

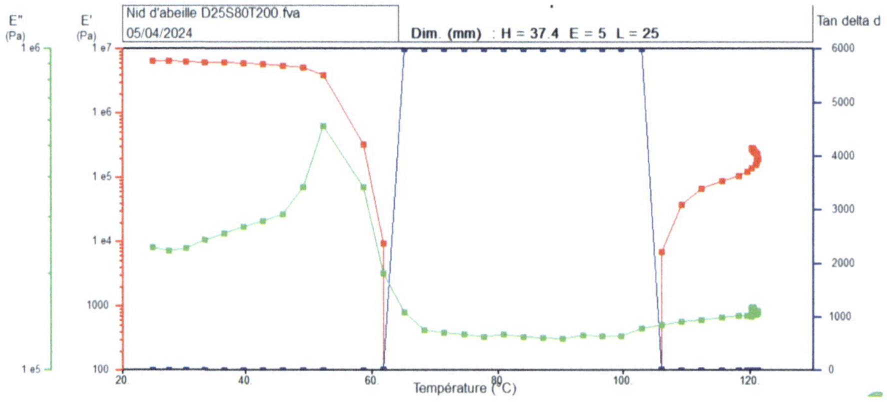

Fig. 7. PLA/Wood honeycomb D25 S80 T200

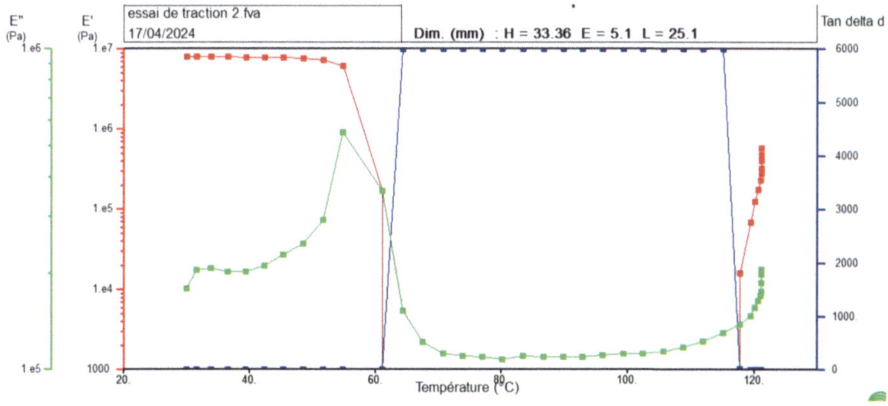

Fig. 8. Pure PLA honeycomb D15 S40 T200

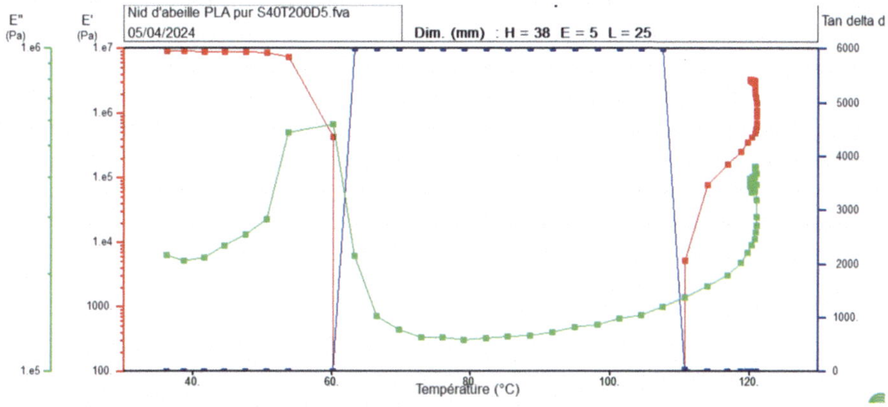

Fig. 9. Pure PLA honeycomb D5 S40 T200

4 Artificial Neural Network Implementation

An Artificial Neural Network (ANN) is a computational model inspired by the human brain's neural networks [12]. It comprises interconnected nodes (neurons) organized in layers. Input data is processed through these layers via weighted connections and activation functions, producing output predictions. ANNs are trained using labeled data and backpropagation, adjusting weights to minimize prediction errors. They're versatile for tasks like classification, regression, and pattern recognition, making them a fundamental tool in machine learning and AI (see Fig. 10).

In the artificial neural network (ANN) modeling code, the activation function used in the hidden layer of the neural network is the Rectified Linear Unit (ReLU) function. The ReLU activation function is defined as:

$$(x) = max(0, x) \quad (1)$$

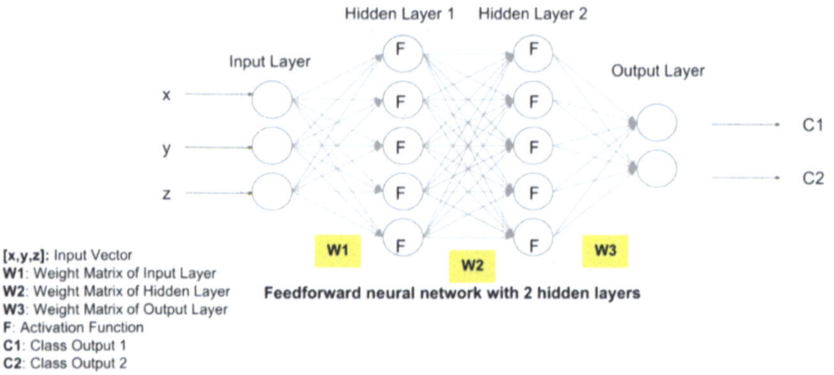

Fig. 10. Artificial Neural Network Modeling

In other words, it returns zero for any negative input and passes the positive input values directly. Mathematically, it's a piecewise linear function that introduces non-linearity to the neural network. This non-linearity is crucial for the network to learn complex relationships in the data and improve its ability to generalize to unseen examples. ReLU has become a popular choice as an activation function for hidden layers in deep learning models due to several advantages, including faster convergence during training (compared to other activation functions like sigmoid or tanh), reduced likelihood of vanishing gradients, and simplicity in computation (no exponentials or logarithms involved).

4.1 Prediction of Elastic Modulus

The best prediction of Elastic modulus by the ANN is illustrated in Fig. 11. The mean squared error (MSE) is a metric used to measure the average squared difference between the actual and predicted values in a regression problem. It is defined as the average of the squares of the differences between the predicted and actual values in Fig. 12.

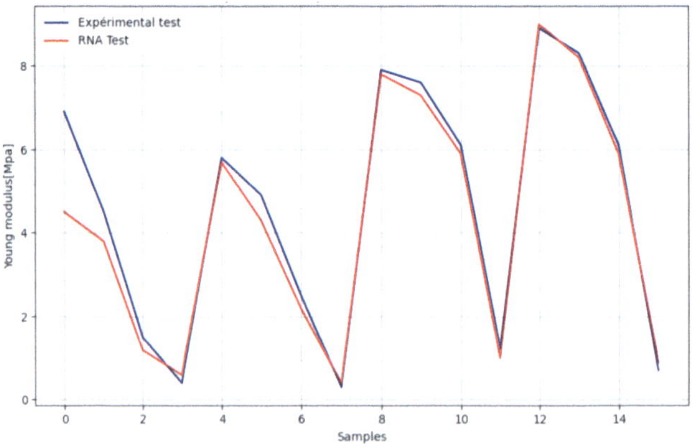

Fig. 11. Comparison of experimental data with RNA test

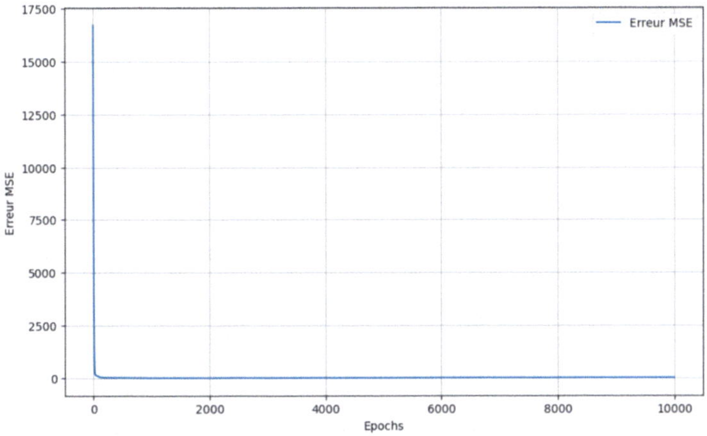

Fig. 12. MSE evolution with Epochs

The formula for MSE is:

$$MSE = \frac{1}{n}\sum_{i=1}^{n}(y_i - \overline{y_i})^2 \tag{2}$$

where:

- n: is the number of samples or data points.
- i: is the actual value for the -th sample.
- $\overline{y_i}$ is the predicted value for the -th sample.

5 Conclusion

In this article, a dynamic experimental dataset was designed to characterize a PLA/ Wood metamaterial and pure PLA. Studies on the influence of specimen fabrication parameters on the transition temperature, as well as on the effect of specimen fabrication parameters (density, filling speed, temperature) on both storage and loss moduli, were conducted. An approach based on artificial neural networks (ANN) is used to predict the Young's modulus of PLA/Wood. Dynamic mechanical analysis data are used as input, and a transformation is established to convert the storage modulus to an elastic modulus over a temperature range.

References

1. Hönig, A., Stronge, W.J.: In-plane dynamic crushing of honeycomb. Part I: crush band initiation and wave trapping. Int. J. Mech. Sci. **44**(8), 1665–1696 (2002)
2. Ju, J., Summers, J.D.: Compliant hexagonal periodic lattice structures having both high shear strength and high shear strain. Mater. Des. **32**(2), 512–524 (2011)
3. Liu, Y.X., Zhang, C.: The influence of cell micro-topology on the in-plane dynamic crushing of honeycombs. Int. J. Impact Eng. **36**(1), 98–109 (2009)

4. Yu, X., Zhou, J., Liang, H., Jiang, Z., Wu, L.: Mechanical metamaterials associated with stiffness, rigidity and compressibility: a brief review. Prog. Mater. Sci. **94**, 114–173 (2018)
5. Palmer, C., RepRap-The Replicating Rapid Prototyper (2009)
6. Lhoste, É., Barbier, M.: FABLABS. L'institutionnalisation de tiers-Lieux du «soft hacking». Revue d'anthropologie des connaissances. **10**(10–1) (2016)
7. Maréchal, P., Bernard, S., & Duflo, H.: Propriétés élastiques effectives d'éprouvettes issues d'une impression 3D, dans un contexte d'optimisation des propriétés–FreeCAD, PrusaSlicer, Scilab. In CNRIUT 2023 Saint Pierre-La Réunion (2023, June)
8. Tlegenov, Y., Hong, G.S., Lu, W.F.: Nozzle condition monitoring in 3D printing. Robot. Comput. Integr. Manuf. **54**, 45–55 (2018)
9. Christiyan, K.J., Chandrasekhar, U., Venkateswarlu, K.: A study on the influence of process parameters on the mechanical properties of 3D printed ABS composite. In: IOP Conference Series: Materials Science and Engineering, vol. 114, p. 012109. IOP Publishing (2016)
10. ASTM Subcommittee D20: 10 on Mechanical Properties. Standard Test Method for Tensile Properties of Plastics. American Society for Testing and Materials (1998)
11. Van Apeldoorn, J., Gilyén, A., Gribling, S., de Wolf, R.: Convex optimization using quantum oracles. Quantum. **4**, 220 (2020)
12. Khaterchi, H., Chamekh, A., BelHadjSalah, H.: Artificial neural network analysis for modeling fibril structure in bone. Int. J. Precis. Eng. Manuf. **16**, 581–587 (2015)

3D Printed Auxetic Structures Based on Re-entrant Hexagon Design for Biomedical Application

Nada Ben Hariz[1,2](✉), Atef Boulila[2], and Mahfoudh Ayadi[3]

[1] Mechanical, Production and Energy Laboratory (LMPE), National Engineering School of Tunis, University of Tunis El Manar, BP 37, Belvédère, 1200 Tunis, Tunisia
nada.benhariz@enit.utm.tn

[2] National Institute of Applied Sciences and Technology, University of Carthage, BP676, 1080 Tunis Cedex, Tunisia

[3] National Engineering School of Bizerte, University of Carthage, BP66, Menzel Abderrahmen, 7035 Bizerte, Tunisia

Abstract. Additive manufacturing, commonly known as 3D printing, has become a pivotal technology in biomedical applications, particularly in the fabrication of patient-specific medical devices. Despite significant advancements, ongoing research is focused on optimizing design methodologies, material properties, and other parameters, including infill density, raster angle, nozzle temperature, and build plate temperature, all of which significantly influence the quality and functionality of the final product. This study investigates the feasibility of employing an auxetic structure, specifically a re-entrant hexagonal design, in fabricating a prosthetic hand. The work encompasses two main phases: first, a systematic review of the existing literature, and second, the modeling and mechanical characterization of the re-entrant structure to assess its applicability in prosthetic design. This research represents a step forward in advancing the use of Fused Deposition Modeling FDM technology for prosthetic development and provides an initial investigation into the mechanical behavior of auxetic structures in biomedical contexts.

Keywords: Fused Deposition Modeling · auxetic structures · prosthetic hand · Polylactic Acid

1 Introduction

The Auxetic structures are defined as materials or configurations that exhibit a negative Poisson's ratio. Unlike conventional materials, which become thinner in directions perpendicular to an applied tensile load, auxetic materials exhibit the opposite behavior: they expand transversely when stretched and contract transversely when compressed [1]. This mechanical property has garnered significant interest due to its potential applications in diverse fields, including engineering, biomedical sciences, and advanced textiles.

M. Soula et al. (Eds.): AIAM 2024, LNME, pp. 69–82, 2025.
https://doi.org/10.1007/978-3-031-94889-3_6

Poisson's ratio is a fundamental property that describes the relationship between transverse and longitudinal deformations of a material under axial loading [2]. While most isotropic materials exhibit a Poisson's ratio between −1 and 0.5, this value can extend these limits in anisotropic materials [3]. Conventional materials typically demonstrate positive Poisson ratios, whereas auxetic materials, due to their negative Poisson ratio, exhibit a remarkable ability to expand laterally under tensile stress. Figure 1 provides a visual comparison of the tensile and compressive behavior of conventional and auxetic materials, illustrating this fundamental distinction [4]. For conventional materials, under tensile loading (ϵ_1), the material elongates in the direction of the applied force while contracting (ϵ_2) in the transverse direction. Conversely, under compressive loading, conventional materials contract in the loading direction (ϵ_1) and expand transversely (ϵ_2). In contrast, auxetic materials behave oppositely. Under tensile loading, they exhibit simultaneous elongation in the loading direction (ϵ_1) and transverse expansion (ϵ_2). Similarly, under compressive loading, auxetic materials contract in both the longitudinal (ϵ_1) and transverse (ϵ_2) directions. This behavior is a direct consequence of their negative Poisson's ratio and unique internal geometry [5].

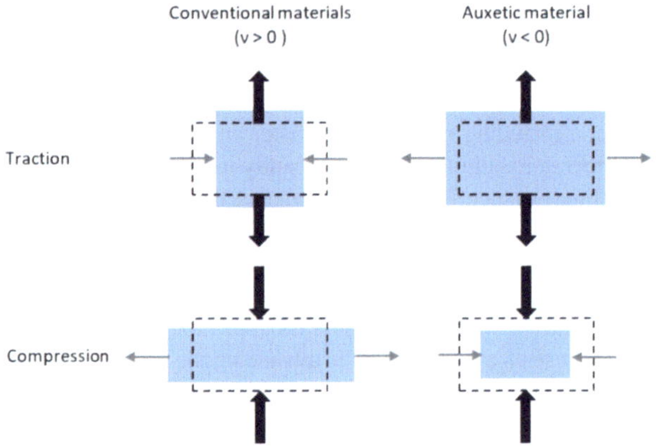

Fig. 1 Compression and tensile test for conventional materials and auxetic material

Re-entrant structures are a prominent category of auxetic materials that derive their auxetic properties from specific geometric designs, such as re-entrant hexagonal patterns. These structures exhibit deformation mechanisms that allow them to expand in two directions under tensile stress, often behaving non-linearly as they reorganize to occupy less space under load [6]. Such properties have made re-entrant structures particularly well-suited for a wide range of applications, including protective equipment, technical textiles, biomedical devices, and advanced structural materials. Their deformation behavior allows for exceptional energy absorption, making them ideal for protective gear such as helmets, bulletproof vests, and shock-absorbing pads [7]. In the biomedical field, re-entrant structures offer significant advantages in the design of prosthetic devices, including joint replacements and endo-prostheses. The ability of these materials to adapt to stress and deformation while maintaining structural integrity improves the performance and comfort of implants, offering a better fit and reducing the risk of complications. For example, in the case of endo-prostheses, the use of re-entrant auxetic materials could enhance the longevity and functionality of the device by imitating the natural movements of bones and tissues, enabling more efficient load distribution and minimizing wear.

Moreover, their adaptability also makes them ideal for designing custom orthopedic devices, such as orthopedic shoes. Re-entrant structures can be engineered to provide optimal support and cushioning, tailored to the specific pressure points of the foot, as shown in Fig. 2. This can alleviate discomfort, correct posture, and improve mobility for individuals suffering from conditions such as flat feet, plantar fasciitis, or arthritis.

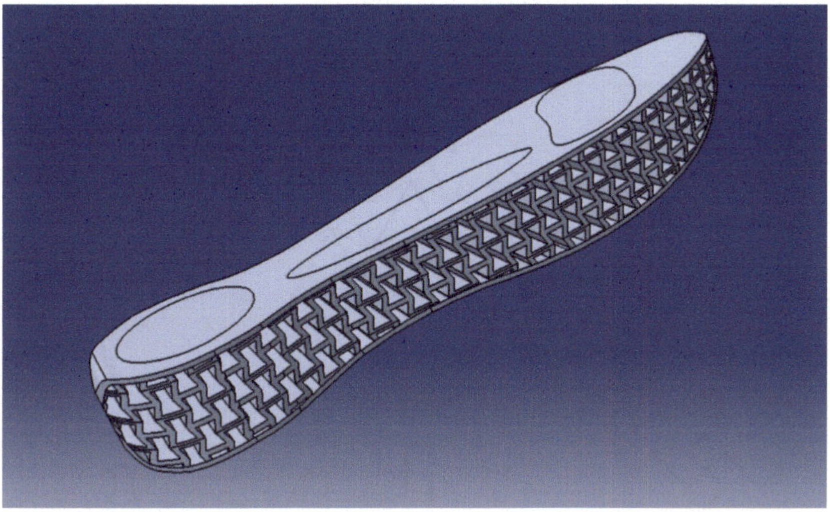

Fig. 2 CAD model of orthopedic shoes based on re-entrant structures

The advent of additive manufacturing, particularly Fused Deposition Modeling (FDM), has enabled the precise fabrication of complex auxetic geometries. Polylactic acid (PLA), a biodegradable and biocompatible thermoplastic, is widely used in FDM due to its favorable properties, such as ease of processing, availability, and suitability for biomedical applications [8]. PLA's biocompatibility allows it to safely interact with biological tissues, making it ideal for medical implants, prosthetics, and tissue engineering scaffolds [9]. Besides, its biodegradability enables it to naturally degrade within the body over time, which is beneficial for temporary implants or devices that aid healing without the need for removal. PLA's mechanical adaptability, combined with its precision in FDM, allows for the creation of custom, patient-specific auxetic structures. These geometries, with their desired mechanical properties, can be tailored for various biomedical applications [10]. Moreover, PLA's cost-effectiveness and environmental sustainability make it an optimal material for advancing biomedical engineering and other innovative fields [11]. This study explores the potential of fabricating re-entrant auxetic structures using FDM technology and PLA, aiming to develop materials with tailored mechanical characteristics for applications in biomedical devices.

2 Modeling and Behavior of the Re-entrant Structure

Auxetic structures encompass a variety of forms, including chiral designs, rotating squares, and bowtie configurations. Additionally, re-entrant structures have many types of forms, as illustrated in Fig. 3. This section aims to analyze the design and behavior of re-entrant structures in the context of 3D printing, with a particular focus on their potential biomedical applications. Re-entrant structures are distinguished by their geometry which causes the structural elements to move inward when compressed. Unlike general auxetic structures, characterized by a negative Poisson's ratio ($\nu < 0$) that allows lateral expansion under tension, re-entrant structures specifically feature inverted internal angles, such as in hexagonal re-entrant structures.

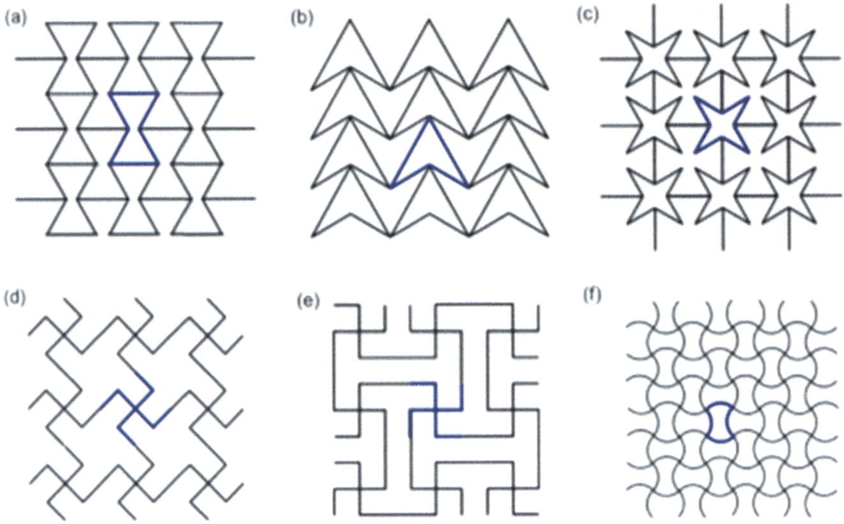

Fig. 3 Re-entrant structures. (a) Re-entrant hexagon, (b) double arrowhead, (c) star honeycomb, (d) structurally hexagonal re-entrant honeycomb, (e) lozenge grids, (f) sinusoidal ligaments [12]

The mechanical behavior of these materials, particularly their enhanced shear resistance, is described by the relationship between shear modulus G and Poisson's ratio ν:

$$G = \frac{E}{2(1 + \nu)} \tag{1}$$

where E is Young's modulus. A decrease in ν increases G, thereby enhancing shear resistance, which is a critical advantage for applications requiring high-impact protection. Energy absorption is another essential property of auxetic materials [13], particularly relevant to biomedical uses. The energy absorbed ω during deformation is calculated as:

$$\omega = \int \sigma(\varepsilon)\, d\varepsilon \tag{2}$$

where σ is the stress and ϵ is the strain. A good energy absorber exhibits a stress-strain curve with a long and high plateau before densification [14].

Geometric parameters, including wall thickness and internal angles, are pivotal in determining the structural performance of re-entrant structures. Varying wall thickness can increase structural strength at the expense of the densification limit because thicker walls provide greater stiffness and resistance to deformation under applied loads, enhancing the material's ability to bear stress. However, this comes with a reduction in the densification limit, as thicker walls lead to a more rapid collapse of the structure during compression. This means that while the structure becomes stronger and better able to withstand initial forces, its ability to absorb energy over an extended deformation period is diminished. This trade-off highlights the need for careful optimization of wall thickness to achieve a balance between strength and energy absorption, depending on the specific requirements of the application [15].

The optimization of these parameters requires rigorous testing and simulation tailored to each specific design. The remarkable energy absorption characteristics and superior resistance to indentation exhibited by re-entrant structures render them particularly well-suited for biomedical applications [16]. These include prosthetic limb sockets, which necessitate a balance of durability and comfort, protective equipment such as helmets and pads, and scaffolds for tissue engineering, which must support cell proliferation while maintaining structural integrity. The combination of mechanical advantages and adaptability offered by re-entrant structures positions them as transformative solutions within the biomedical field, addressing diverse and demanding requirements with unparalleled efficiency.

3 Experimental Study of a Printed Re-entrant Structure via PLA Material

3.1 Geometric Configuration of Re-entrant Structure

The first step of our experimental study is to identify the geometric configuration of our chosen model. The geometric parameters of the re-entrant structure, such as the angles and the distances between the components, are critical in determining its mechanical properties and overall performance. Even small changes in these details can significantly affect the structure's behavior under load, influencing factors such as strength, flexibility, and energy absorption, which are essential for its intended applications, especially in prosthetic devices.

The re-entrant Structure (an assembly of several unit cells) can be distinguished from another by several geometric parameters. The geometry of a re-entrant curtain unit cell is characterized by its neutral axis as shown in Fig. 4. The horizontal neutral axis is labeled H, while the oblique neutral axis is labeled L. The angle formed by the neutral axes H and L is denoted θ. The distance length of one unit cell is labeled d_C. Finally, the wall thickness is identical throughout the structure and is noted Th [17].

Our structure is made up of n cells of height dc in the Y direction, where dc is the height between the two horizontal neutral axes of a unit cell. To determine the total

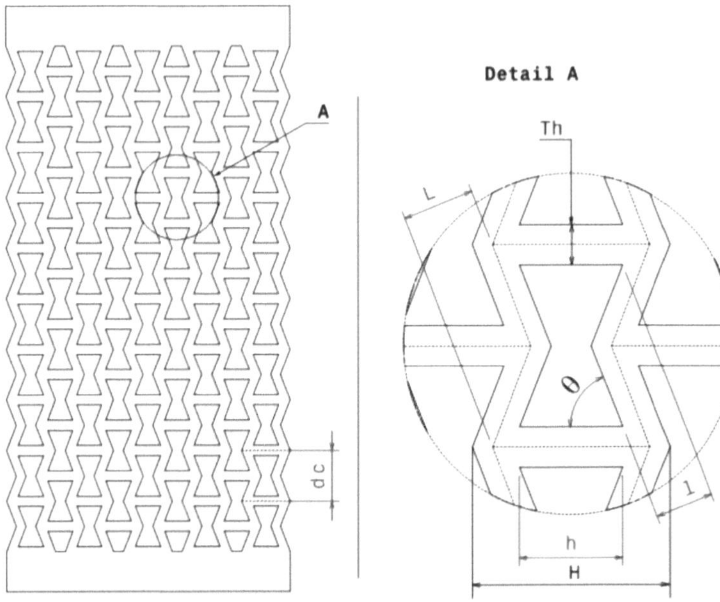

Fig. 4 Re-entrant structure unit cell geometry

height of the structure, it is necessary to add an extra thickness of Th/2 to the extreme cells (top and bottom), so that all walls have the same thickness. D is then defined by:

$$D = n\,dc + Th = 129.54\ \text{mm} \tag{3}$$

All numerical values of geometric parameters are summarized in the Table 2. After identifying all the geometric parameters of the re-entrant structure, the next section will be for the sample fabrication and experimental steps.

3.2 Sample Fabrication

To benefit from the behavior of the re-entrant structure, a sample is fabricated and later tested using the geometric parameters presented in the previous section. For this study, a Creality CR5-Pro 3D printer was employed to fabricate the specimens, utilizing Polylactic Acid (PLA) as the material. PLA, a biodegradable polymer derived from maize starch, is widely recognized for its environmental sustainability and suitability for 3D printing applications. The filament used has a diameter of 1.75 mm, a standard size for Fused Deposition Modeling (FDM) 3D printing. The detailed steps of the fabrication process are presented in Fig. 5, which visually illustrates each stage.

The initial step in the experimental process involves the use of Solidworks software to design the CAD model of the components, as shown in Fig. 6. The CAD model is then converted into a 3D project, which is subsequently sliced into discrete layers. This sliced model is saved in STL format, a widely accepted file type for 3D printing. Subsequently, the Creality Slicer software is utilized to configure the process parameters,

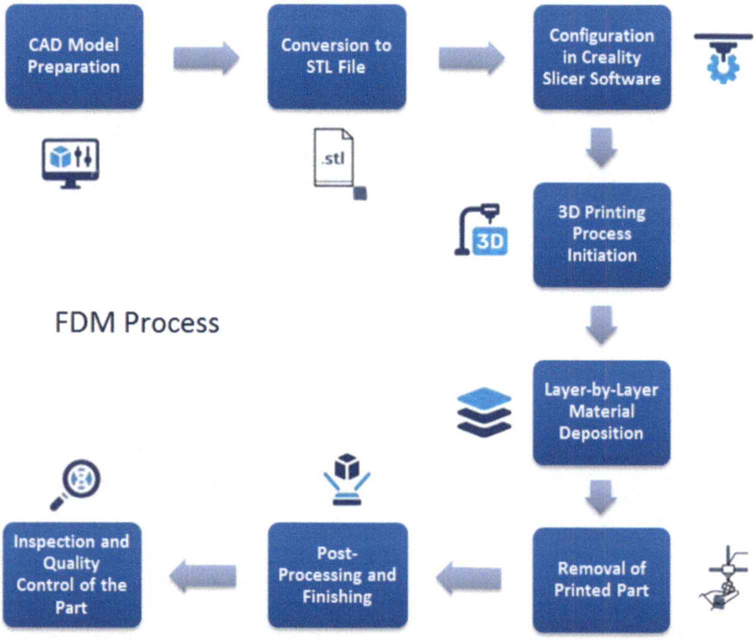

Fig. 5 Detailed steps of the 3D Printing Process of the prosthetic part

including wall thickness, print speed, and infill density. These specific parameters are outlined in Table 1. The selection of these parameters was carried out with careful consideration, drawing upon the findings from our previous work, where we examined the same Prosthetic hand using conventional materials. This previous research provided valuable insights that guided the identification of the appropriate settings to achieve a balance between structural integrity and print quality.

The precise configuration of these parameters is essential to ensuring the success of the printing process, as even minor deviations in these settings can have a substantial impact on the mechanical performance and overall quality of the final product. The objective now is to apply these parameters to create the same structure using PLA material for the prosthetic model, ensuring that the printed parts will meet the required specifications and mechanical properties for prosthetic applications.

Once the settings are determined, the printing process is initiated. The Creality CR5-Pro printer deposits the PLA material layer by layer, gradually constructing the part according to the specifications of the digital model. This method, known as Fused Deposition Modeling (FDM), involves heating the PLA filament until it melts and is extruded through the nozzle. Each layer adheres to the previous one, solidifying upon cooling to create the final structure.

Upon completion of the printing process, the part is carefully removed from the printing platform. Post-processing steps are then conducted to ensure the final product adheres to the desired quality standards. Following the post-processing steps, the part is inspected for any defects, such as printing errors or imperfections, to ensure the quality

Fig. 6 CAD model of the re-entrant parts

Fig. 7 Printed re-entrant *hexagon* structure sample *using PLA material*

Table 1 Geometric dimension of a unit cell of the re-entrant Structure

	θ	Th	l	L	h	H	dc
Unit	°	mm	mm	mm	mm	mm	mm
Value	68	2.54	3.77	4.64	6.6	12.7	12.7

Table 2 Used printing process parameters

Parameters	Unit	Value
Wall thickness	mm	0.9
Top/bottom thickness	mm	0.9
Infill pattern	–	Cubic
Raster angle	°	45°
Build plate temperature	°C	60
Printing speed	mm/s	50
Infill density	%	20
Nuzzle temperature	°	200

and integrity of the final product. Figure 7 presents the final printed sample, illustrating the complete result of the manufacturing process. The finished product, made from PLA, displays precise geometric features, visible layer lines, and a solid structure.

3.3 Mechanical Proprieties

In the following section, we will discuss the mechanical behavior of these structures to better understand their performance. The first property representing the major advantage of cellular structures is their low density compared to conventional materials. As a result of the presence of voids in the structure, a re-entrant strut made of PLA will be lighter at an equivalent volume than a block of material made of the same PLA.

The relative density of a re-entrant structure is the ratio between the density of that structure ρ_c and the density of the raw material ρ_s of which it is made. In the case of an SA, this relative density is expressed as follows:

$$\frac{\rho c}{\rho s} = \frac{Vs}{Vc}. \quad (4)$$

where Vc and Vs are respectively the volumes occupied by the structure and the raw material. Next, we express the volume Vc of the structure as the product of the characteristic length dc and the projected surface area Sc of the structure on the X-Z plane:

$$Vc = dc\ Sc \quad (5)$$

To calculate Sc, the surface projected onto the X-Z plane, we use the following formula:

$$Sc = b\ (H - L \cos\theta) \quad (6)$$

Here, b is the width of the structure, H is the total height, and L is the characteristic length. This expression for Sc gives us the projected surface area of the structure.

Then, to obtain the volume Vs of the raw material, we subtract the volume occupied by the voids (i.e., the empty space inside the structure) from the total volume Vc. This subtraction gives the raw material volume Vs, which is expressed as:

$$Vs = Vc - 2\ b\ \ell\ sin\theta\ (h - \ell\ cos\theta) \quad (7)$$

In this formula, we subtract the material occupying the internal voids of the structure. This is calculated using b, ℓ and the angle θ. This term accounts for the empty spaces within the re-entrant structure.

Finally, by substituting the expressions for Vc and Vs into the initial relative density formula, we obtain the final expression for the relative density. This is written as follows:

$$\frac{\rho c}{\rho s} = 1 - \frac{L \sin\theta - \frac{Th}{2}}{L \sin\theta\ (H - L\cos\theta)}\left[H - L\cos\theta - \frac{Th}{\sin\theta}\left(1 + \frac{\cos\theta}{2}\right)\right] \quad (8)$$

where θ, L, H, and Th are geometric parameters presented in Sect. 3.1.

The density of structured materials can be modulated by varying the size of the voids and the shape of the cells. Not only the choice of cells but also the structure's constituent material can be used to vary the physical properties of cellular materials, such as thermal conductivity, Young's modulus, and compressive strength [16]. The transition between each formula results from substituting the respective volumes into the relative density relation, allowing to integration of the specific geometric characteristics of the re-entrant structure into the final expression.

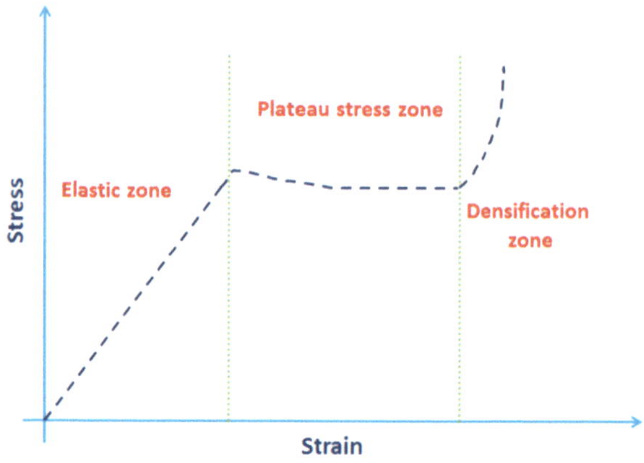

Fig. 8 Typical compressive stress-strain curve [16]

The sample presents a typical compressive stress-strain curve, as shown in Fig. 8, which can be divided into three zones.

The first zone corresponds to the elastic response of the structure and extends from zero strain to a strain at which the stress reaches a particular value called "plateau stress". The second zone corresponds to a long stress plateau. The stress value reached at the end of elastic deformation remains virtually constant until the structure densifies. This stress plateau is caused by the plastic collapse of the edges or solid faces making up the cells. It occurs when the moment in the wall exceeds the maximum plastic moment of these junctions, creating plastic hinges in the cells [18]. It is also during this period that the absorption of mechanical energy, corresponding to the area under the curve, is greatest. In the third zone, when the structure becomes denser, stress increases sharply until the structure is completely crushed. The pores are reduced to a minimum, and the value of the stress-strain slope matches that of Young's modulus of the constituent material [19]. An ideal energy absorber has a long, flat stress-strain curve. Cellular structures are therefore suitable materials for energy absorbers, provided they are sized to absorb the required energy and not exceed the required force.

These mechanical properties, specifically the prolonged stress plateau and energy absorption capability, make the material highly promising for applications such as prosthetics. The ability to absorb significant amounts of energy while maintaining a constant stress level can be critical in prosthetic hands, where comfort and shock absorption are essential. The mechanical behavior of this material allows it to resist repeated impact forces without significant degradation, offering a balance of durability and comfort for users. Furthermore, the tailored compression response of re-entrant structures could be advantageous in protective gear, such as helmets and pads, providing optimal protection during impact while maintaining a lightweight design. The potential applications in the biomedical field are significant, as the material's energy absorption properties align well with the requirements for both prosthetics and protective medical equipment.

4 Applications

Re-entrant materials are already well known for their ability to absorb the energy of a shock or impact while limiting the force transmitted. That's why, we aim to use this structure in an exo-prosthesis to capitalize on these inherent properties. By incorporating re-entrant structures into the design of our prosthetic hand, we intend to enhance its shock absorption capabilities, thereby providing better protection against sudden impacts or external forces. Besides, the use of re-entrant structures can contribute to improving the overall comfort and stability of the exo-prosthesis, allowing users to engage in daily activities with greater ease and confidence.

Figure 9 illustrates the 3D design of prosthetic hands developed using SolidWorks software. This prosthetic is designed as a simple and accessible mechanical solution for individuals with upper limb amputations. It is distinguished by its simplicity, as it lacks sensors or electronic controls, making it both affordable and practical. The design incorporates a three-finger mechanism capable of performing basic tasks [20].

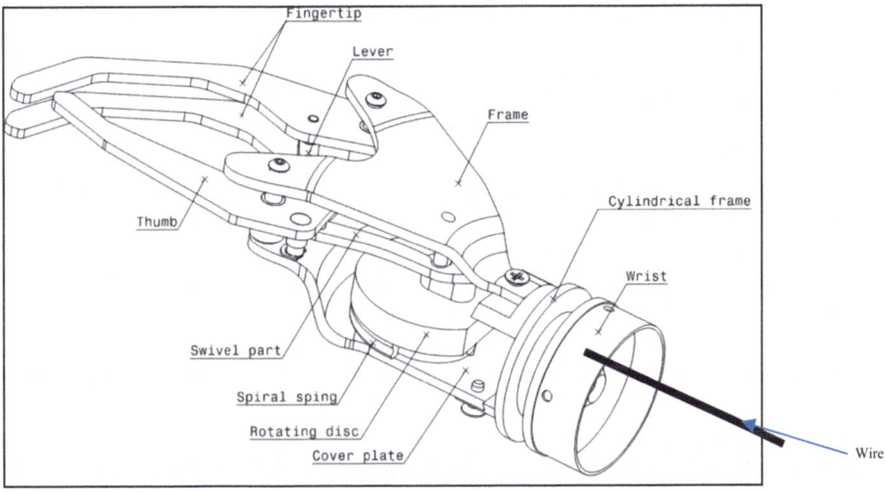

Fig. 9 Prosthetic hand's 3D design is created using Solidworks software

The prosthetic hand features three articulated fingers that are central to its operation. These fingers are actuated through a mechanical cord system, which replicates the natural motion of the hand. When tension is applied to the cord, the fingers flex to grip an object, and upon release, they return to their initial position. This manual mechanism is sufficient for executing fundamental tasks. The system is designed with a modular framework, where each component is intended to be fabricated individually through 3D printing and subsequently assembled. This modularity ensures ease of production and facilitates the replacement of individual parts as needed. Once the components are printed, they are assembled to form a lightweight and functional prosthetic hand. One of the primary advantages of this design is its affordability. By eliminating complex technologies, the prosthetic hand is made accessible to a broader demographic, particularly those in resource-limited settings. Its simplicity further enhances its practicality, as the device requires minimal maintenance and is relatively easy to repair.

The use of a 3D-printed re-entrant structure in hand prostheses provides significant advantages over traditional materials by offering unique properties such as enhanced shock absorption, lightweight design, and adaptability, which contribute to improved comfort and functionality for users. Here are some key applications of this innovative approach:

- Structural support: The re-entrant structure can be used to provide structural support to the hand prosthesis, reinforcing areas where strength and stability are required. For example, it can reinforce joints or load-bearing areas.
- Weight reduction: By using a re-entrant structure, it is possible to reduce the overall weight of the hand prosthesis while maintaining its strength and durability. This can make the prosthesis more comfortable for the user to wear.
- Enhanced functionality: The re-entrant structure can be designed to enhance the functionality of the hand prosthesis. For example, it can be integrated into grasping mechanisms to provide additional support when manipulating objects.
- Component integration: The re-entrant structure can be used to support the integration of electronic components, such as motion sensors or actuators, enabling additional functionality to be added to the hand prosthesis, such as grip control.

In summary, the integration of 3D-printed re-entrant structures in hand prostheses represents a transformative approach to enhancing strength, functionality, and user comfort while offering the flexibility to tailor solutions to individual needs. To validate the robustness and reliability of our design, we will carry out a series of mechanical tests, each addressing a specific aspect of performance. Tensile tests will assess the material's strength under tension, ensuring its durability during stretching motions. Compression tests will evaluate the prosthesis's capacity to endure compressive forces, simulating real-world loads. Three-point bending tests will measure flexibility and resistance to bending forces, which are crucial for dynamic hand movements. Fatigue tests will determine the structure's resilience under repeated loading and unloading cycles, reflecting daily usage conditions. Finally, shear tests will examine performance under lateral forces, ensuring stability during side-to-side movements. These evaluations will provide precise data for detailed finite element analysis, enabling us to refine our designs and ensure optimal performance in practical applications.

5 Conclusion

In conclusion, this study on 3D-printed auxetic structures, focusing on the hexagonal re-entrant design, highlights its potential for biomedical applications. The results demonstrate that these structures can significantly improve the performance of medical devices, particularly by enhancing shock absorption and reducing transmitted forces. For instance, their ability to deform predictably under stress makes them well-suited for applications in prosthetics, where they can provide greater comfort and durability, thereby improving the user experience.

Further research is needed to better understand the performance and limitations of these structures. Future work should focus on assessing their long-term durability under different loading conditions, exploring alternative materials for improved biocompatibility, and optimizing manufacturing processes. These findings underscore the potential of re-entrant structures to drive innovation in prosthetics and other medical devices, paving the way for advanced solutions that enhance functionality and user quality of life.

References

1. Kolken, H.M.A., et al.: Mechanical performance of auxetic meta-biomaterials. J. Mech. Behav. Biomed. Mater. **104**, 103658 (2020). https://doi.org/10.1016/j.jmbbm.2020.103658
2. Li, D., Yin, J., Dong, L., Lakes, R.S.: Strong re-entrant cellular structures with negative Poisson's ratio. J. Mater. Sci. **53**(5), 3493–3499 (2018). https://doi.org/10.1007/s10853-017-1809-8
3. Lvov, V.A., Senatov, F.S., Veveris, A.A., Skrybykina, V.A., Díaz Lantada, A.: Auxetic metamaterials for biomedical devices: current situation, main challenges, and research trends. Materials. **15**(4), 1439 (2022). https://doi.org/10.3390/ma15041439
4. Çelebi, A., İManç, M.M.: Evaluation of mechanical properties of PLA auxetic structures produced by additive manufacturing. J. Mater. Mechatron. A. **4**(2), 384–396 (2023). https://doi.org/10.55546/jmm.1309858
5. Khoshgoftar, M.J., Barkhordari, A., Limuti, M., Buccino, F., Vergani, L., Mirzaali, M.J.: OPEN Bending analysis of sandwich panel composite with a re-entrant lattice core using zig-zag theory. Sci. Rep.
6. Ashby, M.: Designing architectured materials. Scr. Mater. **68**(1), 4–7 (2013). https://doi.org/10.1016/j.scriptamat.2012.04.033
7. Liu, B., et al.: Capability of auxetic femoral stems to reduce stress shielding after total hip arthroplasty. J. Orthop. Transl. **38**, 220–228 (2023). https://doi.org/10.1016/j.jot.2022.11.001
8. Negi, S., Dhiman, S., Sharma, R.K.: Basics, applications and future of additive manufacturing technologies: a review. J. Manuf. Technol. Res. **5**(1/2), 75 (2013)
9. Ben Hariz, N., Boulila, A., Ayadi, M.: Preliminary study of geometric defects and topography of bio-carriers fabricated by the fdm printing process. In: Mabrouki, T., Sahlaoui, H., Sallem, H., Ghanem, F., Benyahya, N. (eds.) Advances in Additive Manufacturing: Materials, Processes and Applications Lecture Notes in Mechanical Engineering, pp. 44–51. Springer Nature Switzerland, Cham (2024). https://doi.org/10.1007/978-3-031-47784-3_6
10. Cheng, X., et al.: Design and mechanical characteristics of auxetic metamaterial with tunable stiffness. Int. J. Mech. Sci. **223**, 107286 (2022). https://doi.org/10.1016/j.ijmecsci.2022.107286
11. Hedayati, R., Yousefi, A., Dezaki, M.L., Bodaghi, M.: Analytical relationships for 2D re-entrant auxetic metamaterials: an application to 3D printing flexible implants. J. Mech. Behav. Biomed. Mater. **143**, 105938 (2023). https://doi.org/10.1016/j.jmbbm.2023.105938
12. Kim, Y., Son, K.H., Lee, J.W.: Auxetic structures for tissue engineering scaffolds and biomedical devices. Materials. **14**(22), 6821 (2021). https://doi.org/10.3390/ma14226821
13. Safikhani Nasim, M., Yaghootian, A., Mosalmani, R.: Energy absorption of the additively manufactured novel re-entrant auxetic structure in comparison with honeycomb structure: experimental and numerical analysis. J. Braz. Soc. Mech. Sci. Eng. **45**(5), 275 (2023). https://doi.org/10.1007/s40430-023-04178-x
14. Kolken, H.M.A., Zadpoor, A.A.: Auxetic mechanical metamaterials. RSC Adv. **7**(9), 5111–5129 (2017). https://doi.org/10.1039/C6RA27333E
15. Simpson, J., Kazancı, Z.: Crushing investigation of crash boxes filled with honeycomb and re-entrant (auxetic) lattices. Thin-Walled Struct. **150**, 106676 (2020). https://doi.org/10.1016/j.tws.2020.106676

16. Petit, A.: "Réponse de Structure auxétique en dynamique rapide : étude théorique et expérimentale," thèse, école doctorale énergie, matériaux, sciences de la terre et de l'univers, université d'orléans, France (2023)
17. Jiang, D., et al.: Advances in additive manufacturing of auxetic structures for biomedical applications. Mater. Today Commun. **40**, 110045 (2024). https://doi.org/10.1016/j.mtcomm.2024.110045
18. Elnasri, I., Pattofatto, S., Zhao, H., Tsitsiris, H., Hild, F., Girard, Y.: Shock enhancement of cellular structures under impact loading: part I experiments. J. Mech. Phys. Solids. **55**(12), 2652–2671 (2007). https://doi.org/10.1016/j.jmps.2007.04.005
19. Acharya, A., Dasgupta, A., Jain, A.: Design and study of novel nested auxetic lattices with tunable and enhanced in-plane elastic properties. Int. J. Solids Struct. **293**, 112749 (2024). https://doi.org/10.1016/j.ijsolstr.2024.112749
20. Ben Hariz, N., Boulila, A., Ayadi, M., Fattouch, S.: Evaluation of a complete characterization of polylactic acid used in additive manufacturing: biomedical application. Ser. Biomech. **36**(3) (2022). https://doi.org/10.7546/SB.02.03.2022

Dynamic Behavior of Honeycomb Composite Structures: Effects of Filling Density Using Split Hopkinson Compression Bars

Manel Chihi[1](✉), Mostapha Tarfaoui[2,3], Sami Chatti[1], and Yumna Qureshi[4]

[1] Mechanical Laboratory of Sousse, ISSATSo, University of Sousse, BP. 264 Erriadh, 4023 Sousse, Tunisia
manel.chihi@eniso.u-sousse.tn
[2] ENSTA Bretagne, IRDL UMR CNRS 6027, F-29200 Brest, France
[3] Green Energy Park (UM6P/GEP), Modelling and Digitalization, Benguerir, Morocco
[4] Department of Mechanical Engineering, Institute of Space Technology, Islamabad, Pakistan

Abstract. This study investigates the impact of filling density (50%, 75%, and 100%) on the dynamic behavior of carbon-fiber-reinforced PETG (CF-PETG) composites fabricated using the cost-effective fused filament fabrication (FFF) technique, commonly employed for producing complex structures. The primary objective is to evaluate how different filling densities influence the material's dynamic characteristics, such as compressive strength, modulus, and damage resistance. Dynamic compression tests were conducted using a split Hopkinson pressure bar (SHPB) at various strain rates, with input pressures of 1.4, 1.7, 2, and 2.4 bars. A high-speed camera was used to monitor and record the damage evolution, offering insights into fracture mechanisms in 3D FFF-printed CF-PETG composites. The results demonstrate that increasing the filling density significantly improves the dynamic characteristics, leading to enhanced compressive strength, higher compressive modulus, and better damage resistance. Higher filling densities result in a more compact and interconnected structure, which increases the stiffness and overall strength of the composite.

Keywords: Carbon-fiber-reinforced PETG (CF-PETG) · fused filament fabrication (FFF) technique · filling density · Split Hopkinson pressure bar (SHPB) · dynamic characteristics · dynamic compressive strength

1 Introduction

Additive manufacturing, often referred to as 3D printing, is revolutionizing the manufacturing industry by enabling the production of complex and customized products through layer-by-layer fabrication from digital models. This innovative technology has seen significant growth, with applications ranging from healthcare to aerospace. One of the most common methods in additive manufacturing is Fused Filament Fabrication (FFF), which uses thermoplastic filaments to create three-dimensional objects. *As an example of additive manufacturing, FFF showcases the flexibility and customization*

M. Soula et al. (Eds.): AIAM 2024, LNME, pp. 83–91, 2025.
https://doi.org/10.1007/978-3-031-94889-3_7

potential that are characteristic of these advanced manufacturing technologies. In the 3D printing industry, several process parameters must be carefully considered to optimize the quality of the printed product. One of the most critical factors is fill density, also known as fill percentage, which determines the amount of material used in creating the internal structure of a printed object. Fill density plays a key role in the mechanical properties, weight, and overall strength of the printed part. Higher fill densities result in a denser internal structure, which improves strength and durability. In contrast, lower fill densities reduce material usage, making the product lighter but potentially compromising its structural integrity. *Finding the right balance in fill density is crucial as it affects both the mechanical performance and manufacturing efficiency, including printing time and material consumption.*

Several studies have investigated the influence of infill density and annealing on the mechanical characteristics of PETG and carbon-fiber-reinforced PETG (CF-PETG) composites manufactured using Fused Deposition Modelling (FDM) [1, 2]. For example, Tarfaoui et al. [3] conducted an experimental and numerical analysis on carbon-fiber-reinforced PETG (CF-PETG) composites manufactured by FFF. They tested dumbbell-shaped specimens with honeycomb (NIDA) and rectilinear (RECT) filling patterns at various filling densities (20%, 50%, 75%, and 100%). The results revealed that increasing the fill density significantly impacted the mechanical behavior, with catastrophic failure occurring in both patterns at 20% fill density. In terms of strength and stiffness, the honeycomb pattern outperformed the rectilinear pattern. Previous studies have primarily focused on the static or quasi-static mechanical behavior of CF-PETG materials under varying fill densities. However, there is limited research on their dynamic behavior under high strain rates, which is critical for practical applications where materials may experience sudden or impact loading. This study aims to bridge this gap by exploring the dynamic response and damage resistance of CF-PETG 3D-printed materials under dynamic conditions, specifically focusing on honeycomb structures.

In this work, dynamic testing is performed using the Split-Hopkinson Pressure Bar (SHPB) at different pressures (1.4, 1.7, 2, and 2.4 bar) to investigate the behavior of CF-PETG materials at high strain rates. The research specifically focuses on how different fill densities (50%, 75%, and 100%) impact the mechanical performance of honeycomb structures under dynamic loading conditions. This research aims to identify the optimal fill density that enhances the honeycomb pattern's ability to withstand dynamic loading without failure or significant deformation. Key parameters such as stiffness and dynamic compressive strength will be evaluated to assess the durability and effectiveness of honeycomb structures under these extreme conditions.

2 Materials

Polyethylene terephthalate glycol (PETG) is a polymer material widely used in 3D printing, distinguished by its transparency, strength, and ease of use. It offers a compromise between the durability of ABS and the simplicity of polylactic acid (PLA). This makes PETG ideal for various applications, particularly in industries such as consumer products, medical devices, and automotive parts, where both strength and ease of use are critical. The use of carbon fiber reinforcements in 3D printing materials, particularly

PETG, is essential for improving mechanical properties, including strength, rigidity, and impact resistance. For this study, carbon fibers with different weight percentages were incorporated into the PETG matrix to significantly enhance the material's overall performance in high-stress applications. This reinforcement improves the durability and structural integrity of 3D-printed items for challenging applications. For this study, CF-PETG filaments from NANOVIA, France, containing carbon fiber reinforcements, were chosen to manufacture samples using the Fused Filament Fabrication (FFF) technique, as shown in Fig. 1. The FFF technique involves the extrusion of a filament through a heated nozzle, layer by layer, which allows for precise control over the printing process. This manufacturing procedure was carried out using a honeycomb structure, which is known for its lightweight yet strong design. The honeycomb pattern provides excellent weight distribution and enhances the structural integrity of the printed samples. Filling density was the only variable modified in the manufacturing of the 3D samples, with all other process parameters remaining constant nozzle temperature, print speed, and layer height), Table 1. Cubic specimens of 13 x 13 x 13 mm^3 were prepared for dynamic testing (Fig. 2). Each figure represents the visualization of respective infill densities i.e. 50%, 75%, and 100%.

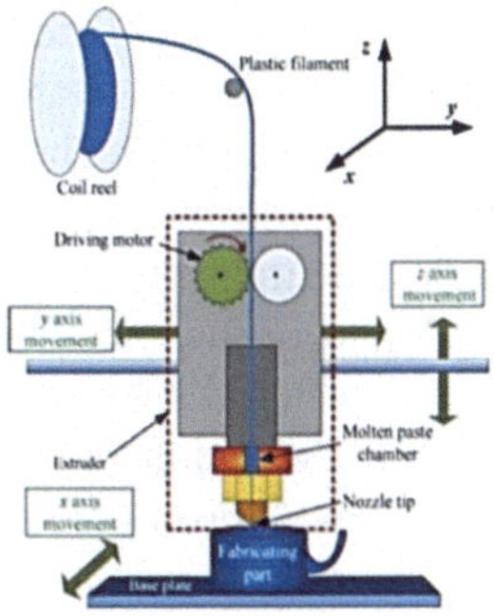

Fig. 1. Process of fused filament fabrication (FFF)

Table 1. Parameters of Fused Filament Fabrication (FFF) technique

Extrusion temperature	220 °C
Heat platform temperature	80 °C
Nozzle	0.3mm
Nozzle moving speed	20 mm/s first layer 40 mm/s other layers
Diameter	1.75/2.85 mm ± 50 μm

(a) 50% (b) 75% (c) 100%

Fig. 2. 3D printed specimens.

3 Dynamic In-Plane Compression Tests

High strain rate testing plays a critical role in material science, as it helps researchers understand how materials behave under rapid loading conditions, which are common in many real-world applications such as automotive and aerospace industries. The Split Hopkinson Pressure Bar (SHPB) is an experimental device commonly used to study the dynamic mechanical characteristics of materials under high strain rates. It consists of three primary components: a striker bar, an incident bar, and a transmission bar (Fig. 3). Each component of the SHPB has a specific role in the generation and measurement of stress waves. The striker bar, which is accelerated towards the incident bar, when creates a compressive stress wave it strikes the incident bar. The incident bar transmits this wave through the material specimen, and as the wave reaches the material, it generates an additional stress wave within the sample. The transmission bar captures the remaining wave that passes through the material. A part of the stress wave is reflected in the incident bar, providing crucial data about the material's resistance to stress. Strain gauges are strategically placed at specific points in the middle of both the incident and transmission bars to accurately measure the strain waves traveling through them. These strain gauges are positioned at the midpoints of the bars to minimize measurement errors from the initial wave impact and ensure precise detection of the strain induced by the stress wave. The signals received from the SHPB experiment—incident, reflected, and transmitted waves—offer detailed insights into the material's behavior under dynamic pressure. These signals are post-processed using the Fast Fourier Transform (FFT) method implemented in Maple software. The FFT is used in this context because it allows for the conversion of time-domain data (the raw strain signals) into frequency-domain data, helping to identify the frequency components of the stress waves. This analysis aids in determining various dynamic parameters, such as wave propagation velocity, strain rate sensitivity, and failure mechanisms in materials subjected to high compression rates.

4 Results and Discussion

Figure 4 illustrates typical signals corresponding to cubic CF-PETG samples with 50% and 100% fill densities subjected to dynamic loading at pressures ranging from 1.4 to 2.4 bar. The signals acquired from a Split Hopkinson Pressure Bar (SHPB) provide useful information on the behavior of materials under high-pressure conditions. In SHPB studies, three distinct waves are frequently observed: the incident wave, the reflected wave,

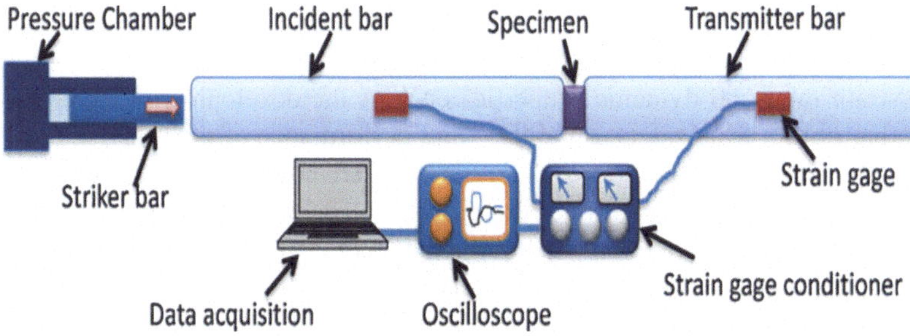

Fig. 3. Split Hopkinson compression bars (SHPB) [4]

and the transmitted wave. These waves provide important information on the dynamic response of materials subjected to dynamic compression. At higher applied pressures (2.4 bar), the presence of damage can be determined by careful analysis of the signals. Specifically, the second peak of the reflected wave in Fig. 4a becomes visible, suggesting material failure or damage. At the same time, an almost negligible transmitted wave highlights the severity of the damage, implying a reduction in structural integrity. When the strain signals for 50% and 100% honeycomb structures are compared, it's clear that 100% honeycomb structures remain macroscopically undamaged, even at the highest pressure (2.4 bar). The term "macroscopically undamaged" refers to the fact that no visible or significant structural failure occurred in the material at the macroscopic level, indicating the material's durability under dynamic loading. This finding demonstrates that the full-density honeycomb pattern is exceptionally resistant to dynamic loading, showcasing its ability to withstand increased pressures without causing significant macroscopic damage. This characteristic makes 100% honeycomb structures potentially valuable in real-world applications such as aerospace and automotive industries, where dynamic loading is a critical factor.

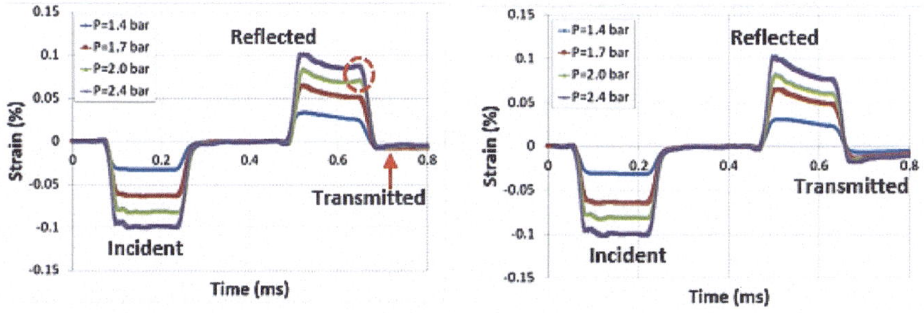

(a) 50% (b) 100%

Fig. 4. Strain vs. time for 50% and 100% CF-PETG specimens subjected to various impact pressures.

To better understand material behavior under dynamic loading conditions, it is necessary to establish a correlation between dynamic characteristics and strain rate. Empirical relationships are effective tools for material characterization, which help in designing and optimizing structures and components subjected to dynamic loads. The ability to correlate strain rates and dynamic properties enables the development of models and solutions that improve the safety, durability, and performance of materials in dynamic situations. For example, the empirical relationship derived from this study highlights the sensitivity of the mechanical properties of 3D-printed CF-PETG to high strain rates, which can aid in material selection for applications requiring high impact resistance.

The maximum stress of 3D printed materials was significantly influenced by any increase in strain rate or filling density, as shown in Fig. 5. An examination of the evolution of maximum stress in 3D CF-PETG composites with different fill densities (50%, 75%, and 100%) shows a consistent trend towards increasing maximum stress with higher fill densities, irrespective of the pressure applied. This finding underlines the beneficial effect of increasing fill density on the mechanical characteristics of composite materials. A denser, better-bonded structure increases load-bearing capacity, resulting in higher maximum stress values under dynamic loading. Increased fill density also contributes to improving stiffness, strength, and mechanical resistance to damage. For material scientists and engineers, these findings suggest that increasing fill density in 3D-printed composites can improve the performance of structures exposed to dynamic stresses, offering a practical guideline for material design in demanding applications.

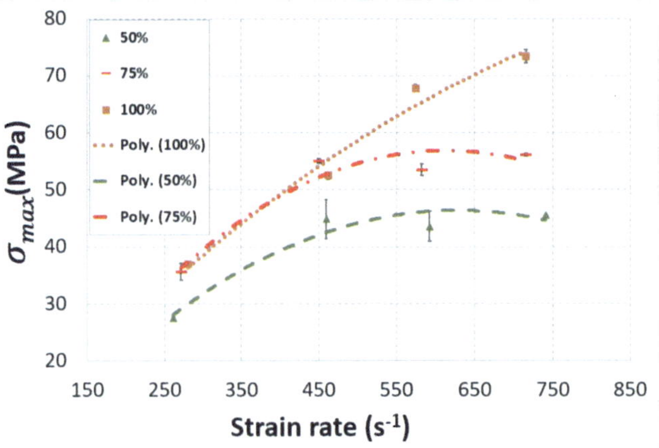

Fig. 5. Maximum stress vs strain rate for different tested 3D specimens.

Furthermore, varying impact pressure or strain rate has a substantial effect on the dynamic compressive strength of 3D-printed CF-PETG materials (Fig. 5). The initial increase in maximum stress as a function of strain rate for 50% and 75% fill densities is most likely due to the materials' sensitivity to strain rate. As strain rate increases, 3D-printed materials respond with higher stress values, indicating a greater ability to absorb and transmit energy under dynamic loading conditions until a threshold is reached, where maximum stress is maintained or increases at a slower rate. This behavior suggests that, at higher strain rates, the material begins to approach a point of saturation, where further strain rate increases have diminishing effects on the stress response. The maximum stress for the composite at 100% fill density increases almost linearly with the strain rate. This trend indicates a more consistent relationship between strain rate and material behavior. The increased density and rigid structure of the 100%-filled composite may result in a more uniform and linear response to changes in strain rate, without immediately reaching saturation as is the case in lower-fill density scenarios.

Similarly, the dynamic compressive modulus of various 3D-printed composites was illustrated (Fig. 6). Trends in compressive modulus as a function of fill density reveal interesting material characteristics under compression. Unexpectedly, the 100% fill density gave a less significant performance, with lower compressive modulus values than those obtained with the 50% and 75% densities. This phenomenon may be due to the extremely rigid and dense structure of the 100%-filled composite, which reduces its flexibility and energy dissipation under compressive loading. In contrast, the configuration with 75% filling density proved the most promising, producing the best values of compressive modulus. This finding implies that a balance between structural density and flexibility is essential to improve compressive modulus under dynamic loading conditions.

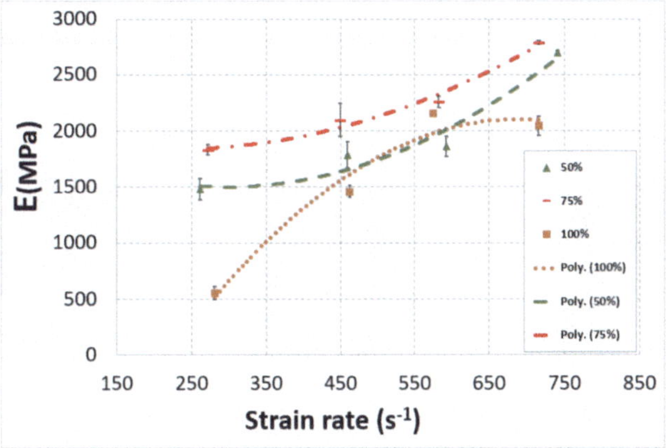

Fig. 6. Compressive modulus vs strain rate for different tested 3D specimens

In addition, in-plane testing of honeycomb-filled composites at different strain rates revealed two fundamental trends in the evolution of compressive modulus as a function of strain rate. For fill densities of 50% and 75%, a convex behavior is observed, indicating immediate softening followed by hardening as a function of increasing strain rates. In contrast, the 100% filling density shows a concave trend, indicating a gradual increase in compressive modulus with higher strain rates, but without the initial softening observed for the other 3D samples. The convex behavior in the lower fill densities suggests that the material softens initially under increased strain, but as the rate continues to increase, the material hardens, likely due to strain rate-induced changes in its microstructure. The concave trend for the 100% fill density could be attributed to the material's higher stiffness, which results in a more gradual and stable response to increasing strain rates.

To assess the damage behavior, a high-speed camera was used to track the damage progression over time, which is crucial for understanding how the material responds to dynamic loading conditions. By capturing the material's response at high strain rates, the camera enabled a detailed observation of damage development under varying pressures, offering valuable insights into how the material's structural integrity evolves during dynamic testing. The results showed significant differences in damage response between different fill densities, as depicted in Fig. 7. Specifically, at the 75% and 100% fill densities, the composites exhibited exceptional damage resistance, with no visible macro-damage, even under the highest applied pressure. The images in Fig. 7 clearly show the difference in damage patterns, where 75% and 100% fill densities demonstrate only slight volume reduction, while the 50% fill density exhibits visible cracking and fracturing at higher pressures. For the 75% and 100% fill densities, the damage was primarily characterized by volume reduction rather than macroscopic cracking or fracturing. This volume reduction suggests a localized, energy-absorbing deformation process that likely involves plastic deformation of the material without catastrophic failure. These observations imply that the higher-density composites have a natural ability to absorb energy through deformation, which can be attributed to microstructural changes such as fiber-matrix interactions and microcracking. The absence of obvious macroscopic damage in these cases indicates that the 75%- and 100%-filled composites exhibit higher damage tolerance, bending without compromising structural integrity under dynamic loading conditions.

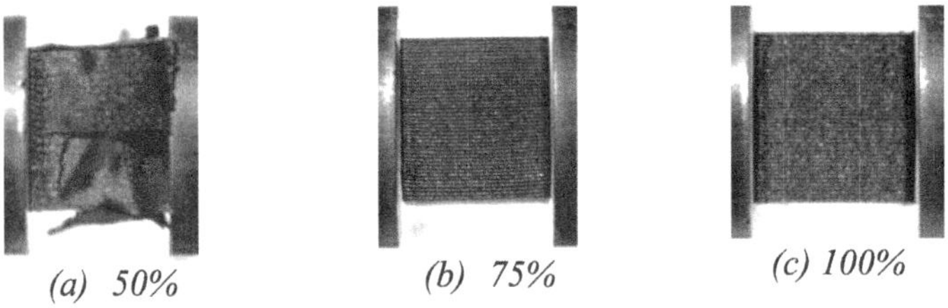

Fig. 7. Each sample after dynamic impact, P=2.4 bar

In contrast, the 50% fill density exhibited macro-damage when subjected to high pressures, which is indicative of a less effective energy absorption mechanism. The observed macro-damage in these samples suggests that the lower fill density lacks the robustness to withstand high dynamic loads, leading to more catastrophic damage such as cracking. This comparison highlights the significant differences in damage mechanisms: while higher densities primarily undergo volume reduction with minimal external damage, lower-density materials experience more visible cracking and fracturing. These findings have important implications for the design and manufacturing of CF-PETG-based composites. Understanding the damage tolerance of 75% and 100% fill densities can inform future material designs, particularly in applications where energy absorption and damage resistance are crucial. This knowledge may lead to improved manufacturing processes that optimize fill density for enhanced dynamic performance in structural applications.

5 Conclusion

The study of 3D CF-PETG composites with varied filling densities revealed an intricate connection between mechanical characteristics, damage resistance, and failure mechanisms. While increasing filling density typically increases maximum stress, surprising results were reported in compressive modulus, with 75% filling density providing the greatest results. Notably, damage resistance and failure behavior differed dramatically, with the 50% filling density failing at the maximum pressure (2.4 bar), indicating a more brittle reaction. In contrast, the 75% and 100% filling densities exhibited elastic-plastic behavior, indicating improved damage resistance and ductility even at high pressures. These results suggest that optimizing filling densities can lead to enhanced composite performance, but further investigation is needed to understand the long-term durability and fatigue behavior of these composites under varying environmental conditions. These findings highlight the complexities of material behavior under dynamic loadings, emphasizing the need for specialized composite designs to provide maximum performance across a wide range of mechanical parameters and suggesting future research into optimizing material configurations for specific applications in high-stress environments.

References

1. Kumar, K.S., Soundararajan, R., Shanthosh, G., Saravanakumar, P., Ratteesh, M.: Augmenting effect of infill density and annealing on mechanical properties of PETG and CFPETG composites fabricated by FDM. Mater. Today: Proc. **45**, 2186–2191 (2021). https://doi.org/10.1016/j.matpr.2020.10.078
2. Shunmugasamy, V.C., Gupta, N., Nguyen, Q., Coelho, P.: Strain rate dependence of damage evolution in syntactic foams. Mater. Sci. Eng. A Struct. Mater. **527**, 6166–6177 (2010). https://doi.org/10.1016/j.msea.2010.06.021
3. Srinidhi, M.S., Soundararajan, R., Satishkumar, K.S., Suresh, S.: Enhancing the FDM infill pattern outcomes of mechanical behavior for as-built and annealed PETG and CFPETG composite parts. Mater. Today: Proc. **45**, 7208–7212 (2021). https://doi.org/10.1016/j.matpr.2021.02.417
4. Tarfaoui, M., Daly, M., Kbaier, R., Chihi, M.: Investigation of 3D printed CF-PETG composites' tensile behaviors: synergizing simulative and real-world explorations. Compos. Sci. Technol. **247**, 110385 (2024). https://doi.org/10.1016/j.compscitech.2023.110385

Influence of Repetitive Recycling of 3D Printing Filament on Mechanical Properties of PLA and PLA/Spent Coffee Grounds Composite

Oumaima Boughanmi, Lamis Allegue, and Haykel Marouani(✉)

National Engineering School of Monastir, Mechanical Engineering Laboratory, University of Monastir, 5019 Monastir, Tunisia
haykel.marouani@enim.rnu.tn

Abstract. Due to its biodegradability, biocompatibility, and mechanical properties, PLA is a leading biomaterial for numerous applications, especially for fused deposition modeling and fused filament fabrication. Nonetheless, the absence of a comprehensive recycling strategy could emerge as a significant source of plastic pollution in the future. Indeed, the polymer undergoes deterioration during melt recycling, resulting in a decrease in some mechanical properties that can compromise the recyclability. To improve the properties of recycled PLA, the utilization of organic fillers coming from renewable materials can be considered as a sustainable solution. The objective of this work is then to evaluate the effect of recycling (reprocessing) on a virgin raw material as well as on biocomposite based on spent coffee grounds. The different types of filaments are extruded and re-extruded and characterized under tensile tests. The results show that the mechanical properties of recycled PLA are lower than those of virgin PLA. The tensile strength of recycled PLA decreases by 14% as compared to virgin PLA with the increase in the number of repetitive recycling. On the other hand, Young's modulus values exhibit fluctuations. The decreasing value of the mechanical properties can be associated with the thermal degradation of the PLA molecular chain during the extrusion process. Regarding the SCG filler, no major enhancement has been observed which is attributed to the poor adhesion between the matrix and the filler. Further investigations on the reprocessing process must be made to optimize the filament extrusion parameters to reduce the mechanical properties degradation.

Keywords: Recycling · re-extrusion · PLA · Coffee grounds · FFF

1 Introduction

In recent years, 3D printing has emerged as a highly appealing additive manufacturing (AM) technique, owing to its versatility and numerous advantages [1]. The increasing use of AM lies in its ability to facilitate the production of intricate designs, reduce manufacturing times, and enable on-demand manufacturing. Among the most prevalent methods of 3D printing are fused deposition modeling (FDM) and fused filament fabrication (FFF). These methods utilize a thermoplastic filament, continuously fed into the

M. Soula et al. (Eds.): AIAM 2024, LNME, pp. 92–101, 2025.
https://doi.org/10.1007/978-3-031-94889-3_8

printer, heated to its melting point, extruded through a heating nozzle, and deposited onto a printing platform. Polylactic acid (PLA) stands out as one of the most utilized thermoplastics in 3D printing. Its widespread adoption is attributed to its biodegradability, biocompatibility, recyclability, and compostability [2]. Additionally, PLA is readily available in large quantities and can be processed using conventional polymer processing techniques such as extrusion and injection molding. With its commendable mechanical, optical, and barrier properties, PLA presents itself as a viable alternative to petroleum-based polymers across various applications [3]. Beyond its inherent properties, PLA is particularly well-suited for 3D printing due to its relatively low melting point (150–160 °C), ease of processing, and satisfactory mechanical characteristics [4]. Furthermore, PLA lends itself well to the production of composite materials, wherein it can be combined with wood [5], metal [6], carbon fiber [7], glass fiber [8], or bio-based fillers such as hemp fibers [9], bamboo fibers [10], banana pseudo-stem fibers [11], etc. to enhance specific properties and expand its applicability even further. However, 3D printing is poised to pose challenges in waste management in the future, encompassing failed prints, support structures, and disposable prototypes. At the same time, AM technology holds promise in enhancing the circular economy framework through the utilization of filaments derived from post-consumer recycled polymers [12] to produce new items. Consequently, within the circular economy paradigm, utilizing PLA or PLA-based biocomposites for 3D printing presents a significant opportunity to adhere to the principles of the 3Rs: reduce, reuse, and recycle. Extensive research has been dedicated to assessing the recyclability and reusability (involving the production of new filaments from 3D printing waste) of such materials despite challenges related to their degradation post-mechanical recycling [13, 14].

Anderson [15] suggested recycling PLA waste directly by grinding and re-extruding the waste materials into filaments. After one recycling cycle, the average mechanical values of the recycled PLA specimens were 5% for tensile modulus and 11% for tensile yield strength, respectively, lower than those of the virgin specimens. Cruz-Sanchez et al. [16] examined the deterioration of PLA mechanical qualities during five recycling cycles while considering four distinct recycling process chains. In a different investigation, Zhao et al. [17] discovered that because the viscosity values of PLA rapidly decline, it can only be reprocessed for two 3D printing cycles. The authors claim that the limited number of reprocessing cycles was caused by the molecular weight reduction that occurred during those cycles, even though mechanical characteristics did not change considerably. Chu et al. investigated [18] the recycling of PLA wastes from university 3D printing labs and recommended that PLA wastes only be recycled once. All these investigations have demonstrated the technical viability of employing recycled PLA for FFF. However, they have also demonstrated that PLA post-consumer waste and processing factors have a significant impact on the mechanical recycling of PLA.

Furthermore, repurposed 3D printed trash finds other uses besides 3D printing. For instance, composites with a recycled PLA matrix supplemented with carbon fiber [19] or silica particles [20] have been made using various leftover PLA from 3D printing. Other authors [21, 22] assessed the PLA wastes from 3D printing and investigated the possibility of turning them into functional films by mechanical recycling. Two types of PLA 3D printing wastes were utilized in the study: wastes derived from a mixture of several PLA grades and wastes from a well-known reference grade. When wastes from the well-known PLA grade were used and put through a rigorous cleaning stage, one extrusion cycle, and compression molding, films with good performance were produced. Furthermore, even if recycled PLA filaments for 3D printers are commercially accessible, there are still certain restrictions associated with using them in terms of the quality of the components that are 3D printed, even with the potential for closed-loop recycling of PLA 3D printing waste.

On the other hand, significant waste is produced by food waste and loss, including spent coffee grounds (SCG) from coffee consumption. By using the same concepts of the circular economy, by incorporating SCG into polymer-based composites for 3D printing, these byproducts can be made more valuable. Composites made of PLA and SCG have potential applications as 3D printing feedstock. The characterization of a PLA/SCG composite feedstock for use in AM pellet extrusion systems was reported. TGA and DSC report that PLA/SCG formulations are thermally robust through 3D printing, with SCG degradation being negligible at PLA 3D printing temperatures of about 190 °C. SCG has a modest impact on the crystallinity of materials that are 3D printed, as their peak intensity ranges are comparable to those of XRD. The formulations behave in a non-Newtonian behavior, and as the weight percentage of SCG increases, their viscosity falls, resulting in enhanced printability and layer adhesion. Tensile tests support these findings by demonstrating greater values of elongation at break and ultimate tensile strength when compared to state-of-the-art methods. Even though micrographs revealed some variation in SCG dispersion and granulometry, the use of pellet feedstock reduced the thermomechanical deterioration brought on by additional raw material processing, such as extruding filaments. Several sample batches, including a sizable furniture product, were successfully 3D printed with intricate geometries or nonplanar slicing techniques.

Nonetheless, biocomposites may require chemical treatments to enhance their mechanical properties, such as strength and durability. While these treatments can be effective in improving material performance, they can also come with significant limitations [23]. Not only can they be financially costly due to the processes and materials involved, but they can also have adverse environmental impacts [24]. Indeed, the chemicals used in these treatments can be toxic and polluting, contributing to environmental degradation and the deterioration of air, water, and soil quality and pose risks to human health [25]. These treatments, such as alkaline treatment, are often associated with environmental pollution during manufacturing, usage, and disposal phases due to the use of chemicals like sodium hydroxide (NaOH) and the generation of waste products previous research found that the pH value of the sodium hydroxide solution was highly alkaline, suggesting adverse effects in the disposal area [26]. Moreover, the cost of chemical treatments, including the procurement of chemicals, treatment processes, and waste management, can be substantial, adding financial burdens to biocomposite production and recycling [27]. Then, in the context of sustainability, the use of chemical treatments in biocomposites contradicts sustainability goals.

This research paper delves into the examination of the mechanical properties of a commercial PLA, subjected to multiple recycling cycles without chemical treatments. Additionally, it explores the impact of integrating 5% SCG on its recycling behavior. By focusing on these aspects, the study aims to provide insights into the viability and sustainability of recycled PLA materials in various applications. Understanding how mechanical properties evolve with recycling, as well as the influence of spent coffee grounds, can inform better practices for recycling PLA and incorporating sustainable additives into its composition. Through meticulous experimentation and analysis, this research contributes to the ongoing efforts to enhance the performance and eco-friendliness of PLA-based materials.

2 Materials and Methods

This section details the characteristics of the PLA investigated, and the sources of coffee grounds. Furthermore, it delineates the methodologies employed for the filament fabrication, the recycling process, the PLA-SCG elaboration, and the characterization of mechanical properties.

2.1 Materials

Investigated PLA, referenced as V PLA, is considered a virgin raw material for this study. It is obtained by grinding a commercial printing filament (Hyper Series PLA 3D Printing Filament from reality, filament diameter of 1.75 mm). The recommended extrusion temperature for this filament ranges between 190 and 230 °C. The SCG is collected from local cafes close to the university (Monastir – Tunisia).

2.2 Methods

PLA-SCG Composite Fabrication

SCG are dried in an oven at 70 °C until the mass-loss is stabilized. The process involves measuring the change in weight of the coffee grounds at regular intervals (every 15 minutes) using a scale throughout the drying process. The examination of each raw material lasted 3.5 h (210 min). The powders were then sieved to achieve a particle size smaller than 180 μm. Subsequently, 5% (in weight) SCG is added to V PLA and mixed in a corotating twin-screw extruder with a temperature profile of 50 °C, 150 °C, and 150 °C (from the hopper to the extruder die) at a speed of 18 rpm (model: CT30/16, China). The extruded material is then pelletized using a professional grinder. Finally, the material is placed again in an oven at 60 °C for 48 h to reduce the moisture content. Figure 1 illustrates the procedure and Fig. 2 the experimental set-up used.

Filament Extrusion and Reprocessing

The methodology used in this study is illustrated in Fig. 3. Firstly, the two types of materials (V PLA, and V PLA / SCG) are dried for 24 h at 60 °C in an oven to remove any moisture. Subsequently, the materials are fed into the hopper of a single-screw extruder (Composer 450 from 3devo) which is equipped with a 4 mm diameter nozzle. The extrusion parameters (Table 1) are adjusted according to the type of material. Once a filament of uniform constant diameter (1.75 ± 0.5 mm) is obtained, the filament is passed through a positioner onto the winding wheel mounted on the winder. The

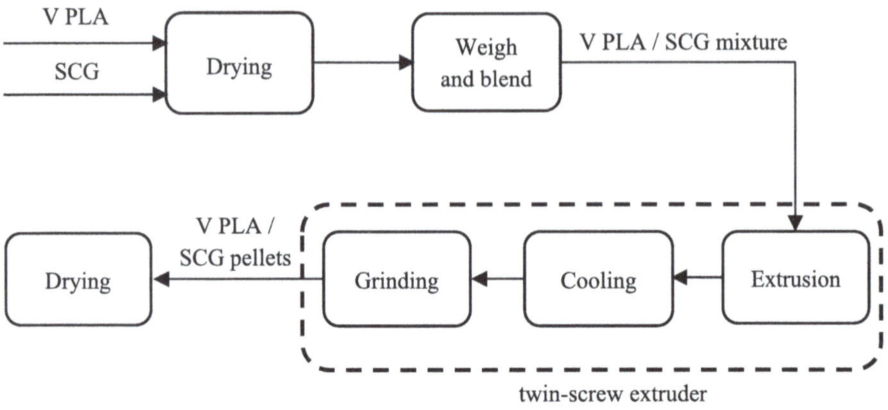

Fig. 1. Flow chart of PLA-SCG composite preparation.

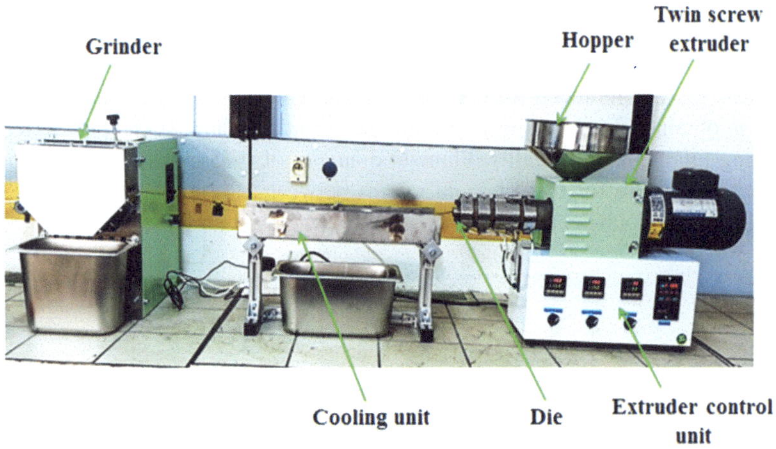

Fig. 2. Experimental set-up for composite preparation (twin screw extruder, cooling unit and grinder).

extruded filament is conveyed through the puller wheels located just below the nozzle. A diameter measurement sensor between the nozzle and the puller wheels is used to measure the diameter (Fig. 4). Based on this measurement and the extrusion rate, the speed of the puller wheel is controlled by the integrated software of the device which generates a controlled tension which aided in pulling the filament and ensuring the diameter uniformity.

Following filament extrusion, PLA experiences reprocessing (recycling) utilizing the same established procedure. Newly produced filaments are designated as R PLA i, wherein i (1 or 2) denotes the number of extrusion cycles undergone during the reprocessing.

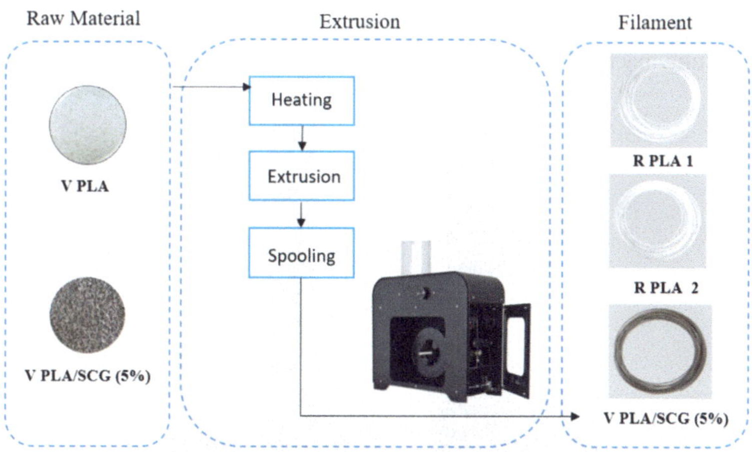

Fig. 3. Filaments fabrication process.

Table 1. The extrusion parameters

Material	Temperature (°C)				Extrusion speed (rpm)	Cooling percentage (%)
	T_1	T_2	T_3	T_4		
V PLA	170	180	190	180	2.5	80
V PLA/SCG	170	185	195	180	4	40

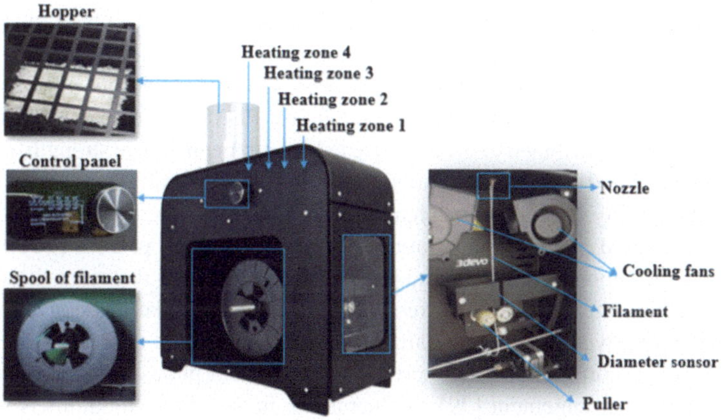

Fig. 4. 3Devo filament maker components' description.

Tensile Tests

The virgin PLA, composite, and recycled filaments are subjected to tensile testing according to ASTM D2256 standard, with a testing speed of 100 mm/min, an initial diameter of 1.75 mm, and a length of 250 mm. The tests were conducted using a Lloyd EZ 20 universal testing machine (Fig. 5). Three replicates were applied for each material.

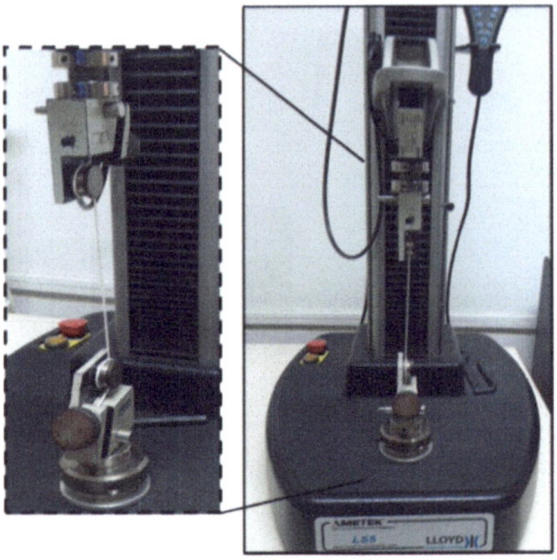

Fig. 5. Tensile test machine.

3 Results and Discussion

Figure 6a exhibits the tensile behavior of V PLA, V PLA / SCG, R PLA 1, and R PLA 2. V PLA has a mean tensile strength of 46.8 MPa (±1.6 MPa). It exhibits a modulus of elasticity equal to 637.59 MPa (±82 MPa) and an elongation value of 0.71%. The reprocessed PLA, as for the composite material (as is manufactured by a reprocessing procedure), denotes a decrease in tensile strength and a dramatic decrease in elongation. Indeed, the tensile strength (Fig. 6b) ranges from 43.28 to 37.56, and the elongation range from 0.16 to 0.11 for R PLA 1, R PLA 2, and V PLA/SCG, respectively (Fig. 6d).

Several factors contributed to the decrease in PLA filament tensile strength, including the recycling cycle and control parameters such as screw speed, barrel temperatures, cooling rate, and humidity. Moreover, mechanical stresses such as shear forces, stresses, and compressions caused polymer degradation during the extrusion process [28]. Biopolymers undergo thermomechanical degradation induced by various factors such as temperature, oxygen, humidity, mechanical stress, and light. This degradation involves reactions such as oxidation, hydrolysis, and chain breakage, leading to a decrease in molecular mass and altering morphological properties including amorphous-crystalline ratio, chirality, and steric rearrangement [29]. A fluctuation in the elastic modulus is shown in Fig. 6c following previous studies [30, 31]. However, elongation shows a decrease after 2 recycling phases with 14%. This decrease is due to the inability of the material to accept progressive deformations.

Tensile strain-stress curves of the PLA-A and PLA-A/ SCG (5%) are illustrated in Fig. 6a. The addition of SCG to the PLA-A matrix resulted in a change in tensile strength, modulus, and elongation. The tensile strength of composite filaments was slightly lower by up to 36% compared to the tensile strength of the PLA filament.

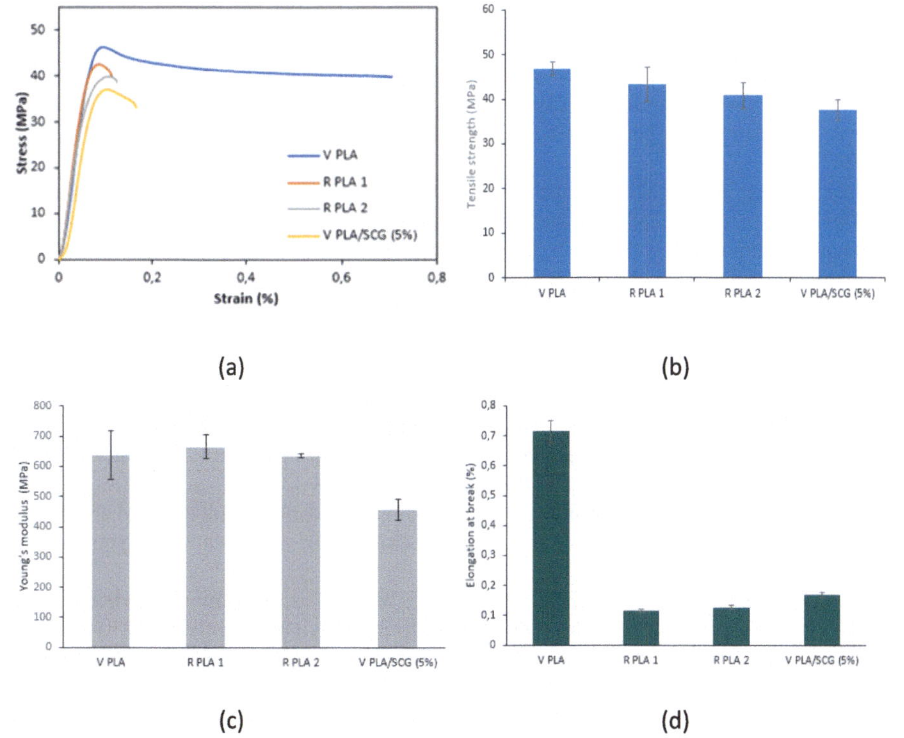

Fig. 6. (a) Tensile stress-strain curve (b) Tensile strength (c) Young's Modulus (d) Elongation at break.

4 Conclusion

This work describes the effect of one and two reprocessing cycles and the incorporation of 5%(wt.) of SCG on the properties of a commercial 3D printing PLA. (Recycled filaments and Composite filaments, consisting of polylactic acid (PLA) spent coffee grounds (with an amount of 5% by weight) were extruded. The filament recycling process begins by initially grinding commercial filament, followed by melting and extruding it. The PLA/SCG composite pellets were successfully prepared via twin-screw extrusion to achieve homogeneity and then extruded once more using the 3devo filament extruder to produce filament. The results show that the mechanical properties of recycled PLA were better when compared with those of the neat PLA. Additionally, the addition of SCG has reduced the tensile properties of the composite, indicating its higher hydrophilic tendency. Overall, our investigations indicate that recycling 3D printing into filament is feasible. However, the mechanical properties are deteriorated compared to the virgin raw material. However, an optimization of the reprocessing process is envisaged to reduce these degradations.

References

1. Praveena, B.A., Lokesh, N., Buradi, A., Santhosh, N., Praveena, B.L., Vignesh, R.: A comprehensive review of emerging additive manufacturing (3D printing technology): methods, materials, applications, challenges, trends, and future potential. Mater. Today Proc. **52**, 1309–1313 (2022)
2. Joseph, T.M., Kallingal, A., Suresh, A.M., Mahapatra, D.K., Hasanin, M.S., Haponiuk, J., Thomas, S.: 3D printing of polylactic acid: recent advances and opportunities. Int. J. Adv. Manuf. Technol. **125**, 1015–1035 (2023)
3. Naser, A.Z., Deiab, I., Darras, B.M.: Poly(lactic acid) (PLA) and polyhydroxyalkanoates (PHAs), green alternatives to petroleum-based plastics: a review. RSC Adv. **11**, 17151–17196 (2021)
4. Kadhum, A.H., Al-Zubaidi, S., AlKareem, S.S.A.: Optimization of mechanical properties and surface characteristics of PLA+ 3D printing materials. Int. J. Chem. Eng. **2023**, 1–15 (2023)
5. Krapež Tomec, D., Balzano, A., Žigon, J., Šernek, M., Kariž, M.: The effect of printing parameters and wood surface preparation on the adhesion of directly 3D-printed PLA on wood. J. Renew. Mater. **10**, 1787–1796 (2022)
6. Ghozali, M., Fahmiati, S., Triwulandari, E., Restu, W.K., Farhan, D., Wulansari, M., Fatriasari, W.: PLA/metal oxide biocomposites for antimicrobial packaging application. Polym. Technol. Mater. **59**, 1332–1342 (2020)
7. Cao, M., Cui, T., Yue, Y., Li, C., Guo, X., Jia, X., Wang, B.: Investigation of carbon fiber on the tensile property of FDM-produced PLA specimen. Polymers (Basel). **14**, 5230 (2022)
8. Ismail, K.I., Pang, R., Ahmed, R., Yap, T.C.: Tensile properties of in situ 3D printed glass fiber-reinforced PLA. Polymers (Basel). **15**, 3436 (2023)
9. Dogru, A., Sozen, A., Neser, G., Seydibeyoglu, M.O.: Effects of aging and infill pattern on mechanical properties of hemp reinforced pla composite produced by Fused Filament Fabrication (FFF). Appl. Sci. Eng. Prog. (2021)
10. Ye, W., Dou, H., Zhang, D., Yang, F., Cheng, Y., Cai, W.: Effects of process parameters on mechanical properties and interface of 3D printed bamboo-inspired CCFR-PLA/TPU composites. Polym. Compos. **45**, 227–242 (2024)
11. Pei, P., Zou, R., Zhang, C., Yu, M., Chang, S., Tan, J., Li, J., Li, X., Li, S.: Optimization of alkali-treated banana pseudo-stem fiber/PBAT/PLA bio-composite for packaging application using response surface methodology. Bioresources. **18**, 39–59 (2022)
12. Shanmugam, V., Das, O., Neisiany, R.E., Babu, K., Singh, S., Hedenqvist, M.S., Berto, F., Ramakrishna, S.: Polymer recycling in additive manufacturing: an opportunity for the circular economy. Mater. Circ. Econ. **2**, 11 (2020)
13. Chawla, K., Singh, R., Singh, J.: On recyclability of thermoplastic ABS polymer as fused filament for FDM technique of additive manufacturing. World J. Eng. **19**, 352–360 (2022)
14. Vidakis, N., Petousis, M., Tzounis, L., Maniadi, A., Velidakis, E., Mountakis, N., Papageorgiou, D., Liebscher, M., Mechtcherine, V.: Sustainable additive manufacturing: mechanical response of polypropylene over multiple recycling processes. Sustain. For. **13**, 159 (2020)
15. Nagengast, N., Bay, C., Döpper, F., Schmidt, H.-W., Neuber, C.: Thermo-mechanical recyclability of additively manufactured polypropylene and Polylactic acid parts and polypropylene support structures. Polymers (Basel). **15**, 2291 (2023)
16. Vidakis, N., Petousis, M., Maniadi, A.: Sustainable additive manufacturing: mechanical response of high-density polyethylene over multiple recycling processes. Recycling. **6**, 4 (2021)
17. Anderson, I.: Mechanical properties of specimens 3D printed with virgin and recycled Polylactic acid. 3D print. Addit. Manuf. **4**, 110–115 (2017)

18. Cruz Sanchez, F.A., Boudaoud, H., Hoppe, S., Camargo, M.: Polymer recycling in an open-source additive manufacturing context: mechanical issues. Addit. Manuf. **17**, 87–105 (2017)
19. Zhao, P., Rao, C., Gu, F., Sharmin, N., Fu, J.: Close-looped recycling of polylactic acid used in 3D printing: an experimental investigation and life cycle assessment. J. Clean. Prod. **197**, 1046–1055 (2018)
20. Chu, J.S., Koay, S.C., Chan, M.Y., Choo, H.L., Ong, T.K.: Recycled plastic filament made from post-consumer expanded polystyrene and polypropylene for fused filamant fabrication. Polym. Eng. Sci. **62**, 3786–3795 (2022)
21. Tosto, C., Pergolizzi, E., Vitiello, L., Filippone, G., Cicala, G.: Production of PLA-milled carbon fibers (MCF) filled filaments for fused filament fabrication (FFF) printing. Macromol. Symp. **404** (2022)
22. Vidakis, N., Petousis, M., Velidakis, E., Mountakis, N., Tzounis, L., Liebscher, M., Grammatikos, S.A.: Enhanced mechanical, thermal and antimicrobial properties of additively manufactured Polylactic acid with optimized Nano silica content. Nano. **11**, 1012 (2021)
23. Kuo, C.-C., Chen, J.-Y., Chang, Y.-H.: Optimization of process parameters for fabricating Polylactic acid filaments using design of experiments approach. Polymers (Basel). **13**, 1222 (2021)
24. Beltrán, F.R., Arrieta, M.P., Moreno, E., Gaspar, G., Muneta, L.M., Carrasco-Gallego, R., Yáñez, S., Hidalgo-Carvajal, D., de la Orden, M.U., Martínez Urreaga, J.: Evaluation of the technical viability of distributed mechanical recycling of PLA 3D printing wastes. Polymers (Basel). **13**, 1247 (2021)
25. Maga, D., Hiebel, M., Thonemann, N.: Life cycle assessment of recycling options for polylactic acid. Resour. Conserv. Recycle. **149**, 86–96 (2019)
26. Kalia, S., Thakur, K., Celli, A., Kiechel, M.A., Schauer, C.L.: Surface modification of plant fibers using environment-friendly methods for their application in polymer composites, textile industry and antimicrobial activities: a review. J. Environ. Chem. Eng. **1**, 97–112 (2013)
27. Vattathurvalappil, S.H., Shaukat, M.M., Theravalappil, R., Shuja, S.Z., Abdelrahman, W.G.: Techno-environmental evaluation of alkaline treatment in flax reinforced thermoplastics. Polymers (Basel). **16**, 662 (2024)
28. Bashir, I., Lone, F.A., Bhat, R.A., Mir, S.A., Dar, Z.A., Dar, S.A.: Concerns and threats of contamination on aquatic ecosystems. In: Bioremediation and Biotechnology, pp. 1–26. Springer, Cham (2020)
29. Das, P.P., Manral, A., Ahmad, F., Sharma, B., Chaudhary, V., Gupta, S., Gupta, P.: Environmentally sustainable chemical treatment of plant fibers for improved performance of polymeric composites. Polym. Compos. **43**, 7155–7169 (2022)
30. Khoshnava, S.M., Rostami, R., Ismail, M., Valipour, A.: The using fungi treatment as green and environmentally process for surface modification of natural Fibres. Appl. Mech. Mater. **554**, 116–122 (2014)
31. Briassoulis, D.: Analysis of the mechanical and degradation performances of optimized agricultural biodegradable films. Polym. Degrad. Stab. **92**, 1115–1132 (2007)

Mechanical Behavior for 3D Printing Processes

Use of the RDPP-SF Method to Analyze Rheology Variation in an Am-Cement-Based Process

Marwen Amdouni(✉), Helmi Nasraoui, Mohamed Ali Rezgui, and Ali Trabelsi

Laboratory of Mechanics, Production, and Energy (LR18ES01), National Higher School of Engineers of Tunis (ENSIT), Tunis, Tunisia
marwan.amdouni@gmail.com

Abstract. The process of 3D Cementitious Material Printing (3DCMP) is widely used to create complex geometries in buildings and structures of various sizes. Printing through material deposition is the basis of extrusion technology and it is highly productive, labor-efficient, and cost-effective. Unlike polymeric materials, the behavior of cement-based materials has very little sensitivity to temperature, and it necessitates adjustment of the printing speed and geometry to the solidification rate of the cement material, which is controlled by the cement hydration. However, even though many rheological models are devised to address the mechanical properties of cement-based materials, still the 3DCMP process is laced with many variations and faces structural failures. In this work, we employ a case study approach to examine natural and non-natural variations occurring in two process responses, i.e., static and dynamic yield stresses. The static RDPP-SF method is combined with the GRA technique to find out the most robust combination of the input factors leading to the lesser variation of the process responses.

Keywords: Additive manufacturing · cement-based materials · random and non-random variation · stochastic frontier · multi-objective processes · GRA method

1 Introduction

One promising technology in the civil engineering industry is additive manufacturing (AM), also known as digital fabrication of cementitious material [1–3]. The technology allows for the creation of complex geometric structures from CAD models by layering materials using the extrusion technique. Moreover, the cementitious AM technology is advantageous since it allows a gain in automation, eliminates the use of formworks, reduces construction waste, and improves geometric precision. Many research works have focused on the properties of materials and the applicability of the 3DCMP process [4, 5], however, challenges pertaining to environmental, and economic matters remain to be resolved. In 2009, the ASTM and the SME created the committee "F42 for Additive Manufacturing Technology". They organized the 3DCMP into six major categories; (a) material extrusion, (b) photo-polymerization, (c) fusing, and direct application of

M. Soula et al. (Eds.): AIAM 2024, LNME, pp. 105–114, 2025.
https://doi.org/10.1007/978-3-031-94889-3_9

energy, (d) binder jetting, (e) injection of materials, and (f) lamination of sheets. Because the AM-FDM process requires fast solidification to achieve a rapid printing rate [6], the application of external energy sources (laser, UV rays, etc.) for phase changing or softening the material is a common use. However, these techniques do not apply to cementitious-based materials.

1.1 The Digital Fabrication of Cement-Based Materials

In civil engineering, the extrusion/deposition method is extensively utilized for AM processes. This method enables the production of structures at various scales, from individual components to whole buildings. However, the method using injection into a bed of particles is also used in the field of concrete-based materials and has made it possible to obtain small and big structures [7]. Less frequently, alternative techniques like projection onto mobile supports [8], controlled slip-forming processes [9], and injection into a mesh screen made by robots (permeable formwork) [10] are also employed. All processes require the material to have enough fluidity to be easily pumped and buildable (to assure the stability of the structure), regardless of the complexity of the model or the consistency of the mortar or concrete to be printed [10–14]. Hence, the determination of a limited set of rheological properties is necessary to account for these contradictory qualities. This applies to the four steps of the printing process. i.e., (i) the production of the concrete, (ii) transportation of the concrete optionally mixed with a hydration accelerator, (iii) extrusion of the concrete, and deposition of a single layer, (iv) and the production of a structure that must be stable throughout the printing process. Based on this breakdown, it is possible to determine the various rheological characteristics that the concrete-based material should possess at each printing stage [7–15]. The size of the structure is determined by the elastoplastic parameters of the material's mechanical behavior; additionally, the quality of the layer-to-layer interface needs to be characterized to prevent the development of structural weaknesses referred to as cold joints. The strength gradient across the structure's height should be considered when analyzing the printed structure's stability at a particular point in time [9]. This gradient is rooted in varying ages, resting times, and even the levels of hydration of the material for each layer after it is deposited.

1.2 Rheology of Cement-Based Materials

Understanding the material's transition from liquid-like to solid-like behavior is essential [6–16]. The elastoplastic behavior will control and scale the "static" phases of the printing process that is after layers have been deposited. According to previous work [6], the five groups of compounds that make up the cementitious material utilized in 3DCMP are sieved sand (SS), fly ash (FA. Class F), water, silica fume (SF, unidentified, Grade 940, Elkem company), and ordinary Portland cement (OPC, ASTM type I, Grade 42.5). To save water, a 6 g superplasticizer (SP, MasterPozzolithR168 from BASF) is employed in each experiment. Figure 1 illustrates the particle size distribution of cement, SF, and FA, which are analyzed by Master Sizer 2000 [13].

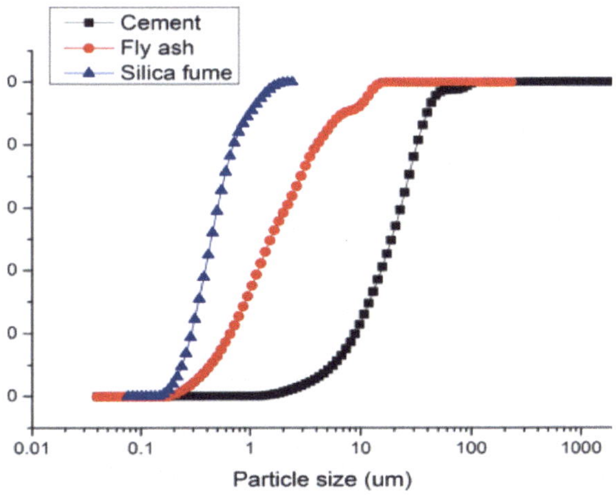

Fig. 1. Particle size distribution of OPC. FA and SF [13]

1.3 Robust Design for Products and Processes Using the Stochastic Frontier Method

Because cement materials have a wide range of properties, it is essential to plan appropriate experiments to create a workable compound with different ingredients and see how those ingredients affect the rheological properties. To have the robust design of a 3DCMP process, it is important to reduce assignable sources of variation. This problem has been addressed by the RDPP-SF (robust design for process and products using stochastic frontier) method [7, 8], which aids in choosing a combination of input parameter levels that produce the least variation in the output(s). The RDPP-SF method is employed for both static and dynamic systems and it is based on the composed error technique, i.e., the total variability in a process into random and non-random variations as shown in Fig. 2.

The math model accounting for the composed error is stated in Eq. (1).

$$yi = xi\beta + (\mathrm{vi} - \mathrm{ui}) \quad \text{for } \mathrm{i} = 1, \ldots, \mathrm{n} \tag{1}$$

The output values are bound above by exp.($x_i\beta$ + v_i+ v_i), which is the deterministic frontier;

- Y_i and X_i are logged;
- $X_i = [1, \ln(X_{1i}), \ln(X_{2i}),\ldots,\ln(X_{ki})]$ and $\beta = [\beta_0, \beta_1,\ldots, \beta_k]'$;
- $V_i \sim N(0, {\sigma_v}^2)$ represents random variation (part-to-part, measurement, etc.).
- $U_i \sim N^+(0, {\sigma_u}^2)$ represents nonrandom variation (calibration, tools, deterioration, batch, etc.).

$Y_i = f(x_i.\ \beta)\ exp\ (v_i)$: stochastic
$Y_i = f(x_i.\ \beta)\ exp(v_i\text{-}u_i)$: observed error

The stochastic frontier model is largely used in econometric studies, and for an output-oriented technical inefficiency, the model presented in Eq. (1) explains why a given firm

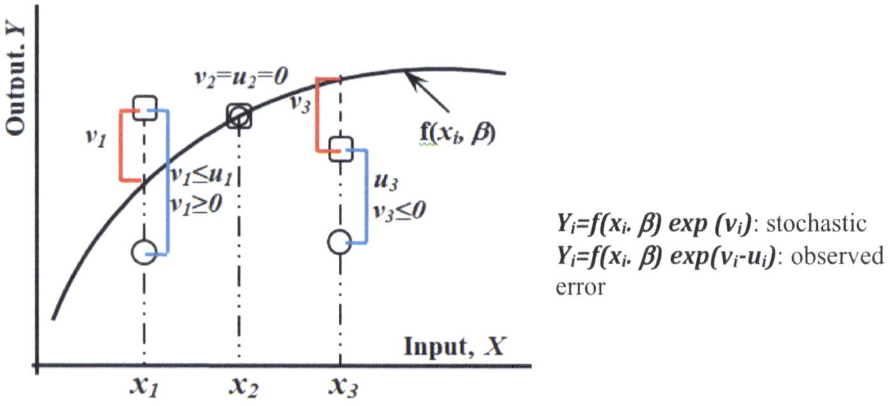

Fig. 2. Technical inefficiency using the SF model from an output-oriented point of view standpoint [7, 8]

(DMUi) is unable to achieve the maximum feasible output performance beyond noise sources. By following this principle, the RDPP-SF method [11, 12] has aligned the lack of robustness (robust design point of view) with the inefficiency (econometric point of view).

The RDPP-SF approach [11, 12] can be used for multi-objective problems and has been applied to both static and dynamic processes. Figure 3 shows the procedural scheme of the RDPP-SF method.

2 Case Study for Method

The study is retrieved from the work of Liu et al. [13]. We apply the RDPP-SF method to find out the robust design solution of the 3DCMP process, that is, the combination levels of the inputs, which yield minimum variation of the process outputs in the presence of environmental noises. The experiments are conducted according to a mixture design strategy. The input parameters are the proportions (in %) of five ingredients; x1: ordinary Portland cement (OPC, ASTM Type I, Grade 42.5), x2: sieved sand (SS), x3: fly ash (FA. Class F), x4: water and x5: silica fume (SF, unidentified, Grade 940, Elkem company). 6 g Superplasticizer (SP, MasterPozzolithR168 from BASF) is added to reduce water consumption in each experiment. The process outputs are the fresh rheological properties of the cementitious materials, that is, Y1: static yield stress (in Pa), and Y2: dynamic yield stress (in Pa). Table 1 shows the levels of the input parameters and Table 2 shows the results of the mixed design experiments.

Because of the lack of degrees of freedom for the designed experiment, only a linear model has been adopted. This corresponds to a Cobb Douglas (C.D) regression model for the input-output transfer function in the RDPP-SF method. The software FRONTIER 4.1® is used to generate the least sensitive design solution.

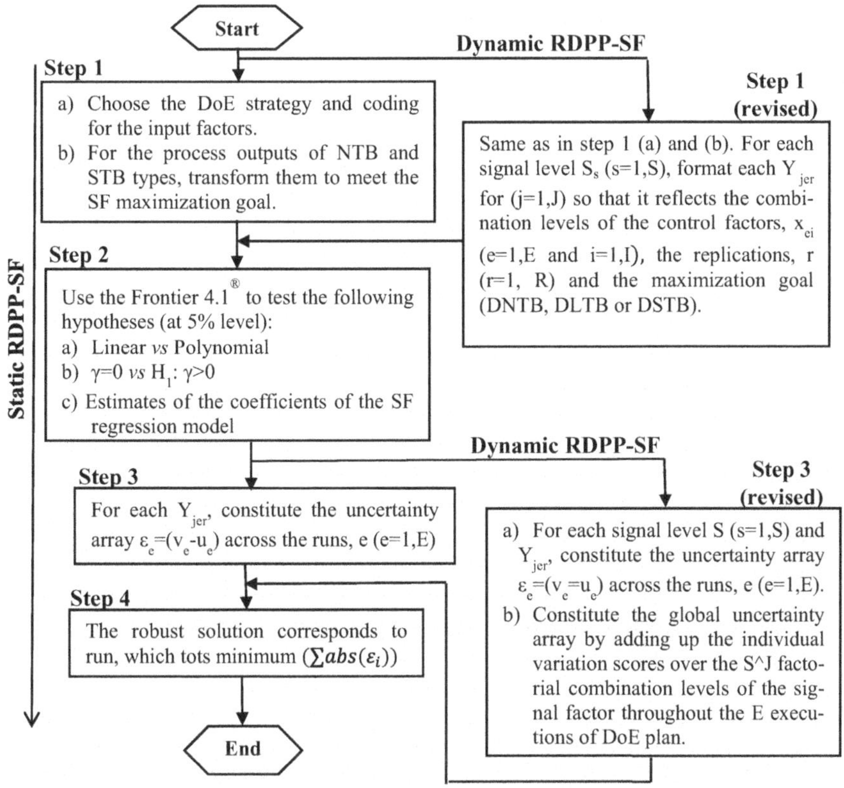

Fig. 3. Functional scheme of the RDPP-SF method for multi-objective problems [12]

Table 1. Factors of 3D cementitious material printing

Factors	Symbols	Levels
OPC (%)	x1	12; 14; 15; 16
SS (%)	x2	21; 22; 22.6; 23; 24; 26
FA (%)	x3	25; 26; 26.6; 27; 28; 30
Water (%)	x4	33; 33.9; 34; 35
SF (%)	x5	2; 2.9; 3; 4

3 Results and Discussion

Using FRONTIER 4.1® at 95% C.I., the regression model for the stochastic frontier is linear. Table 3 shows the model coefficients for Y1, and Y2 as well as the significant test using the t-ratio metrics.

Table 2. Design of experiments of 3D cementitious material printing

Run	Input Parameters					3DCMP Responses	
	x1	x2	x3	x4	x5	Y1 (Pa)	Y2 (Pa)
1	0.16	0.21	0.28	0.33	0.02	2363.13	244.81
2	0.12	0.23	0.3	0.33	0.02	913.88	153.16
3	0.12	0.21	0.3	0.35	0.02	517.19	70.53
…	…	…	…	…	…	…	…
10	0.14	0.21	0.3	0.33	0.02	572.11	256.26
11	0.12	0.26	0.25	0.35	0.02	969.86	126.55
12	0.16	0.21	0.26	0.33	0.04	5952.11	248.16
13	0.12	0.26	0.25	0.33	0.04	1687.37	208.8
…	…	…	…	…	…	…	…
18	0.12	0.26	0.27	0.33	0.02	497.46	72.67
19	0.16	0.22	0.25	0.35	0.02	526.25	95.19
20	0.15	0.21	0.25	0.35	0.04	1775.26	226.39
21	0.14	0.226	0.266	0.339	0.029	1082.26	168.61
22	0.14	0.226	0.266	0.339	0.029	1117.42	179.49

According to Table 3, the γ-values for the outputs Y1 and Y2 are lesser than 5%, arguing that only natural variations are accountable for the 3DCMP process variation at the 5% level. To obtain the robust solution, the variation array, $\varepsilon i = |vi—ui|$ should be constituted for Y1 and Y2 in relation to each of the 22 runs.

As our problem focuses on the rheological aspect of the mixture specific to the 3D printing of cement, the behavior of the static and dynamic elasticity have the same importance. Similarly for the error terms. Moreover, the problem is multi-objective, and a robust solution is obtained when applying the Grey Relational Analysis technique.

For the stochastic frontier method, Y1 and Y2 should be maximized and the $\varepsilon i1$ and $\varepsilon i2$ should be minimized. Table 4 shows the variation scores of each run for the 3DCMP process variation.

When applying the RDPP-SF method, it is shown that the robust solution corresponds to run 12 (refer to Table 4). Therefore, the best setting of the input parameter are; x1 (cement) = 16%; x2 (sand) = 21%; x3 (fly ash) = 26%; x4 (water) = 33%, and x4 (silica fume) =4%. For economic consideration, other runs (runs 14 and 16) could be chosen as highlighted in Fig. 4.

Table 3. CD regression models for the static (Y1) and dynamic (Y2) yield stresses

Variables	Param.	Y1 in (Pa) ($\gamma \neq 0; \mu = \eta = 0$)		
		Coef.	Std Error	t-ratio
Cte	β_0	−100.199	−1.040	96.264
Ln(x1)	β_1	−7.692	−0.626	12.288
Ln(x2)	β_2	−17.454	0.816	−21.384
Ln(x3)	β_3	−18.878	0.981	−19.229
Ln(x4)	β_4	−36.203	1.709	−21.183
Ln(x5)	β_5	−0.521	0.244	−2.128
$\sigma^2 = \sigma_v{}^2 + \sigma_u{}^2$		0.134	0.040	3.314
γ		0.916E-04	0.058	0.001
μ. η		–	–	–
Log (likelihood) = −9.129				
Variables	Param.	Y2 in (Pa) ($\gamma \neq 0; \mu = \eta = 0$)		
		Coef.	Std. Error	t-ratio
Cte	β_0	18.811	3.333	5.652
Ln(x1)	β_1	2.707	0.451	6.007
Ln(x2)	β_2	1.793	0.676	2.6602
Ln(x3)	β_3	3.902	0.797	4.898
Ln(x4)	β_4	−3.347	1.189	−2.808
Ln(x5)	β_5	1.153	0.168	6.844
$\sigma^2 = \sigma_v{}^2 + \sigma_u{}^2$		0.057	0.0173	3.335
γ		0.267E-04	0.029	0.902E-03
μ. η		–	–	–
Log (likeli hood) = 0.113				

Table 4. Robust design for the 3DCMP process using the RDPP-SF method

Run	$\varepsilon_{1i}=\|v_{i1}-u_{i1})\|$	$\varepsilon_{2i}=\|v_{i2}-u_{i2}\|$	Total variation	Ranking
1	0.599	0.548	1.147	9
2	0.865	0.659	1.524	22
3	0.891	0.534	1.425	16
4	0.503	0.437	0.94	4
5	0.816	0.434	1.25	13
6	0.908	0.451	1.359	15
7	0.787	0.711	1.498	19
8	0.96	0.507	1.467	18
9	0.991	0.322	1.313	14
10	0.306	0.719	1.025	6
11	0	1	1	5
12	0.71	0	0.71	1
13	0.835	0.322	1.157	11
14	0.736	0.091	0.827	2
15	0.744	0.408	1.152	10
16	0.529	0.336	0.865	3
17	0.787	0.309	1.096	8
18	0.708	0.341	1.049	7
19	0.738	0.472	1.21	12
20	1	0.429	1.429	17
21	0.918	0.591	1.509	20
22	0.886	0.624	1.51	21

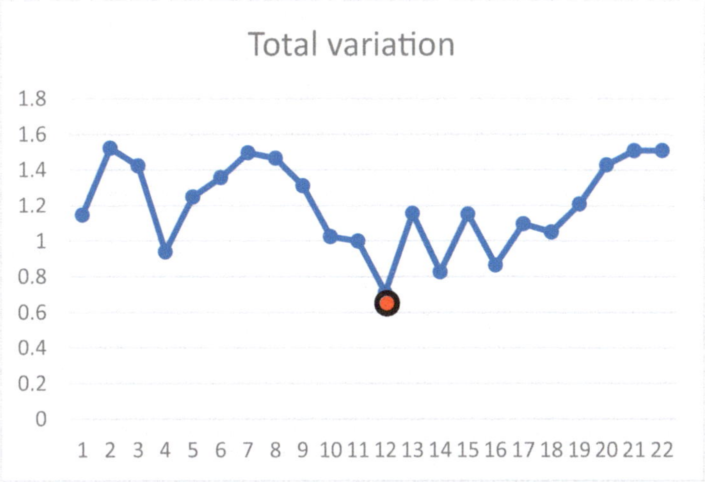

Fig. 4. Robust design solution based on the GRA method

4 Conclusion

The article has applied the static RDPP-SF method for the robust design of the 3DCMP process [8]. For a given process, the RDPP-SF method can decompose the total variation that is lacing the process/system into two neoclassic error components, i.e., pure random/natural and assignable/non-natural variations. For the case being, run 12 is the robust design solution and the best setting of the input factors is; x1 (cement) = 16%; x2 (sand) = 21%; x3 (fly ash) = 26%; x4 (water) = 33%, and x4 (silica fume) =4%. Other potential settings, such as runs 14 and 16, can be chosen in light of economic considerations.

References

1. Roussel, N., Oussel, N., Ovarlez, G., Garaults.: The origins of thixotropy of fresh cement pastes. Cem. Concr. Res. **42**(1), 148–157 (2012)
2. Nerella, V.: Micro-and macroscopic investigations of the interface between layers on the interface of 3D-printed cementitious elements. J. Mater. Civ. Eng. **29**(7) (2017)
3. Pegna, J.: Application of cementitious bulk materials to site processed solid freeform construction. SFF Symposium, Austin, TX (1995)
4. Weger, D., Lowke, D., Gehlen, C.: 3D printing of concrete structures using the selective binding method - effect of concrete technology on contour precision and compressive strength. In: 11th Fib International PhD Symposium in Civil Engineering, pp. 403–410. The University of Tokyo Edition, Tokyo (2016)
5. Lowek, D., Dini, E., Perrot, A.: Particle-bed 3D printing in concrete construction—possibilities and challenges. Cem. Concr. Res. **112**, 50–65 (2018)
6. Yiwei, W., Li, M., Ming, J.T., Shunzhi, Q.: Design 3D printing cementitious materials via Fuller Thompson theory and Marson-Percy model. Constr. Build. Mater. **163**, 600–610 (2018)

7. Zhixin, L., Mingyang, L., Yiwei, W., Teck, N.W.: Mixture design approach to optimize the rheological properties of the material used in 3D cementitious material printing. Constr. Build. Mater. **198**, 245–255 (2019)
8. Delphine, M., Shiho, K., Bessaies-Bey, H., Mantellato, S., Serina, N.: Hydration and rheology control of concrete for digital fabrication: Potential admixtures and cement chemistry. Cem. Concr. Res. **112**, 96–110 (2018)
9. Le, S.A., Austin, S., Lim, R.A., Buswell, R.L., Gibb, T.: Hardened properties of high-performance printing concrete. Cem. Concr. Res. **42**(3), 558–566 (2012)
10. Sanjayan, J.G., Nematollahi, B., Xia, M., Marchment, T.: Effect of surface moisture on inter-layer strength of 3D printed concrete. Swinburne Res Bank. **172**, 468–475 (2018). https://doi.org/10.1016/j.conbuildmat.2018.03.232
11. Rezgui, M.A., Trabelsi, A.: Use of the stochastic frontier and the grey relational analysis in robust design of multi-objective problems. Concurrent Engineering Research and Applications. **28**, 1–20 (2020). https://doi.org/10.1177/1063293X20908317
12. Trabelsi, A., Rezgui, M.A.: Robust design of processes and products using the mathematics of the stochastic frontier methodology (SF). J. Adv. Manuf. Technol. **106**, 2829–2841 (2020). https://doi.org/10.1007/s00170-019-04503-6
13. Zhixin, L., Mingyang, L., Yiwei, W., Teck, N.W., Ming, J.T.: Mixture design approach to optimize the rheological properties of the material used in 3D cementitious material printing. Constr. Build. Mater. **198**, 245–255 (2019)
14. Bouasker, M., Mounanga, P., Turcry, P., Loukili, A., Khelidj, A.: Chemical shrinkage of cement pastes and mortars at very early age: effect of limestone filler and granular inclusions. Cem. Concr. Compos. **30**(1), 13–22 (2008). https://doi.org/10.1016/j.cemconcomp.2007.06.004
15. Barret, P., Bertrandie, D.: Fundamental hydration kinetic characteristics of the major cement constituents: Ca3SiO5 and βCa2SiO4. J. Chim. Phys. **83**, 765–775 (1986)
16. Scrivener, K.L., Nonat, A.: Hydration of cementitious materials, present and future. Cem. Concr. Res. **41**(7), 651–665 (2011). https://doi.org/10.1016/j.cemconres.2011.03.026

MSLA 3D Printing of Wax Pattern for Rapid Investment Casting Inconel 718-Superalloy Aeronautical Part

Chams Eddine Dakhli[1,2](✉), Brahim Ben Fathallah[1,2], Mohamed Soula[3], and Badreddine Aouadi[3,4]

[1] Laboratoire de Mécanique, Matériaux Et Procédés (LMMP), LR 99ES05, École Nationale Supérieure d'Ingénieurs de Tunis (ENSIT), Université de Tunis (UT), 5 Avenue Taha Hussien, B.P 56, Bab Menara, 1008 Tunis, Tunisia
chamseddine.dakhli@gmail.com

[2] ENIT, University of Tunis El Manar, BP 37, Le Belvédère, 1002 Tunis, Tunisia

[3] Applied Mechanics and Engineering Laboratory (LR-11-ES19), University of Tunis El Manar, ENIT, BP 37, Le Belvédère, 1002 Tunis, Tunisia

[4] EUROCAST (Pursuit Aerospace), GP1, Route de Sousse KM7, 2033 Megrine, Tunis, Tunisia

Abstract. The precision-machining of injection molds for disposable wax patterns intended to be used in investment casting of complex superalloy aerospace components is a costly and time-consuming procedure; thus, making this method particularly unlucrative for unit or limited production runs. Yet, additive manufacturing of casting models can be more suitable in this situation. Our work, then, is to explore the feasibility of using the Masked Stereo-lithography Apparatus (MSLA)-based 3D printing technique to make a unitary experimental investment casting pattern of a jet fuel distributor unit made of Inconel 718. The model is first designed with the aid of numerical casting simulation. Then, the approved model design is 3D-printed, and the operation outcomes are discussed and analyzed. The simulation-assisted pattern-design iteration has granted to a porosity-free cast with a microporosity index and a shrinkage deformation less than 0.54% and 6.6×10^{-2} mm, respectively, altogether with a satisfactory yield of 69.17%. MSLA 3D printing has led to a cost-effective, high-quality pattern with a yield of 88.6% thanks to proper preprocessing settings including calibrations, and proper slicing design and refinements. The implemented process was found, however, to be perfectible through some technical adjustments we discussed briefly at the end of this work.

Keywords: Additive Manufacturing · Lost-wax casting · Mold design · Nickel-alloys · Stereolithography Apparatus

1 Introduction

Investment casting (IC) is widely utilized in the manufacturing of complex and high-performance aerospace components made with superalloys meant to withstand extreme functional conditions [1]. The procedure of obtaining precision components with IC

M. Soula et al. (Eds.): AIAM 2024, LNME, pp. 115–132, 2025.
https://doi.org/10.1007/978-3-031-94889-3_10

is generally managed as follows [2, 3]: (1) the pattern molds' making using precision machining, (2) the wax injection into the pattern molds, (3) the manual wax pattern assembly to a wax gating system, (4) the shell making with multilayer ceramic slurry coatings of the assembled pattern and curing, (5) the dewaxing using rapid heating, (6) the metal pouring into the resulting investment mold, (7) the mold breakage after cooling down and (8) the cut-off of the investment cast from the gating system.

Despite the rise of metal additive manufacturing (AM) as an alternative, IC remains the first choice for mass production of high-performance components due to its cost-effectiveness [4–6]. In addition, AM-made parts suffer from key issues like porosity, residual stresses, poor fatigue endurance [7, 8], and lack of accuracy and repeatability [9].

However, the IC turns out to be costly in case of small production runs [2]. Complex-geometry wax injection mold making is time and resource-consuming [10], on the one hand, and limits design flexibility on the other hand. Opting, however, for a Rapid Investment Casting (RIC) technique based on AM to make investment patterns with a castable material, could be a solution to overcome this problem. The main AM technologies utilized and studied for this purpose are Fused Filament Fabrication (FFF), Stereolithography (SLA), and Multi-Jet Modeling (MJM) [11–16]. This hybrid approach offers the AM benefits of producing intricate geometries with limitless design freedom and flexibility in addition to the IC consistency [6] and accuracy when producing thin-walled and complex parts, in a lucrative way, which is unreachable with traditional manufacturing techniques [10, 17]. Thanks to these advantages, the AM-based RIC could be suitable for making superalloy precision aerospace parts [18–20].

Nevertheless, failures and poor-quality casts weigh heavily on the overall production cost of superalloy precision casting [21, 22], the same as the meager yield caused by oversized gating and rising systems [23, 24]. The use of casting simulation to predict and minimize cast defects, therefore, sounds like an efficient aiding design tool that enables adjustments on the casting system and the process parameters before launching a limited series production. Consequently, the costs of conventional trial-and-error methods based on experiments could be cut down largely [25, 26]. The cost-effectiveness and design freedom of AM-based RIC joined with accurate casting simulation could be, then, a convenient solution to make successful IC superalloy parts in small production runs.

This work is an attempt to implement RIC in a single experimental production run of a jet fuel distributor made of Inconel 718, by performing Masked-SLA (MSLA) 3D printing of its cast pattern with a UV-curable castable resin. Thus, we will start with the part's lost wax-pattern design based on the evaluation of the casting defects via numerical simulation. Then, the validated casting pattern's AM will be implemented using proper settings. The results will be commented on, and areas for improvement will be discussed. This study is carried out at the request of Pursuit Aerospace Tunisia.

2 Materials and Methods

2.1 The Casting Material

The studied part (Fig. 1) is made from Inconel 718 Nickel-based superalloy characterized by its high performance and yield strength at high temperatures as well as its excellent corrosion and fatigue resistance, making it one of the widely used materials for critical get-engine components [27]. Its typical chemical composition limits according to Aerospace Material Specification (AMS) 5383 standards are shown in Table 1.

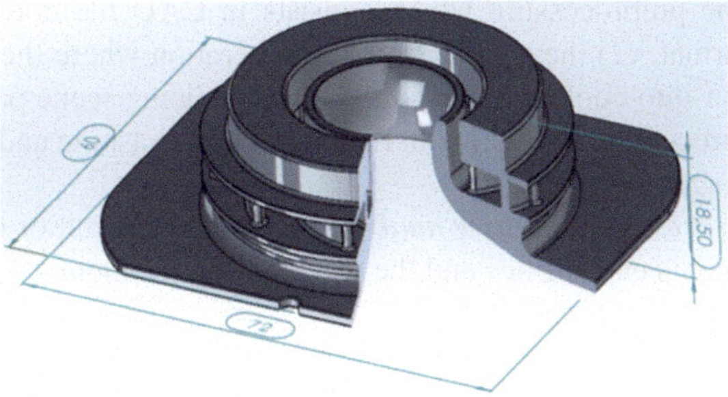

Fig. 1. The part's CAD model with zonal section view (Dimensions in mm).

Table 1. Typical Inconel chemical composition (wt. %) [28, 29].

Ni + Co	Cr	Fe	Nb + Ta	Mo	Ti	Al	Co	C	Mn	Si	Cu
50.0 55.0	17.0 21.0	Balance	4.75 5.50	2.803.30	0.65 1.15	0.20 0.80	1.0 max	0.08 max	0.35 max	0.35 max	0.3 max

2.2 The Process and the Work Methodology

As shown in Fig. 1, the part has an intricate design that includes complex internal multi-cavities and thin-walled blades, making its fabrication using common manufacturing methods other than investment casting irrelevant or even impossible. Molding its casting wax pattern, however, needs a complex and compound injection mold tooling which is non-profitable for testing samples or small production runs; That's why it becomes necessary to think of using an AM solution to produce this pattern in a way that no tooling is required, and a design freedom is granted.

In this study, the making of the 3D printed lost wax pattern design is based on the casting simulation results (Fig. 2). The Casting parameters are given in Table 2, and the investigation methodology is carried out as follows:

- *(Stage 1) The iterative numerical-simulation-based design:* The casting pattern design project, which includes the main part, the gating, the runner, and the riser systems, is verified using the Altair Inspire Cast Manufacturing Solver to simulate the metal pouring and the thermomechanical behavior of the metal at its cooling phase and the resulted cast defects. Based on the outcomes of simulations, the pattern design is readjusted depending on the defect's nature and locations, then reevaluated, and so on until the obtention of the "defect-free" one.
- *(Stage 2) The wax pattern's AM fabrication:* The validated pattern design is then 3D-printed using the Masked Stereolithography Apparatus (MSLA) technology with "Prizma" castable resin provided by MakertechLabs. To do so, three steps are involved: (1) the preprocessing which consists in CAD file conversion into STL 3D-printable format, (2) the very 3D printing operation where the 3D-print is built up layer by layer into completion according to the slicing scene parameters and (3) the model postprocessing consisting in IPA99-Alcohol washing and post-curing with 405 nm Wavelength UV light.
- *(Stage 3) The process's qualitative analyses:* Finally, the result of the 3D printing is evaluated to gage its capabilities and the ways of improvement.

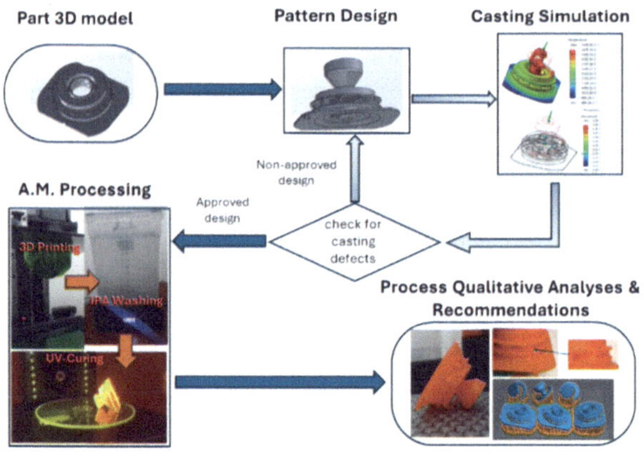

Fig. 2. The Investigation Methodology.

Table 2. Investment casting parameters.

Casting material	Inconel 718
Shell-mold material	Ceramic
Pouring temperature (°C)	1450
Initial mold temperature (°C)	750
Shell thickness (mm)	5–6
Flow rate (kg/s)	0.045

2.3 Investment Casting Simulation

The purpose of the iterative numerical-simulation-based design is to attain a shell-mold design apt to produce a cast with the minimum defects possible. The defects intended to be assessed are the shrinkage porosity and microporosity known to have the most damaging effect on cast-parts' performances [30, 31]. The first one (also called Macro porosity) refers to the volume of macroscale voids mainly caused by shrinkage and gas formation during the solidification, while the latter is the volume fraction within the cast of microscale pores (porosity index) caused by gas formation and retention during the solidification phase as well [34]. This index is considered excellent, below 1% and deemed bad above 5% [36]. The Altair software utilizes the Niyama Criterion to predict these two defects [37, 38] which are calculated as follows:

$$N_y = \frac{G}{\sqrt{R}} < C_{Ni(macro,micro)} \tag{1}$$

where G is the local thermal gradient (K/m), R *is* the local cooling rate (K/s), and C_{Ni} *is* the critical value depending on the type of porosity, above each defect is likely to occur in a specific area of the cast.

Additionally, we have evaluated the displacements (deformations) due to the metal retraction and solidification-induced stresses during and cooling down, which affect the geometry and the dimensional accuracy.

2.4 The Utilized 3D Printing Technique and the Major Settings

The second stage, involving the wax pattern AM fabrication, is carried out once the design has been validated. In the following subsections, we will describe the principle of the MSLA technology used and the major settings we managed to apply to accomplish its implementation.

The MSLA 3D Printing Principle. The utilized MSLA technique (Fig. 3) progressively builds a 3D object using a reflected UV light which polymerizes, at every Z-incremental rise of the built-plate, a thin layer of a UV-sensitive resin poured into a vat having a transparent base. The curing light shines through very specific regions of the bottom of the vat which are defined according to the cross-sectional shape of each layer by a selectively transparent LCD screen. This technology allows an "instant" flash curing with a high resolution of an entire layer of resin regardless of its size or shape complexity, leading to a faster printing speed compared to the conventional SLA technique based on a moving UV-laser beam. The STL CAD file of the model must be, therefore, introduced to the 3D printer in a "sliced" format that defines the cross-sectional shape of each single layer. So, the LCD screen can control where exactly the UV light impacts the bottom of the resin tank through an arrangement of deactivated pixels matching the shape of each drawn layer.

Procedure and Major Settings

Parameters Calibration. A UV overexposure during the polymerization of each drawn layer leads to losses in quality and accuracy, delamination, and a drop in productivity [39,

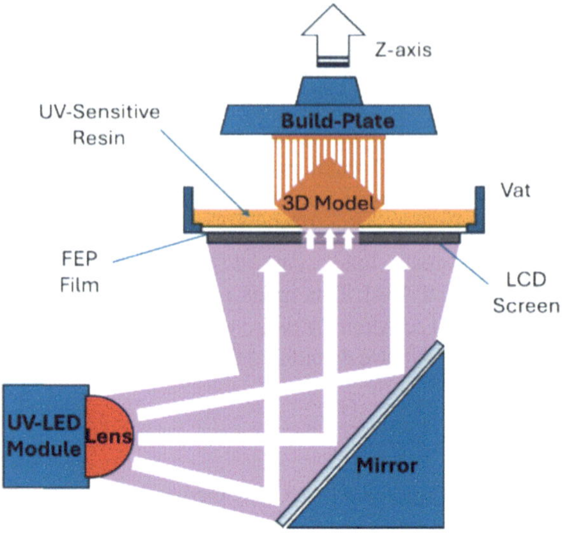

Fig. 3. MSLA 3D printing principle.

40]; while underexposure as well as high built-plate's rising speed causes deformation and collapse. Therefore, performing a series of qualitative calibration tests is needed to optimize the 3D printing parameters (UV light intensity, exposure times, speeds, etc.) to make the MSLA 3D printing succeed.

The 3D-print Design in the Slicing Phase. A bad 3D-print orientation relative to the build-plate could be a source of shrinkages, poor surface finish, and longer build time. With this in mind, we have ensured that the part is tilted in such a way as to minimize large overhangs and abrupt changes in cross-sections between successive layers, altogether avoiding as far as possible the excessive number of drawn layers and the need for a large support structure. The proper design of the support which allows the 3D print to be attached to the build-plate, at its part, plays a crucial role in the success of the operation. The inadequate design of this structure not only alters its adhesion to the plate, structural integrity, quality, and precision but also leads to unnecessary waste of material and production time. Figure 4 illustrates a 3D-print collapse due to low density and thin supports associated with under-exposure and unfitting 3D-print model tilting. In our support design, we utilized, at the first attempt, the automatic part orientation and support generation with medium preset; then, we managed to check manually the model layer by layer to proceed, if necessary, for reorientation to balance between fabrication lead time and progressive cross-sectional changes, and to proceed, also, for reinforcements in critical sections (bottom edges and large islands) prone to engage failure or damages when needed. Additionally, for a perfect adhesion of the support structure to the build-plate, we have added a raft at the bottom with burn-in layers being exposed to UV-curing in a duration over ten-fold as normal layers. All these settings combined could give our model 3D printing the best chance of success.

Fig. 4. 3D-printing collapse due to poor support design structure and model orientation.

3 Results and Discussion

3.1 Casting Simulation

Analysis of the Initial Pattern Design Attempt. In the first design attempt (Fig. 5), we used an end-gating system with side ingates connected to the inner ring and arranged in a cross form to guarantee an even and stable supply of the molten metal. We have designed a rising system composed of four risers connected to the sprue cup to ensure, in addition to the metal gapes' compensation, the role of a venting system. This configuration has also been proven efficient in suppressing the cooling rate of the molten metal [25]. This design led to a yield of 68.47% with filling and solidification times estimated to 5 and 101 s respectively.

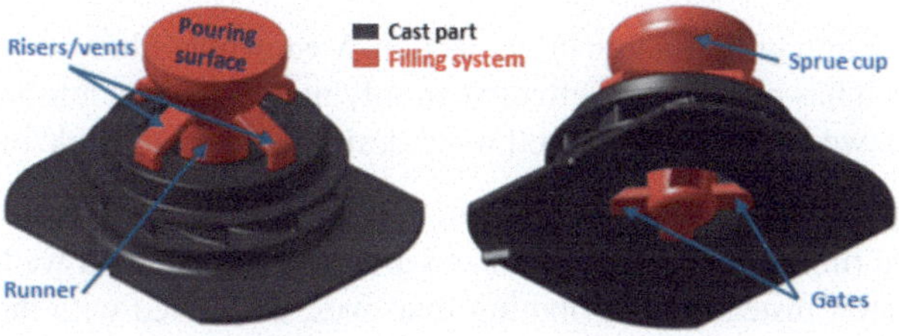

Fig. 5. Initial pattern design.

Figure 6a and b show respectively the distribution of macro- and microporosity in the casting model. A shrinkage porosity zone with a volume of 17.3 mm^3 is located at the very top of the inner ring. The defect formation in the concerned zone is due to its isolation, its level near the top position, and its thin-walled nature, making it the last area to fill with the molten Inconel 718 alloy (Fig. 7) and more likely to form and retain bubbles. Regarding microporosity, it also appears at the upper surface of the inner ring with, however, a low ratio (0.63%), and a relatively higher index of 2.44% in a tiny, isolated location at the upper ring of the outlet deflector, meaning no consequential effect on the overall integrity of the cast.

To address the shrinkage porosity issue, it seems that adding a riser or a vent at the top of the inner ring to refill the molten metal voids or conduct the gases out of the cast

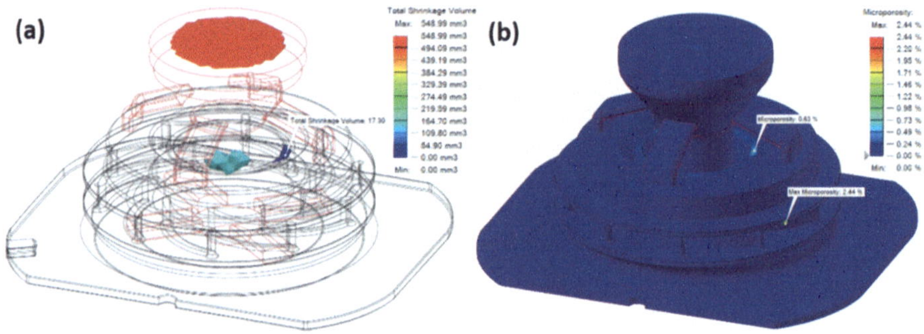

Fig. 6. Porosities in the initial pattern design: (a) Shrinkage porosity, (b) Microporosity.

Fig. 7. Last areas to fill in the initial pattern design.

is a non-viable option given its thin cross-section. Meanwhile, tilting the mold could be an alternative as it has been proven effective in solving gas-related defects in investment casting [23]. So, we have studied a shell mold design tilted at an angle of 10°.

Analysis of the 10°-Tilted Pattern Design. We have designed our new tilted pattern with a preserved runner, and gating sizes and disposition. Although we have kept only one riser/ventilator, instead of distributing four ones, connected with the highest edge and modified the sprue cup in a way to keep the pouring surface at a horizontal position to guarantee a complete filling up of the tilted-mold cavity (Fig. 8). The resulted yield has become 69.17%, while the filling and solidification times have remained approximately at 5 s and 101 s respectively.

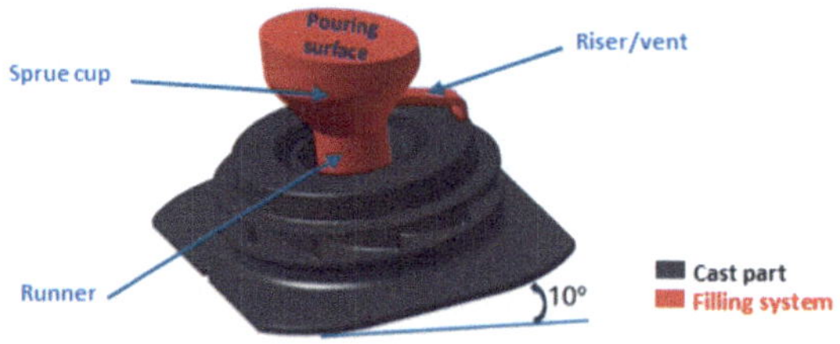

Fig. 8. 10° tilted pattern design.

When tilting the shell-mold at 10°, the pattern's main part becomes shrinkage-free (Fig. 9a). Meanwhile, only a tiny volume is located at the center of the riser, which doesn't affect the integrity of the main part, anyway. Regarding microporosity, Fig. 9b shows a very inconsequential index as the maximum value displayed does not exceed 0.54%. This enhancement is due to the improved filling session.

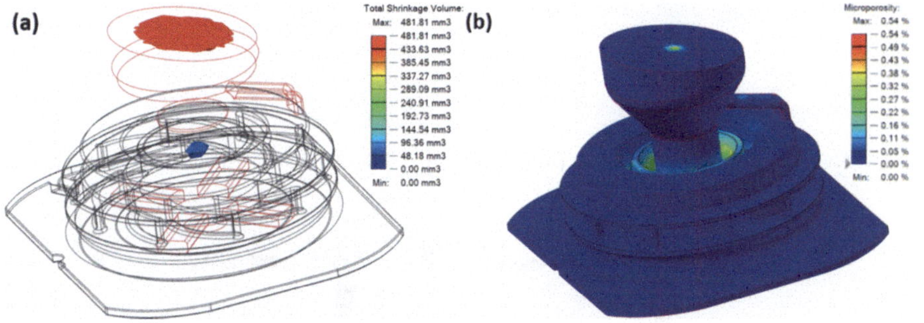

Fig. 9. Porosities in the 10° tilted pattern design: (a) Shrinkage porosity, (b) Microporosity.

First, the tilting has permitted a complete filling of the inner ring. Figure 10 shows a neat reduction in the "last air" zones at the upper edge of the ring due to its shifting down from the near top of the cast, which means a reduction of the possibility of retaining trapped gases or gaps from metal retraction.

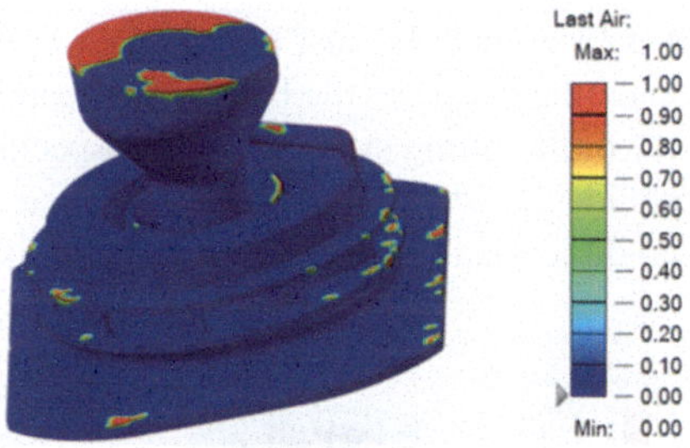

Fig. 10. Last areas to fill in the 10° tilted pattern design.

In addition, as shown in Fig. 11 representing the velocities' vectors profiles, the flow front of the molten metal touches down the bottom of the runner in a way that it streams down first on the peripheral runner's surface, in the case of tilted mold, instead of sharply impacting the bottom. Consequently, less turbulence and splashes occur when filling the main cast part, which means less casting voids and porosity formation.

As the filling progresses (Fig. 12), a steady flow is observed. This situation reduces the bubbles' formation and permits an easy escape of the trapped gas and inclusions through the riser/ventilator tube.

With regards to displacements (deformations) caused by metal retraction and stresses during the cooling phase, Fig. 13 shows a very low deformation amplitude after demolding, with a maximum displacement of 6.6×10^{-2} mm located at the runner. At the

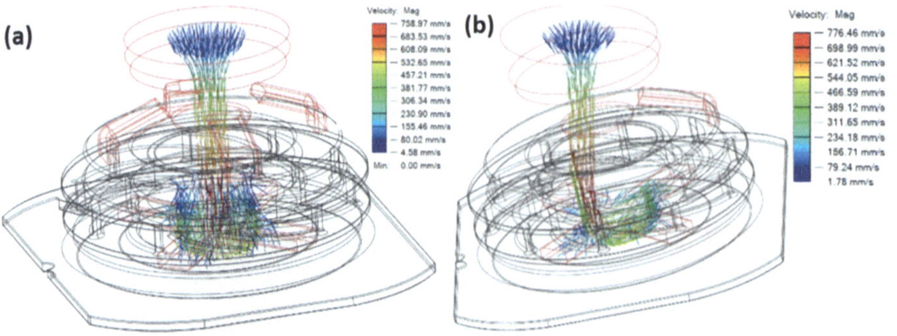

Fig. 11. Velocities' vectors profile: (a) first design at t=0.22s, (b) 10° tilted design at t=0.23s.

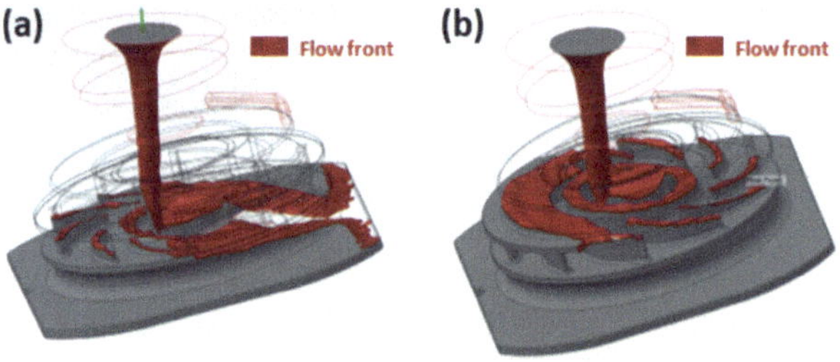

Fig. 12. Flow progression at the filling session of the 10° tilted mold: (a) t=1.88s, (b) t=2.85s.

critical and thin-walled locations, the cast has displayed very descent displacement at the selected checkpoints (the flat base, the blowing deflector blades, and the inlet ring), thanks to the efficiency of the filling system design, especially the location of the gates and the riser which provides constant feeding of metal to compensate the retracted metal as well as uniform and directional cooling down progression (Fig. 14).

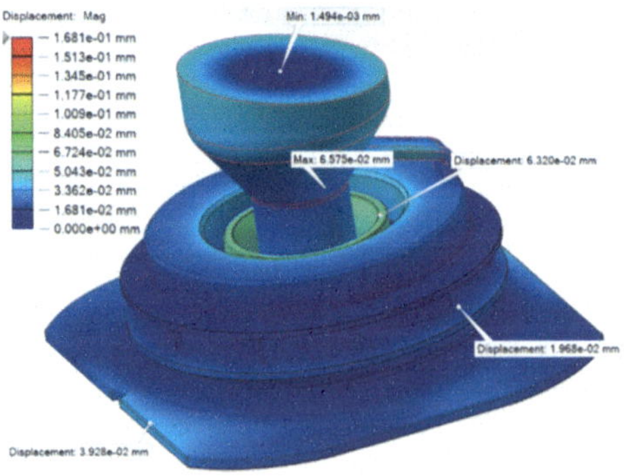

Fig. 13. Displacement (deformation) magnitude distribution after cooling (10° tilted mold).

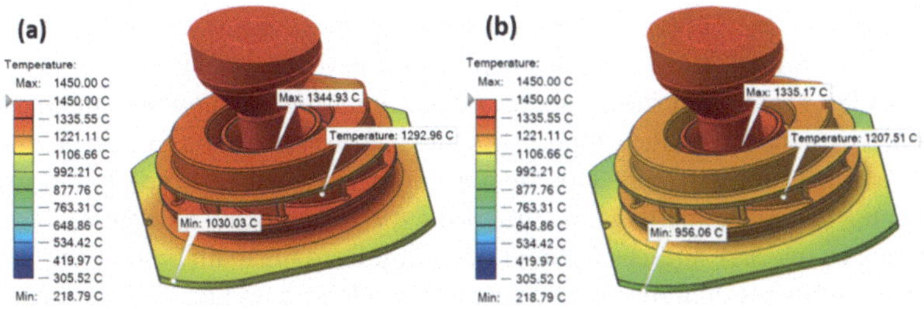

Fig. 14. Cooling temperature progression in 10° tilted mold: (a) t=12,36s, (b) t=38.37s.

3.2 Model 3D Printing

3D Printing Calibration. To calibrate the machine for the utilized resin, we have applied qualitative calibration tests using the "XP finder" 3D-printing test model. We set as constants the layer height to 0.05 mm and the light intensity to 100%; then, we experimented with a series of 3D printing tests with different exposure times and lift speeds to spot their impact on the integrity of shapes and details of the test model. Figure 15 represents the resulting optimal 3D test-print where neither signs of underexposure like vanishing boundaries or missing details nor signs of overexposure like expansion, flattening, and inconsistency in pattern's spacings are portrayed. The related optimal 3D printing conditions are shown in Table 3.

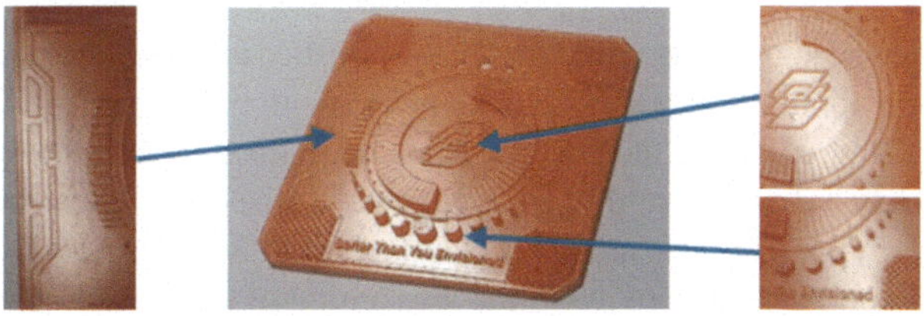

Fig. 15. XP finder 3D-printing test model (optimal 3D test-print).

Table 3. The main optimal 3D printing parameters (50 μm layer thick. & 100% light int.)

Parameters	Burn In layers	Normal Layers
Layers count	6	–
Exposure time (s)	55	5.5
Transition layers count	8	–
Lift speed (mm/min)	45	45
Retract speed (mm/min)	55	65

CAD STL-Conversion and Slicing. The approved CAD model has been converted to a highly defined resolution STL mesh file with cordal and angular deviations of 5 microns and 0.5° respectively to obtain as far as possible the most accurate dimensions and the smoothest transition between triangles particularly in the small, curved surfaces of the blowing blades. Figure 16a shows the prepared layout for slicing where the model has been properly oriented, and the support structure has been added on. As the layout of the 3D print shows, we have managed to orient the model in a way that the support structure is mainly attached to the flat base and the sprue; so, we avoid any alteration or roughing of functional or internal surfaces during its removal from the main pattern and we make the polishing of its residual tips as easy and accessible as possible. The prepared model slicing has been processed using the obtained optimal 3D printing parameters from calibration.

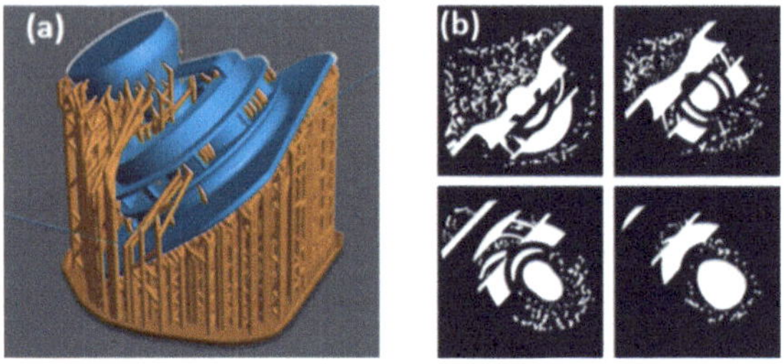

Fig. 16. Slicing of the model: (a) 3D model layout ready for slicing, (b) Slice-images captures from the 3D printing scene.

To smooth out the jagged edges and obtain the polished effect of the curved surfaces of 3D-print, we have applied slight anti-aliasing refinement, before performing the slicing, to blur the pixelated edges of the image slices to be displayed digitally on the LCD screen for each layer. To achieve this goal, we have utilized law intensity settings with a Blur Spread Radius of 2 pixels and a Gray Offset of 30%. The smoothing hasn't been applied to the support structure to avoid the weakening of their junctions with the main pattern model. Figure 16b depicts some slice-image captures generated from the slicing of the prepared and smoothed-out 3D printing model being utilized afterward to build up the 3D-printing scene via LCD flashing.

The 3D Printing Result Analysis. The file containing the slice arrangement and the print settings has been sent to the MSLA machine to perform the 3D printing. After the achievement of the operation, the 3D print was post-processed as follows: IPA Alcohol washing, support removal, UV-post-curing for 1 minute, and residual support tip's polishing with ceramic auto-generative file (Fig. 17).

As a result, the 3D printing has generated very smooth surfaces and reproduced the finest details namely the fillets and all the curvatures of the deflector's blowing blades (Fig. 18). The process has cost circa 36.9 ml of resin consumption, including 4.2 ml for the support structure -meaning a yield of 88.6%, with a fair electricity consumption of 0.84 kWh.

However, despite the quality and flexibility offered when implementing MSLA 3D printing to fabricate our investment casting pattern, the method is not without some downsides. In the following part, we will discuss some of the issues encountered and

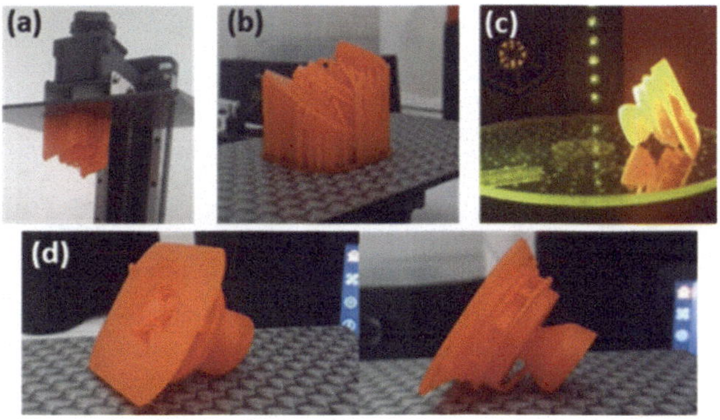

Fig. 17. 3D-print model at the different stages of fabrication: (a) MSLA 3D printing-achieved, (b) Drying up after IPA-washing, (c) During the UV-post-curing after support removal, (d) Finished state after polishing the residual support tips.

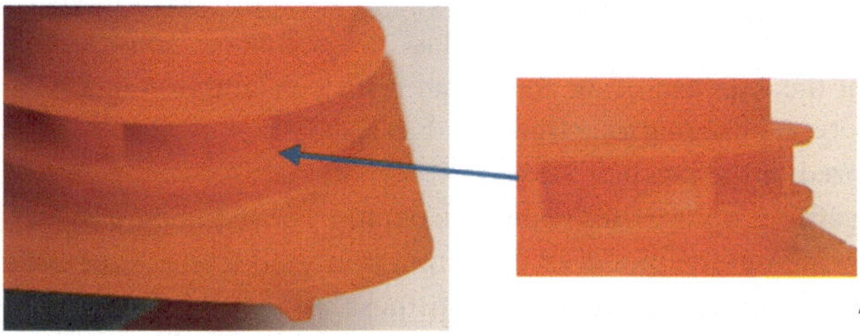

Fig. 18. 3D-print surface quality and details' reproduction.

possible ways to address them. One of its problems is the effective printing time (around 7 h 02 minutes in the case studied). Moreover, the material used can no longer be recycled after melting, due to the irreversibility of the solidification process based on UV polymerization. Another very important aspect not to be overlooked, when using this technology, is the hazardous impact on human health and the environment if safety rules are not respected when handling resin or managing production waste.

A thoughtful use of this process, however, aids in benefiting from this technology while limiting the technical constraints mentioned above. In terms of 3D printing time, the intrinsic advantage related to the MSLA technology over other 3D printing technologies, including the conventional SLA, can be exploited. As it allows a simultaneous flash-printing of all the shapes of each layer in a single LCD-masked projection, a batch of identical or different parts can be printed in the same amount of time. Figure 19a shows a proposed solution for a printable arrangement of three identical copies of the part in question, in case, by way of example, of studying the same model design behavior under three different casting conditions. Here the printing time per model will be cut to one-third.

Another commonly utilized solution is possible if only one part is needed to be 3D-printed. This one involves splitting the model into parts while exploiting the space allotted on the build plate; in this way, we can reduce the height of the 3D print, which

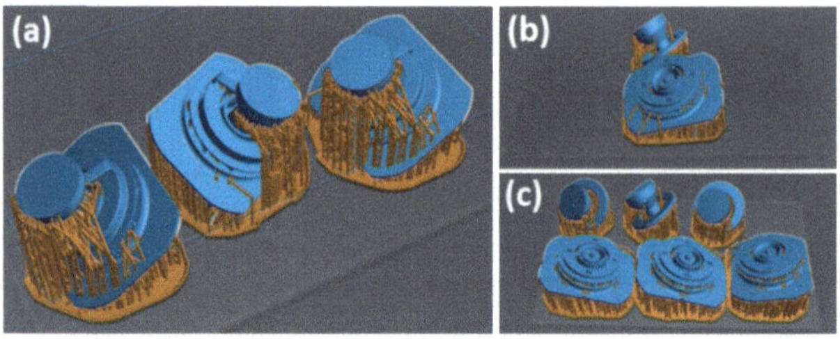

Fig. 19. Time-saving arrangement solutions for the studied pattern: (a) three-copies batch printing, (b) Single-part splitting, (c) combination of multiple-parts batch 3D printing with model splitting.

will lead to a significant reduction in the time required. Figure 19b portrays a feasible solution where our model is split into two sub-parts. The estimated printing time for this configuration will be reduced to 5 h 3 min, which means a time saving of around 30%. If combined with batch printing (Fig. 19c), the 3D printing time per model will be cut down further. However, regarding the last two solutions, considerable time saving is achievable at the expense of an additional resin consumption and a lower yield cut to 52%. In this type of scenario, a trade-off has to be, then, made based on whether the yield is prioritized or the productivity.

Finally, regarding optimizing resin consumption or reducing unnecessary material wastage, hollowing out the model or designing it in shell format proves useful but only on condition that it does not compromise stiffness, and on condition that the necessary measures are taken to avoid suctions or cracking, such as good orientation or the provision of sealable holes with lids to allow reaction gases to exhaust during the 3D printing process. In our case study, for instance, this option applies to the feeding system. Managing and optimizing the support's thicknesses, density and distribution could also be in concert with the above-mentioned solutions which could further reduce the amount of resin consumed. However, this option also needs to be handled with care, as a weakly built support structure can lead to the collapse and failure of the 3D-print.

4 Conclusion

This work aims to study the feasibility of implementing AM-based RIC to cost-effectively produce an investment casting pattern of an Inconel 718 get axial fuel distributor, for which the design is aided by numerical simulation of the casting process. The choice of an AM solution has been set to avoid investing in sacrificial experimental wax injection molds intended for single-part production and to profit from the flexibility and the unlimited design freedom that it offers as well.

The simulation-aided design consists of evaluating the macro- and microporosities as well as the deformations resulting from the solidification phase while keeping in mind the need to maintain an optimum casting yield. This phase has resulted in a porosity-free model design with a microporosity index of no more than 0.54% and a very satisfactory shrinkage deformation of no more than 6.6×10^{-2} mm, even in critical curved and small-section areas, thanks to an appropriate design and a 10° tilt angle guaranteeing steady filling and directional cooling. All this is achieved with a satisfactory yield of 69.17%.

The cast pattern has been produced using 3D printing based on MSLA technology with a qualitative low-viscosity castable resin. The study of 3D printing has been approached, in this work, with a technical perspective in which we have reported the steps and rules followed from the 3D printing model design phase to the obtaining of the finished product. These rules have led to the success of the operation. As a result, we have been finally able to produce a high-quality model at a very reasonable cost and with an excellent yield of 88.6%.

Nevertheless, we have found that the implementation of this AM process has come with some drawbacks that need to be addressed to take full advantage of this promising technique. One of these problems is the relatively long time taken to print the subject model, for which we have proposed two practical solutions, either to choose one of them or to combine depending on the context, namely, the model splitting or batch printing. Another problem is that of the wastage of non-reusable solidified resin, especially in the support structure, where there are viable solutions such as model hollowing out, shell printing of the model, and optimization of the support structure, which can be combined as well, but on condition that they do not adversely affect the operation because of stiffness losses, cracking and suction problems caused by a poor 3D-print design. In our exploration of solutions, we have also found that in some cases adopting one solution to a problem can create or exacerbate another, so a trade-off needs to be made depending on priorities.

In summary, RIC based on the MLSA 3D printing can take full advantage of this AM's unmatched design freedom flexibility, and quality, provided that all necessary measures in the design and the slicing phase are taken to lead the operation toward success; therefore, we could work on a lucrative pattern-making method for investment casting of high-precision and complex aerospace small parts in unit or small production runs.

This study opens up prospects for further in-depth study of this approach; accordingly, it could be complemented by studies that can be spread over two main strands: (1) Optimization of the 3D printing process and study of the model in terms of accuracy and repeatability; (2) Performing the investment casting from the MSLA 3D-printed models and analyzing the casting results and defects regarding simulation results.

Acknowledgments. Authors sincerely express their gratitude to the industrial "Pursuit Aerospace" EUROCAST, Tunisia: an investment casting company specializing in complex aerospace machined components. The authors also thankfully acknowledge the reviewers' helpful comments and suggestions, which have improved this work.

References

1. Kanyo, J.E., Schafföner, S., Uwanyuze, R.S., Leary, K.S.: An overview of ceramic molds for investment casting of nickel superalloys. J. Eur. Ceram. Soc. **40**(15), 4955–4973 (2020)
2. Aouadi, B., Ghannem, S., Ben Fathallah, B., Yallese, M.A.: Additive manufacturing and investment casting comparison of Superalloys: aerospace industry. In: Mabrouki, T., Sahlaoui, H., Sallem, H., Ghanem, F., Benyahya, N. (eds.) Advances in Additive Manufacturing: Materials, Processes and Applications. AIAM 2023 Lecture Notes in Mechanical Engineering, pp. 125–131. Springer, Cham (2024)
3. Tewo, R.K., Rutto, H.L., Focke, W., Seodigeng, T., Koech, L.K.: Formulations, development and characterization techniques of investment casting patterns. Rev. Chem. Eng. **35**(3), 335–349 (2019)
4. Aboulkhair, N.T., Simonelli, M., Parry, L., Ashcroft, I., Tuck, C., Hague, R.: 3D printing of Aluminium alloys: additive manufacturing of Aluminium alloys using selective laser melting. Prog. Mater. Sci. **106**, 100578 (2019)
5. Maconachie, T., Leary, M., Lozanovski, B., Zhang, X., Qian, M., Faruque, O., Brandt, M.: SLM lattice structures: properties, performance, applications and challenges. Mater. Des. **183**, 108137 (2019)
6. Richard, C.T., Kwok, T.H.: Analysis and design of lattice structures for rapid-investment casting. Materials. **14**(17), 4867 (2021)
7. Leuders, S., Meiners, S., Wu, L., Taube, A., Tröster, T., Niendorf, T.: Structural components manufactured by selective laser melting and investment casting—impact of the process route on the damage mechanism under cyclic loading. J. Mater. Process. Technol. **248**, 130–142 (2017)
8. Li, C., Liu, Z.Y., Fang, X.Y., Guo, Y.B.: Residual stress in metal additive manufacturing. Proc. Cirp. **71**, 348–353 (2018)
9. Revilla-León, M., Özcan, M.: Additive manufacturing technologies used for 3D metal printing in dentistry. Curr. Oral Health Rep. **4**, 201–208 (2017)
10. Cygan, R., Rakoczy, Ł., Szybicki, D.: 3D printing vane rings models used in the investment casting process. In: 7th International Conference INTEGRITY-RELIABILITY-FAILURE, pp. 59–464. INEGI-FEUP, Funchal-Portugal (2020)
11. Singh, D., Singh, R., Boparai, K.S.: Development and surface improvement of FDM pattern based investment casting of biomedical implants: a state of the art review. J. Manuf. Process. **31**, 80–95 (2018)
12. Singh, R., Singh, S., Kapoor, P.: Investigating the surface roughness of implant prepared by combining fused deposition modeling and investment casting. Proc. Instit. Mech. Eng. Part E J. Proc. Mech. Eng. **230**(5), 403–410 (2016)
13. Carneiro, V.H., Rawson, S.D., Puga, H., Meireles, J., Withers, P.J.: Additive manufacturing assisted investment casting: a low-cost method to fabricate periodic metallic cellular lattices. Addit. Manuf. **33**, 101085 (2020)
14. Huang, Y., Xue, Y., Wang, X., Han, F.: Effect of the cross-sectional shape of struts on the mechanical properties of aluminum-based pyramidal lattice structures. Mater. Lett. **202**, 55–58 (2017)
15. Puga, H., Carneiro, V.H., Correira, P., Vieira, V., Barbosa, J., Meireles, J.: Mechanical behavior of honeycomb lattices manufactured by investment casting for scaffolding applications. Proc. Instit. Mech. Eng. Part L J. Mater. Design Appl. **231**(1–2), 73–81 (2017)
16. Mukhtarkhanov, M., Perveen, A., Talamona, D.: Application of stereolithography based 3D printing technology in investment casting. Micromachines. **11**(10), 946 (2020)
17. Al-Ketan, O., Rowshan, R., Al-Rub, R.K.A.: Topology-mechanical property relationship of 3D printed strut, skeletal, and sheet based periodic metallic cellular materials. Addit. Manuf. **19**, 167–183 (2018)

18. Wu, H., Li, D., Tang, Y., Guo, N., Cui, F., Sun, B.: Rapid casting of hollow turbine blades using integral ceramic molds. Proc. Inst. Mech. Eng. B J. Eng. Manuf. **223**(6), 695–702 (2009)
19. Wu, H., Li, D., Chen, X., Sun, B., Xu, D.: Rapid casting of turbine blades with abnormal film cooling holes using integral ceramic casting molds. Int. J. Adv. Manuf. Technol. **50**, 13–19 (2010)
20. Aguilar, J., Schievenbusch, A., Kättlitz, O.: Investment casting technology for production of TiAl low-pressure turbine blades–process engineering and parameter analysis. Intermetallics. **19**(6), 757–761 (2011)
21. Nawrocki, J., Motyka, M., Szeliga, D., Ziaja, W., Cygan, R., Sieniawski, J.: Effect of cooling rate on macro-and microstructure of thin-walled nickel superalloy precision castings. J. Manuf. Process. **49**, 153–161 (2020)
22. Tao, P., Shao, H., Ji, Z., Nan, H., Xu, Q.: Numerical simulation for the investment casting process of a large-size titanium alloy thin-wall casing. Prog. Nat. Sci. Materials Int. **28**(4), 520–528 (2018)
23. Kim, Y.C., et al.: Improvement of quality and yield for investment casting of centrifugal pump impeller by tilting mold and optimizing runner/riser system. Int. J. Adv. Manuf. Technol. **130**(5), 2369–2379 (2024)
24. Sheikh, A.K., Khan, M.A.A.: Mold design optimization and quality assessment of steel castings through integrated simulations and experiments. J. Mech. Sci. Technol. **34**(7), 2975–2983 (2020)
25. Huang, P.H., Shih, L.K.L., Lin, H.M., Chu, C.I., Chou, C.S.: Novel approach to investment casting of heat-resistant steel turbine blades for aircraft engines. Int. J. Adv. Manuf. Technol. **104**, 2911–2923 (2019)
26. Donghong, W., Yu, J., Yang, C., Hao, X., Zhang, L., Peng, Y.: Dimensional control of ring-to-ring casting with a data-driven approach during investment casting. Int. J. Adv. Manuf. Technol. **119**(1), 691–704 (2022)
27. Li, X., Wang, Y., Guo, K., Miao, L.: A simulation analysis of the cutting process for Inconel718 nickel-based superalloy. Int. J. Adv. Manuf. Technol. **114**, 2721–2738 (2021)
28. Deng, D.: On the microstructures and anisotropic mechanical behaviors of additively manufactured IN718 (Doctoral dissertation, Linköping University Electronic Press) (2019).
29. Inconel® 718 (Alloy 718) Material Data - UNS N07718 | Ulbrich, https://www.ulbrich.com/alloys/inconel-718-uns-n07718, last accessed 21 Mar 2024
30. Hardin, R.A., Beckermann, C.: Effect of porosity on deformation, damage, and fracture of cast steel. Metall. Mater. Trans. A. **44**, 5316–5332 (2013)
31. Hardin, R.A., Beckermann, C.: Prediction of the fatigue life of cast steel containing shrinkage porosity. Metall. Mater. Trans. A. **40**, 581–597 (2009)
32. Blondheim Jr., D., Monroe, A.: Macro porosity formation: a study in high pressure die casting. Int. J. Met. **16**(1), 330–341 (2022)
33. Caceres, C.H., Selling, B.I.: Casting defects and the tensile properties of an AlSiMg alloy. Mater. Sci. Eng. A. **220**(1–2), 109–116 (1996)
34. Kang, M., Gao, H., Wang, J., Ling, L., Sun, B.: Prediction of microporosity in complex thin-wall castings with the dimensionless Niyama criterion. Materials. **6**(5), 1789–1802 (2013)
35. Guo, J., Beckermann, C., Carlson, K., Hirvo, D., Bell, K., Moreland, T., Gu, J., Clews, J., Scott, S., Couturier, G., Backman, D.: Microporosity prediction and validation for Ni-based superalloy castings. In: IOP Conference Series: Materials Science and Engineering, vol. 84, p. 012003. IOP Publishing (2015)
36. Kuyumcu, A.M., Haktanir, Y.: Design and optimization of runner and gating systems, in high-pressure casting and sand casting, for an aluminum alloy valve cover. Int. J. Automat. Eng. Technol. **12**(4), 154–164 (2023)

37. Puspitasari, P., Amar, A., Tsamroh, D.I., Karim, S.A.A., Achyarsyah, M.: Simulation for oil pan production against its porosity, shrinkage, and Niyama criterion. In: Nanotechnologies in Green Chemistry and Environmental Sustainability, pp. 27–44. CRC Press (2022)
38. Carlson, K.D., Beckermann, C.: Development of thermophysical property datasets, benchmark niyama results, and a simulation qualification procedure. In: Proceedings of the 64th SFSA Technical and Operating Conference, paper (No. 5.5). Steel Founder's Society of America, Chicago, IL (2010)
39. Yao, W., Li, D., Zhao, Y., Zhan, Z., Jin, G., Liang, H., Yang, R.: 3D printed multi-functional hydrogel microneedles based on high-precision digital light processing. Micromachines. **11**(1), 17 (2019)
40. Schittecatte, L., Geertsen, V., Bonamy, D., Nguyen, T., Guenoun, P.: From resin formulation and process parameters to the final mechanical properties of 3D printed acrylate materials. MRS Commun. **13**(3), 357–377 (2023)

Taguchi Optimization of FDM Process Parameters on Ultimate Tensile Strength of 3D-Printed PLA Parts

Boulahem Khaled[1(✉)], Souli Imen[2], Badreddine Larbi[3], Mohamed Toumi Nasri[3], and Ayadi Mahfoudh[4]

[1] Laboratory of Applied Mechanics and Engineering, National Engineering School of Tunis, Tunis, Tunisia
khaled.boulahem@enit.utm.tn

[2] Research Laboratory Structures and Mechanics Applied, Polyethnic School of Tunis, Tunis, Tunisia
imen.souli@enib.ucar.tn

[3] Laboratory of Mechanics, Production, and Energy, National High School of Engineering of Tunis, Tunis, Tunisia
{badreddine.larbi,mohamedtoumi.nasri}@enib.ucar.tn

[4] Department of Mechanical Engineering, National Engineering School of Bizerte, Bizerte, Tunisia
mafoudh.ayadi@esstt.rnu.tn

Abstract. Fused deposition modeling (FDM) is the most common method for 3D printing polymers and is gaining popularity in engineering due to its ability to easily create complex parts. However, the mechanical properties of 3D-printed parts heavily rely on the optimal selection of process parameters. This study aims to investigate the optimum 3D printing process parameters for achieving the highest ultimate tensile strength using polylactic acid material. The FDM input variables include raster angle, layer thickness, nozzle temperature, print speed, and bed temperature. The experiments were performed using the Taguchi method. Additionally, linear regression analysis was employed to establish the relationship between the input variables and the output variable. The analysis of variance was used to assess the significance of the mathematical models and to measure the percentage contribution of control factors. The optimal parameter combination was selected based on the signal-to-noise ratio and the main effects plot. The results revealed a model summary with a coefficient of determination value of 97.52%, indicating successful experimentation. The experimental results indicated that the optimum process parameters for maximum ultimate tensile strength were a raster angle of 0°, a layer thickness of 0.1 mm, a nozzle temperature of 215 °C, a print speed of 40 mm/s, and a bed temperature of 60 °C. Finally, the effects of the process parameters on the ultimate tensile strength were studied.

Keywords: Fused deposition modeling · Taguchi method · Polylactic acid · Ultimate tensile strength · Optimization

M. Soula et al. (Eds.): AIAM 2024, LNME, pp. 133–148, 2025.
https://doi.org/10.1007/978-3-031-94889-3_11

1 Introduction

Fused deposition modeling (FDM) is an additive manufacturing (AM) technique that enables the production of various complex shapes and structures from 3D models. It involves layer-by-layer printing of materials to create the desired object. Compared to traditional processing technology, 3D printing offers numerous advantages, such as cost-effective production, reduced component count, minimal waste production, and the ability to produce intricate geometries [1]. In recent years, researchers have consistently explored methods to enhance the quality of FDM products, with a particular focus on their mechanical properties as a key evaluation criterion. Therefore, it is crucial to choose and optimize the FDM process parameters, which include layer thickness, infill density, nozzle temperature, build direction, raster angle, bed temperature, print speed, raster width, and others, to obtain the desired quality characteristics of the printed parts [2]. Regarding this issue, researchers aim to investigate the impact of controlled printing parameters on the mechanical properties of 3D-printed parts. To examine this correlation, researchers have employed a range of experimental methods. Numerous studies have employed artificial intelligence techniques, design of experiments (DOE) [3], and evolutionary algorithms [4–6] to investigate the impact of FDM parameters on build quality. Among the conventional optimization approaches used, the Taguchi method can be found. Widely used in manufacturing, it offers multiple solutions for optimization problems with numerous variables [7, 8]. This method can help develop models that take into account the interactions between printing parameters and mechanical properties, which allows for optimization. Besides, Hasdiansah et al. [9] conducted an optimization study on FDM 3D printing for TPU surface finish. The study focused on optimizing various parameters including flow rate, layer thickness, nozzle temperature, print speed, overlap, and fan speed. Using the Taguchi method, they determined that layer thickness had the greatest impact on surface roughness variation, accounting, for 65.11% of the variation. Accordingly, Nagendra and Prasad [10] performed FDM parameter optimization to enhance the mechanical properties of nylon parts, including tensile strength, flexural strength, impact strength, and compression strength. The optimum settings identified through a Grey Taguchi design of experiments were a layer height of 0.4 mm, an extruder temperature of 300 °C, a density of 90%, a raster angle of 90°, and a rectilinear filling model. Also, Rarani et al. [11] utilized the Taguchi method to optimize FDM printing parameters for PLA specimens. Through their study, they determined that an infill density of 80%, a print speed of 40 mm/s, and a layer thickness of 0.1 mm were the optimum for achieving the desired modulus of elasticity and ultimate tensile strength. For the failure strain, the optimum settings were an infill density of 80%, a printing speed of 40 mm/s, and a layer thickness of 0.2 mm. Moreover, Auffray et al. [12] conducted a study on the mechanical properties of PLA specimens produced using FDM technology. By utilizing the Taguchi design of experiments approach, the research identified that the most influential parameters were infill density, infill pattern, printing velocity, and printing orientation. However, layer thickness, nozzle temperature, and outline overlap had minimal impact on Young's modulus and the yield strength of the samples. Furthermore, Rasheed et al. [13] conducted FDM parameter optimization using a Taguchi orthogonal array to investigate the influence of varying FDM process parameters on the mechanical properties and to identify the optimum set of parameters for achieving better tensile strength in a bi-layered composite of PLA-ABS. They discovered that the optimum parameters for better tensile strength were an infill density of 75%, 30 layers per part, a print speed of 20 mm/s, and a bed temperature of 100 °C.

In light of recent inquiries regarding the materials, surface roughness, and mechanical properties of 3D printed parts, it is imperative to emphasize the significance of studying the relationship between the mechanical properties of FDM-3D printed specimens and the FDM process parameters. This emphasis is further compounded by the diversity in printed parameters, complex shapes, design of experiments, and workpiece materials. For this reason, the main goal of this research is to develop a mathematical model for predicting the optimum ultimate tensile strength (UTS) of polylactic acid (PLA). This model takes into account several factors, such as layer thickness (H), nozzle temperature (T), raster angle (O), print speed (V), and bed temperature (Tb). The Taguchi method and analysis of various process parameters are used to achieve this objective. Additionally, the impact of different input process parameters on the response parameter of FDM is demonstrated. Ultimately, the optimum printing parameters needed to achieve the desired ultimate tensile strength of PLA material have been identified.

2 Materials and Methods

2.1 Materials

The experimental study was carried out by measuring the tensile strength of the 3D-printed parts using the FDM machine. The testing material used for sample printing was polylactic acid (PLA) filament with a density of 1.25 g/cm^3. The mechanical properties of PLA materials are given in Table 1.

Table 1. Mechanical and specifications properties of PLA

Properties	Values	Units
Tensile strength	72	MPa
Flexural modulus	2100	MPa
Flexural strength	69	MPa
Rockwell hardness	120	MPa

The Tronxy 2 Pro, a 3D printer with a nozzle diameter of 0.4 mm, was used to manufacture the specimens for the tensile tests. The specimen designs were created using SolidWorks CAD software. The tensile specimens were printed according to the ISO 527-2 standard method with a crosshead speed of 5 mm/min. The UltiMaker Cura5 package was used to generate G-code files and control the printing parameters throughout the printing process of the test samples, Fig. 1. The parameters are kept constant, and their respective values are shown in Table 2.

A universal tensile testing machine (Ibertest, TESTCOM-10, Spain) was used to conduct the tensile test. A tensometer (Mini MFA 2 extensometer, Ibertest, Spain) with

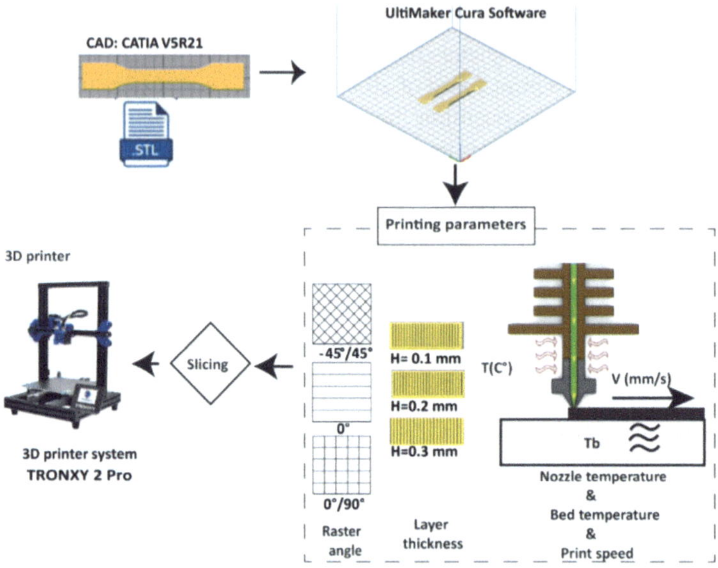

Fig. 1. 3D printing process

Table 2. Constant process parameters during 3D printing of PLA parts

Parameters	Values	Parameters	Values
Filament diameter	1.75 mm	Overlap	30%
Nozzle diameter	0.4 mm	Outer shell speed	60 mm/s
Shells	3 Shell	Inner shell speed	30 mm/s
Contour lines	0.6 mm	Infill pattern	Rectilinear
Infill density	80%	Build orientation	Flat X direction

a 50 mm gauge length was attached to the side layer of the specimens. Experimental results involving tensile tests were collected and recorded using desktop WinTest 32 software. The tests were carried out at a temperature of 25 °C and 50% relative humidity, in accordance with ISO R291:1977, Fig. 2.

2.2 Methods

This section describes the research methodology employed in this study, which is summarized in the flowchart presented in Fig. 3.

Design of Experiments. The Taguchi method is known as a robust tool for the experimental optimization of a model. This method was chosen to examine the interaction and the effect of five FDM process parameters on the tensile strength properties of the material on the printed parts shown in Table 3. This design of experiments is available

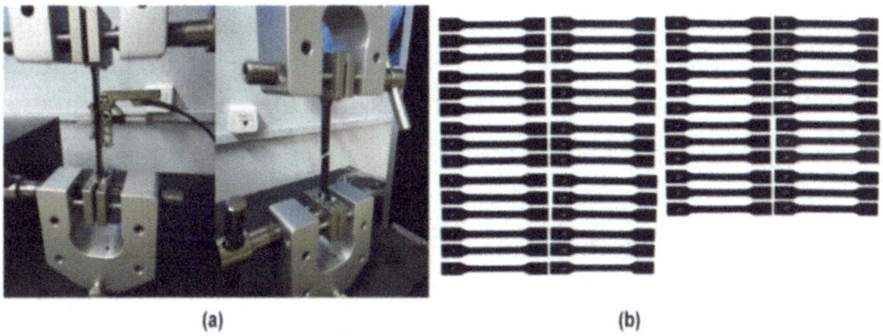

Fig. 2. a) Tensile test b) Tensile specimens

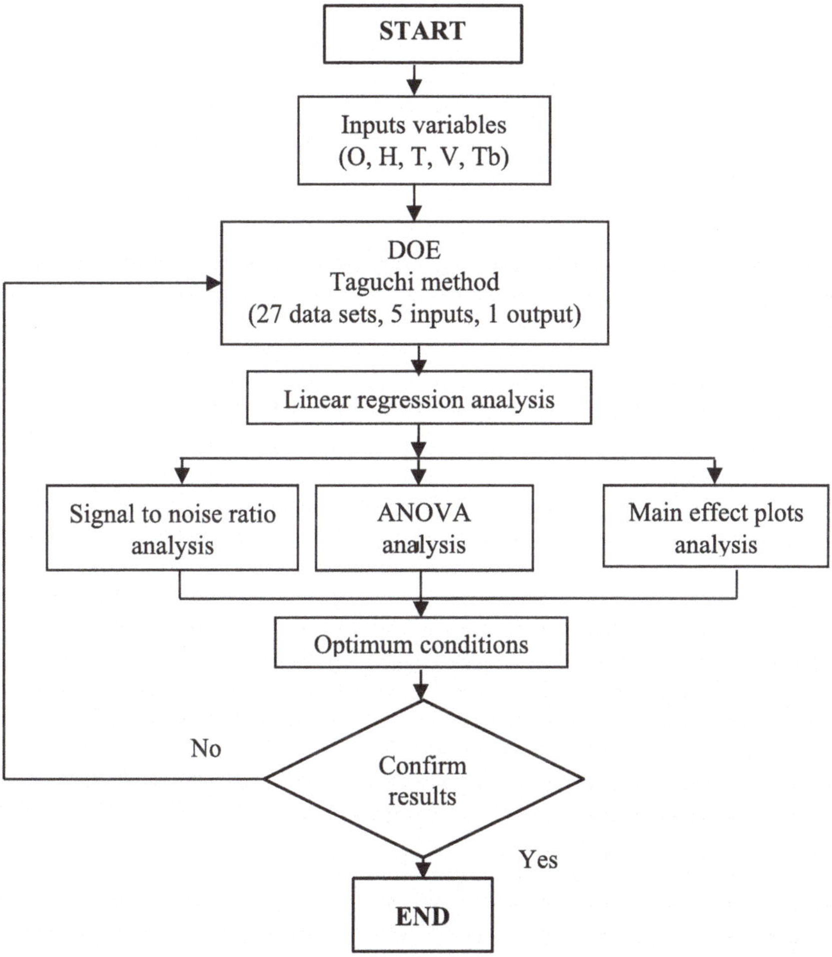

Fig. 3. Flowchart of Taguchi methodology for enhancement of ultimate tensile strength

in the Minitab 19 software package and includes twenty-seven experiments. For each experiment two specimens are printed. The FDM parameters set are layer thickness (H), nozzle temperature (T), raster angle (O), print speed (V), and bed temperature (Tb). Ultimate tensile strength (UTS) is selected as a response characteristic, Table 3.

Table 3. Levels of process parameters

Controlling factors	Symbols	Units	Range		
			Low Level	Centre point	High Level
Nozzle temperature	T	°C	210	215	230
Layer thickness	H	mm	0.1	0.2	0.3
Raster angle	O	°	0°	−45°/45°	0°/90°
Print speed	V	mm/s	40	60	80
Bed temperature	Tb	°C	50	55	60

The experimental results obtained for the trials, along with the S/N ratio and standard deviation (SD) of ultimate tensile strength based on the printing parameters, are presented in Table 4.

Table 4. Experimental plan using Taguchi design for experimentation

N°	Factor levels						Responses	
	H	O	T	V	Tb	Mean UTS (MPa)	SD	S/N ratio (db)
1	0.1	0	200	40	50	31.05	1.34	30.11
2	0.1	0	200	40	55	33.70	1.83	30.55
3	0.1	0	200	40	60	39.59	1.66	31.95
4	0.1	−45/45	215	60	50	31.05	7.14	29.55
5	0.1	−45/45	215	60	55	32.75	9.65	30.30
6	0.1	−45/45	215	60	60	33.55	4.64	30.51
7	0.1	0/90	230	80	50	33.65	2.70	30.54
8	0.1	0/90	230	80	55	28.39	2.58	29.06
9	0.1	0/90	230	80	60	24.84	4.04	27.90
10	0.2	0	215	80	50	31.64	1.70	30.00
11	0.2	0	215	80	55	33.14	2.99	30.40
12	0.2	0	215	80	60	36.44	1.28	31.69

(continued)

Table 4. (*continued*)

N°	Factor levels						Responses	
	H	O	T	V	Tb	Mean UTS (MPa)	SD	S/N ratio (db)
13	0.2	−45/45	230	40	50	29.71	4.38	29.45
14	0.2	−45/45	230	40	55	31.93	3.28	30.35
15	0.2	−45/45	230	40	60	32.73	6.22	30.29
16	0.2	0/90	200	60	50	28.04	2.71	28.95
17	0.2	0/90	200	60	55	27.77	2.89	28.55
18	0.2	0/90	200	60	60	30.14	4.22	30.40
19	0.3	0	230	60	50	30.07	1.62	29.56
20	0.3	0	230	60	55	32.39	2.02	30.20
21	0.3	0	230	60	60	35.67	1.89	31.04
22	0.3	−45/45	200	80	50	25.32	3.02	28.07
23	0.3	−45/45	200	80	55	27.26	4.35	28.71
24	0.3	−45/45	200	80	60	29.32	2.55	29.04
25	0.3	0/90	215	40	50	29.62	4.23	29.43
26	0.3	0/90	215	40	55	29.44	2.33	29.07
27	0.3	0/90	215	40	60	32.99	3.54	30.37

Linear Regression Model. In this study, a quadratic regression model is used to determine the relationship between the output response of ultimate tensile strength and the variables of layer thickness (H), nozzle temperature (T), raster angle (O), print speed (V), and bed temperature (Tb). This relationship is represented by a second order polynomial equation, which can be expressed as follows:

$$\mathrm{Y_i} = \mathrm{b_0} + \sum_{i=1}^{5} \mathrm{b_i.X_i} + \sum_{i=1}^{5} \mathrm{b_{ii}.X_i}^2 + \sum_{i=1}^{4}\sum_{i=i+1}^{5} \mathrm{b_{ij}.X_i.X_j} + \mathrm{e_i} \tag{1}$$

where Y_i is the response, b_0, b_i, b_{ii}, and b_{ij} are the mean values of responses, linear, quadratic, and interaction constant coefficients, respectively, and e_i is the experimental error.

To investigate the adequacy of the regression model and to check the underlying assumptions, residual plots were used. The residual is defined as the difference between the observed and fitted values of the response variable [14].

Signal to Noise Ratio. To estimate and optimize printing parameters for better responses, the obtained outcomes are further analyzed using the signal-to-noise ratio (S/N ratio) [9, 15]. The factor levels with the highest S/N ratios aim to minimize noise sensitivity and achieve optimized levels. To evaluate the impact of input parameters (H, O, T, V, and Tb) on the response (UTS), the performance indicator adopted for this study is the S/N ratio with the greater-the-better characteristic developed by the Taguchi method, as shown in Table 5. The S/N ratio is calculated using the following equation:

$$S/N = -10\log_{10}\left(\frac{1}{n}\sum_{j=1}^{n}\frac{1}{{y_i}^2}\right) \tag{2}$$

where y_i is the measured ultimate tensile strength for each test specimen, and n denotes the number of specimens for each test trial.

Table 5. Regression equations for UTS

Raster angle	Equations
0°	UTS = −138.4 - 43.5 H + 0.846 T + 1.156 V + 1.66 Tb + 0.156 H*T - 1.648 H*V + 1.790 H*Tb - 0.0009 T*V- 0.016 T*Tb - 0.012 V*Tb + 0.026 Tb^2
−45°/45°	UTS = −124.5 - 43.5 H + 0.846 T + 1.156 V + 1.35 Tb + 0.156 H*T - 1.648 H*V + 1.790 H*Tb - 0.0009 T*V - 0.016 T*Tb - 0.012 V*Tb + 0.02 Tb^2
0°/90°	UTS = −104.9 - 43.5 H + 0.846 T + 1.156 V + 0.92 Tb + 0.156 H*T - 1.648 H*V + 1.790 H*Tb - 0.0009 T*V - 0.016 T*Tb - 0.012 V*Tb + 0.026 Tb^2

Analysis of Variance and Main Effect Plots. The analysis of results was conducted using the analysis of variance (ANOVA) technique and the main effects plot, implemented with the statistical software Minitab 19. The ANOVA is a statistical technique employed to evaluate the influence of processing parameters and their interactions on the ultimate tensile strength. It utilizes a Fisher ratio test and a P-value test, with a significance level set at P-value<0.05 [16, 17]. The contribution of each parameter and their interactions were examined to determine their impact and influence on the results. The main effects plot, on the other hand, displays the average mean response values for each level of a design parameter or process variable. This plot enables the comparison of the relative impact of the effects of different factors. These graphical representations make it easier to analyze individual factors, identify significant effects or trends, and determine the key drivers or influences on the studied outcome [12, 18]. In this study, the main effects plot of the mean of means and SN ratio were utilized. This analysis identified the parameters that had the greatest effect on the outcomes.

Optimization of Process Parameters. The Taguchi design-based formula was applied for optimization, as follows [19–21]:

$$\eta_{op} = \eta_m + \sum_{i=0}^{o}(\bar{\eta}_i - \eta_m) \tag{3}$$

where η_{op} is the optimum S/N ratio, η_m is the total mean S/N ratio, $\bar{\eta}_i$ is the mean S/N ratio at the optimum level, and o is the number of the main design parameter that affects the quality of the outcome.

The prediction of the best value of the ultimate tensile strength for the optimum parameters selected can be predicted as stated below:

$$y_{op}^2 = \sqrt{(10)^{\eta_{op}/10}} \text{ (Larger is better)} \quad (4)$$

where y_{op} is the optimum predicted value of ultimate tensile strength.

3 Results

3.1 Linear Regression Models

To establish a correlation between process inputs and response, regression models were developed. Based on this estimation, the equation for deriving the model of ultimate tensile strength can be found in Table 5.

3.2 ANOVA Analysis

Table 6 presents the percentage contribution and P-values of each printing parameter. It is evident that the interaction (O*Tb) has the most significant on the UTS printed results at 28.37%. Following this, the (O) factor contributes 18.37%, while the (V*Tb) interaction contributes 12.66%, the (T*Tb) interaction contributes 12.47%, and the (H*V) interaction contributes 11.09%. The process parameters (H*Tb) and (T) have a moderate impact with 6.54% and 5.77%, respectively. Factors (H), (V), and (Tb), the interaction term (HT), and the quadratic term (Tb^2) have negligible impact, with P-values exceeding 0.05.

4 Discussion

4.1 Checking Model Adequacy and Residual Analysis

Based on the results of the ANOVA, a probability (Prob) value of less than 0.05 indicates a statistically significant and effective model, as evidenced by a high coefficient of determination ($R^2 = 97.52\%$), Table 6. These findings support the applicability of the developed model for future engineering applications.

Figure 4 reveals a close alignment between the point clouds and the first bisector, indicating a convergence between the experimental results and the model output. This result demonstrates the strong correlation between the desired response and the predicted values. Figure 4 displays the confidence interval, which serves to gauge the accuracy of the regression outcomes.

The plots of normal residuals indicate a nearly aligned distribution for the UTS response, suggesting that the errors are normally distributed, Fig. 5a. Furthermore, the residual plot demonstrates a random distribution of points according to the order of the tests, confirming the assumption of independence of the residual, Fig. 5b. Consequently, the validation of the considered models is confirmed.

Table 6. ANOVA results of UTS response

Source	DOF	Adj SS	Adj MS	F Value	P Value	Contribution %
H	1	0.123	0.12	0.29	0.19	0.08
T	1	8.486	8.48	12.10	13.29	5.77
V	1	3.566	3.56	5.66	5.59	2.43
Tb	1	0.432	0.43	0.61	0.68	0.29
O	2	27.002	13.50	19.57	21.15	18.37
H*T	1	0.082	0.08	0.19	0.13	0.06
H*V	1	16.302	16.30	25.64	25.54	11.09
H*Tb	1	9.612	9.61	16.01	15.06	6.54
T*V	1	0.109	0.11	0.24	0.17	0.07
T*Tb	1	18.327	18.32	26.03	28.71	12.47
V*Tb	1	18.600	18.60	26.43	29.14	12.66
Tb*O	2	41.695	20.84	30.39	32.66	28.37
Tb^2	1	2.614	2.6	3.65	4.09	1.78
Error	11	7.022	0.63			
Total	26	283.446				
R^2	97.52					

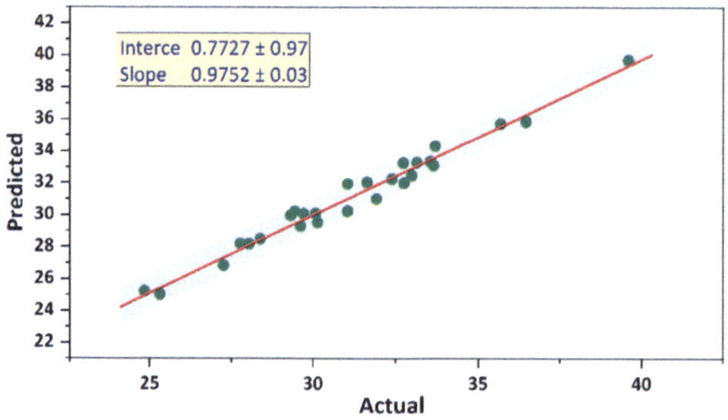

Fig. 4. Predicted and actual values plot of UTS

4.2 Effects Print Parameters

Figure 6 illustrates the analysis of the mean effects of various input parameters (H, T, O, V, and Tb) on the response UTS. In Fig. 6a, the UTS demonstrates a decrease as the layer thickness increases from 0.1 mm to 0.3 mm. The results indicate that an increase in layer thickness leads to enhanced thermal diffusion properties and the production of fewer layers. This, in turn, results in an increasing stair-stepping effect, which can lead to the formation of voids and cracks in the microstructure. The strength between fibers

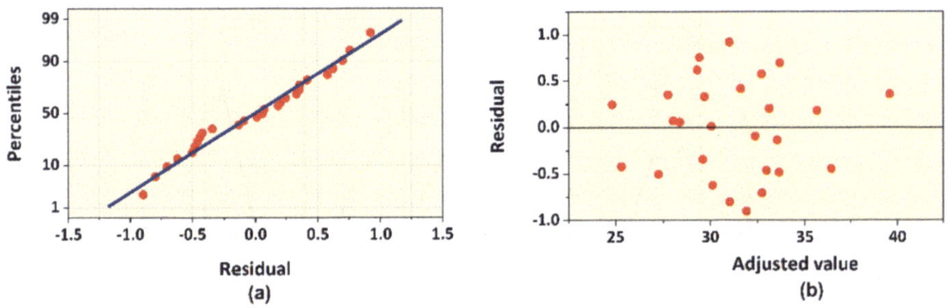

Fig. 5. (a) Normal probability plot for the standardized residual distribution and (b) residual plot versus adjusted value for UTS

is enhanced by reducing the deformability of each layer. This allows the semi-melted material to flow within the expanded space between the fibers, ultimately leading to a decrease in strength [22, 23]. Conversely, Alafaghani et al. [24] observed contrasting results. Their study examined the influence of layer thickness on the mechanical properties of PLA samples fabricated. Layer thicknesses of 0.1, 0.25, 0.3, and 0.4 mm were investigated. The findings indicated that increasing the layer thickness generally resulted in improved mechanical properties of the specimens.

Figure 6 also illustrates that the UTS can be significantly reduced by increasing the raster angle beyond 0° (X-axis). This is because higher raster angles will result in fewer rasters aligned with the tensile stress direction, thereby reducing distortion. When the material is processed with a part print direction at a flat X-axis (0°), the deposited rasters provide improved creep resistance load as their length is parallel to the applied load. In other words, the raster angle (0°/90°) only has 50% of its structure capable of supporting higher tensile loads (layers at 0°) [25]. Therefore, increasing the raster angle to (−45°/45°) and (0°/90°) causes misalignment between the two axes and leads to the accumulation of residual stress, resulting in increased distortion and weaker parts in terms of UTS [26].

The results indicate that the UTS increases when the nozzle temperature is raised from 210 to 215 °C and decreases thereon [27]. Higher temperatures enhance the fluidity of the molten material, promoting adhesion between the freshly deposited PLA fiber and the preexisting extruded material. This leads to improved interlayer welding, and stronger bonds [28]. However, raising the nozzle temperature to 215 °C can result in material deterioration, sub-optimal layer adhesion, and structural challenges. Moreover, the viscosity of the molten PLA decreases, reducing heat transfer from the deposited layers and potentially increasing porosity in the printed parts. This can be achieved by maximizing the cooling time during the printing process, although this may adversely affect the tensile properties.

Decreasing the print speed in FDM can improve the ultimate tensile strength of the printed parts. This is because thinner layers are deposited, resulting in a smoother surface finish and fewer defects [22, 29]. As a result, stress concentrations are reduced leading to improved overall strength of the part. Slower printing speeds allow the molten material more time to flow and bond with the previous layer, creating stronger interlayer bonds and enhancing the overall strength of the parts.

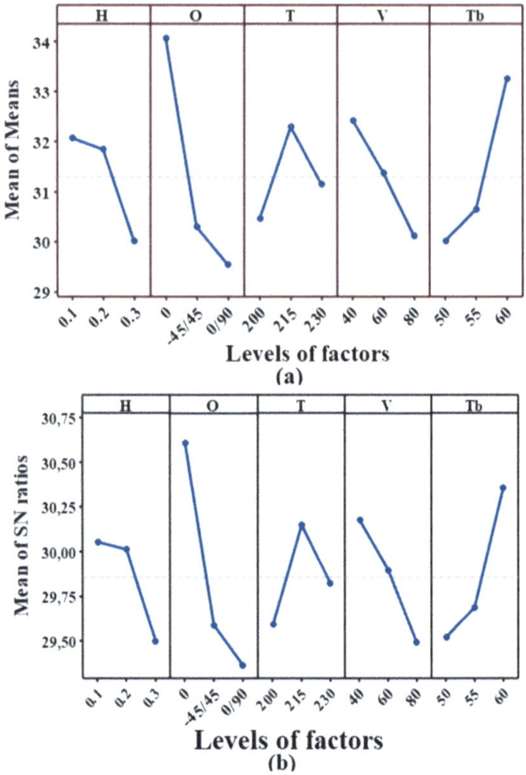

Fig. 6. Main effects plot of (a) mean of means and (b) mean of SN ratio for UTS

As observed in Fig. 6a, the UTS can be enhanced at the highest bed temperatures. This can be attributed to the fact that higher bed temperatures facilitate superior interlayer bonding and minimize warping of the printed part. Conversely, lower bed temperatures can result in weak layer adhesion and reduced overall part strength [30]. Higher bed temperatures can also improve initial layer adhesion, reducing the chances of warping or delamination during printing. These findings contradict previous research conducted by Rasheed et al. [13], who concluded that the tensile strength initially decreases until it reaches a minimum at 95 °C and, then increases with further increase in bed temperature. The disparity in research outcomes may arise from variations in experimental designs.

4.3 Optimization of Process Parameters

According to Fig. 6 and Table 7 of S/N ratios, the optimum process parameters are H1O1T2V1Tb3, where H1, O1, T2, V2, and Tb3 represent the means of S/N ratio with layer thickness at level 1, raster angle at level 2, nozzle temperature at level 2, bed temperature at level 3, respectively, and T is the overall mean, Table 7. By utilizing Eqs. 3 and 4, the prediction of S/N ratio values for UTS can be evaluated for the optimal levels of the process parameters.

$$\eta_{op} = \mathrm{H1} + \mathrm{O1} + \mathrm{T2} + \mathrm{V1} + \mathrm{Tb3} - 4\mathrm{T} = 30.06 + 30.62 + 30.15 + 30.18 + 30.36 - 4^{*}(29.8 \tag{5}$$

Now, the optimum predicted value of ultimate tensile strength is

$$y_{op} = \sqrt{(10)^{32.17/10}} = 40.59\ \mathrm{MPa} \tag{6}$$

Based on the mean and SN ratio values presented in Table 7, the optimum achievement of ultimate tensile strength can be obtained by using the following combination: layer thickness of 0.1 mm, raster angle of 0°, nozzle temperature of 215 °C, print speed of 40 mm/s, and bed temperature of 60 °C.

Table 7. Response table for Means and signal-to-noise ratios for ultimate tensile strength

Process parameters	Levels	Means					S/N ratio				
		H	O	T	V	Tb	H	O	T	V	Tb
Average value	L1	32.06	34.08	30.47	32.42	30.02	30.06	30.62	29.60	30.18	29.52
	L2	31.84	30.29	32.29	31.38	30.64	30.01	29.59	30.15	29.90	29.69
	L3	30.01	29.54	31.15	30.11	33.25	29.50	29.37	29.83	29.49	30.36
Delta	1.83	2.05	4.53	1.83	2.31	3.24	0.55	1.25	0.56	0.69	0.84
Rang	5	4	1	5	3	2	5	1	4	3	2

4.4 Confirmation Run

To validate the results of the optimization process, a series of tests were conducted by setting the process parameters to their optimum levels, Table 8. The predicted response value was 40.59 MPa, while the experimental value obtained was 39.63 MPa. These results indicate a strong agreement between the experimental and predicted values. Furthermore, the UTS response model fulfills all the necessary criteria for validation, with only a 2.4% margin of error.

Table 8. Comparison of confirmation tests with the results

Responses	Experimental Responses				Optimization response	Error (%)
UTS (MPa)	UTS_1	UTS_2	UTS_3	$UTS_{average}$	UTS_{opt}	2.4
	40.06	39.12	39.72	39.63	40.59	

5 Conclusion

In this study, the Taguchi method was utilized to determine the optimum FDM process parameters on the ultimate tensile strength of polylactic acid material. Based on the obtained results, the conclusions drawn from this study are as follows:

- The regression model for ultimate tensile strength with an (R^2) value of 97.52%, demonstrates a high level of agreement with the experimental data.
- The ANOVA analysis reveals that the interaction between raster angle and bed temperature has the most significant impact on the UTS printed results. Additionally, raster angle is identified as the most influential factor affecting the ultimate tensile strength.
- The ultimate tensile strength increases as the nozzle temperature rises to the intermediate level of 215 °C, thereafter, it begins to decline.
- Increasing the layer thickness, raster angle, and print speed leads to a decrease in the ultimate tensile strength. Conversely, the ultimate tensile strength shows an upward trend with an increase in bed temperature.
- The optimum values for the input parameters are as follows: a layer thickness of 0.1 mm, a raster angle of 0°, a nozzle temperature of 215 °C, a print speed of 40 mm/s, and a bed temperature of 60 °C.

References

1. Penumakala, P.K., Santo, J., Thomas, A.: A critical review on the fused deposition modeling of thermoplastic polymer composites. Compos. Part B. **201**, 108336 (2020). https://doi.org/10.1016/j.compositesb.2020.108336
2. Mohamed, O.A., Masood, S.H., Bhowmik, J.L.: Optimization of fused deposition modeling process parameters: a review of current research and prospects. Adv. Manuf. **3**(1), 42–53 (2015). https://doi.org/10.1007/s40436-014-0097-7
3. Moradi, M., Aminzadeh, A., Rahmatabadi, D., Rasouli, S.A.: Statistical and experimental analysis of process parameters of 3D nylon printed parts by fused deposition modeling: response surface modeling and optimization. J. Mater. Eng. Perform. **30**(7), 5441–5454 (2021). https://doi.org/10.1007/s11665-021-05848-4
4. Saad, M.S., Nor, A.M., Baharudin, M.E., Zakaria, M.Z., Aiman, A.F.: Optimization of surface roughness in FDM 3D printer using response surface methodology, particle swarm optimization, and symbiotic organism search algorithms. Int. J. Adv. Manuf. Technol. **105**(12), 5121–5137 (2019). https://doi.org/10.1007/s00170-019-04568-3
5. Enemuoh, E.U., Asante-Okyere, S.: Impact of feature selection on neural network prediction of fused deposition modeling (FDM) print part properties. Int. J. Interact. Des. Manuf. **18**, 7413–7427 (2023). https://doi.org/10.1007/s12008-023-01598-w
6. Alhaddad, W., He, M., Halabi, Y., Yahya Mohammed Almajhali, K.: Optimizing the material and printing parameters of the additively manufactured fiber-reinforced polymer composites using an artificial neural network model and artificial bee colony algorithm. Structure. **46**, 1781–1795 (2022). https://doi.org/10.1016/j.istruc.2022.10.134
7. Dong, G., Wijaya, G., Tang, Y., Zhao, Y.F.: Optimizing process parameters of fused deposition modeling by Taguchi method for the fabrication of lattice structures. Addit. Manuf. **19**, 62–72 (2018). https://doi.org/10.1016/j.addma.2017.11.004

8. Zeng, Y.S., Hsueh, M.H., Lai, C.J., Hsiao, T.C., Pan, C.Y., Huang, W.C., et al.: An investigation on the hardness of Polylactic acid parts fabricated via fused deposition modeling. Polymers (Basel). **14**(14) (2022). https://doi.org/10.3390/polym14142789
9. Hasdiansah, H., Yaqin, R.I., Pristiansyah, P., Umar, M.L., Priyambodo, B.H.: FDM-3D printing parameter optimization using Taguchi approach on surface roughness of thermoplastic polyurethane parts. Int. J. Interact. Des. Manuf. **17**(6), 3011–3024 (2023). https://doi.org/10.1007/s12008-023-01304-w
10. Nagendra, J., Prasad, M.S.G.: FDM process parameter optimization by taguchi technique for augmenting the mechanical properties of Nylon–Aramid composite used as filament material. J. Inst. Eng. India Series C. **101**(2), 313–322 (2020). https://doi.org/10.1007/s40032-019-00538-6
11. Heidari-Rarani, M., Ezati, N., Sadeghi, P., Badrossamay, M.R.: Optimization of FDM process parameters for tensile properties of polylactic acid specimens using Taguchi design of experiment method. J. Thermoplast. Compos. Mater. **35**(12), 2435–2452 (2020). https://doi.org/10.1177/0892705720964560
12. Auffray, L., Gouge, P.-A., Hattali, L.: Design of experiment analysis on tensile properties of PLA samples produced by fused filament fabrication. Int. J. Adv. Manuf. Technol. **118**(11), 4123–4137 (2022). https://doi.org/10.1007/s00170-021-08216-7
13. Rasheed, A., Hussain, M., Ullah, S., Ahmad, Z., Kakakhail, H., Riaz, A.A., et al.: Experimental investigation and Taguchi optimization of FDM process parameters for the enhancement of tensile properties of Bi-layered printed PLA-ABS. Mater. Res. Express. **10**(9), 095307 (2023). https://doi.org/10.1088/2053-1591/acf1e7
14. Boulahem, K., Ben Salem, S., Bessrour, J.: Prediction model of ultimate tensile strength and investigation on microstructural characterization of friction stir welded AA2024-T3. Int. J. Adv. Manuf. Technol. **95**(1), 1473–1486 (2018). https://doi.org/10.1007/s00170-017-1215-z
15. Heidari-Rarani, M., Ezati, N., Sadeghi, P., Badrossamay, M.: Optimization of FDM process parameters for tensile properties of polylactic acid specimens using Taguchi design of experiment method. J. Thermoplast. Compos. Mater. **35**(12), 2435–2452 (2022). https://doi.org/10.1177/0892705720964560
16. Mishra, A., Kumar, R., Sharma, A.K., Gupta, N.K., Somani, N.: A statistical analysis on the effect of process parameters on tensile, flexural, and hardness characteristics of wood–polylactic acid composites using FDM 3D printing. Int. J. Interact. Des. Manuf. **18**, 1303–1315 (2023). https://doi.org/10.1007/s12008-023-01658-1
17. Boulahem, K., Salem, S.B., Shiri, S., Bessrour, J.: Experimental modeling and multi-response optimization in friction stir welding process parameters of AA2024-T3 using response surface methodology and desirability approach. Exp. Tech. **48**, 833–849 (2024). https://doi.org/10.1007/s40799-023-00691-9
18. Vicente, C.M.S., Martins, T.S., Leite, M., Ribeiro, A., Reis, L.: Influence of fused deposition modeling parameters on the mechanical properties of ABS parts. Polym. Adv. Technol. **31**(3), 501–507 (2020). https://doi.org/10.1002/pat.4787
19. Tiwary, V.K., Padmakumar, A., Malik, V.: Adhesive bonding of similar/dissimilar three-dimensional printed parts (ABS/PLA) considering joint design, surface treatments, and adhesive types. Proc. Inst. Mech. Eng. C J. Mech. Eng. Sci. **236**(16), 8991–9002 (2022). https://doi.org/10.1177/09544062221089849
20. Tiwary, V.K., Ravi, N., Arunkumar, P., Shivakumar, S., Deshpande, A.S., Malik, V.R.: Investigations on friction stir joining of 3D printed parts to overcome bed size limitation and enhance joint quality for unmanned aircraft systems. Proc. Inst. Mech. Eng. C J. Mech. Eng. Sci. **234**(24), 4857–4871 (2020). https://doi.org/10.1177/0954406220930049

21. Tyagi, R., Singh, G., Kumar, R., Kumar, V., Singh, S.: 3D-printed sandwiched acrylonitrile butadiene styrene/carbon fiber composites: investigating mechanical, morphological, and fractural properties. J. Mater. Eng. Perform. **33**, 5061–5074 (2023). https://doi.org/10.1007/s11665-023-08292-8
22. Shergill, K., Chen, Y., Bull, S.: An investigation into the layer thickness effect on the mechanical properties of additively manufactured polymers: PLA and ABS. Int. J. Adv. Manuf. Technol. **126**(7), 3651–3665 (2023). https://doi.org/10.1007/s00170-023-11270-y
23. Sharma, M., Sharma, V., Kala, P.: Optimization of process variables to improve the mechanical properties of FDM structures. J. Phys. Conf. Ser. **1240**(1), 012061 (2019). https://doi.org/10.1088/1742-6596/1240/1/012061
24. Alafaghani, A., Qattawi, A., Alrawi, B., Guzman, A.: Experimental optimization of fused deposition modelling processing parameters: a design-for-manufacturing approach. Proc. Manuf. **10**, 791–803 (2017). https://doi.org/10.1016/j.promfg.2017.07.079
25. Gordelier, T.J., Thies, P.R., Turner, L., Johanning, L.: Optimising the FDM additive manufacturing process to achieve maximum tensile strength: a state-of-the-art review. Rapid Prototyp. J. **25**(6), 953–971 (2019). https://doi.org/10.1108/RPJ-07-2018-0183
26. Çakan, B.G.: Effects of raster angle on tensile and surface roughness properties of various FDM filaments. J. Mech. Sci. Technol. **35**(8), 3347–3353 (2021). https://doi.org/10.1007/s12206-021-0708-8
27. Guessasma, S., Belhabib, S., Nouri, H.: Thermal cycling, microstructure and tensile performance of PLA-PHA polymer printed using fused deposition modelling technique. Rapid Prototyp. J. **26**(1), 122–133 (2020). https://doi.org/10.1108/RPJ-06-2019-0151
28. Hsueh, M.-H., Lai, C.-J., Wang, S.-H., Zeng, Y.-S., Hsieh, C.-H., Pan, C.-Y., et al.: Effect of printing parameters on the thermal and mechanical properties of 3D-printed PLA and PETG, using fused deposition modeling. Polymers. **13**(11), 1758 (2021)
29. Miazio, Ł.: Impact of print speed on strength of samples printed in FDM technology. Agric. Eng. **23**(2), 33–38 (2019). https://doi.org/10.1515/agriceng-2019-0014
30. Nathaphan, S., Trutassanawin, W.: Effects of process parameters on compressive property of FDM with ABS. Rapid Prototyp. J. **27**(5), 905–917 (2021). https://doi.org/10.1108/RPJ-12-2019-0309

Overall Equipment Effectiveness Modelling and Assessment of Additive Manufacturing Printing Machines

Amel Hajji[1,2], Karem Dhouib[3](✉), and Ali Gharbi[4]

[1] LMPE Laboratory, ENSIT, University of Tunis, Tunis, Tunisia
amel.hajji.12@gmail.com
[2] Industrial Engineering Department, ENIT, University of Tunis-El Manar, Tunis, Tunisia
[3] RIFTSI Laboratory, ENSIT, University of Tunis, Tunis, Tunisia
karemdhouib@yahoo.fr
[4] C2SP Laboratory, École de Technologie Supérieure (ÉTS), Montréal, Canada
Ali.Gharbi@etsmtl.ca

Abstract. The advent of additive manufacturing, also known as 3D printing, within the framework of Industry 4.0 has significantly transformed traditional production processes by enabling highly customizable and efficient manufacturing solutions. Unlike conventional methods, additive manufacturing supports decentralized production, which reduces the need for large-scale, centralized factories and enhances the flexibility and adaptability of the manufacturing sector. This decentralization allows manufacturers to produce closer to demand points, respond swiftly to customer requirements, and minimize inventory costs. This paper focuses on evaluating the performance of 3D Additive Manufacturing Printers by analyzing the impact of random equipment failures, repair times, and scrap and rework probabilities of non-conforming parts on the Overall Equipment Effectiveness of these machines. This is accomplished by developing analytical models that assess the effectiveness of 3D printing machines under different industrial scenarios, guided by the specific policies of production and quality management departments. Additionally, simulation models are constructed to mimic the real-world dynamic and stochastic behavior of these Additive Manufacturing Printers, providing a robust validation of the proposed analytical approaches. According to specific industrial contexts, and based on several configurations of Additive Manufacturing Printers, the results show that the proposed analytical models generate a negligible relative error, inferior to 1%, compared to simulation ones, providing insights for improving the efficiency and reliability of additive manufacturing processes across various industrial contexts.

Keywords: Additive manufacturing · 3D printers · Overall Equipment Effectiveness · Random disturbances · Analytical-Simulation modeling

M. Soula et al. (Eds.): AIAM 2024, LNME, pp. 149–162, 2025.
https://doi.org/10.1007/978-3-031-94889-3_12

1 Introduction

To improve operational efficiency and product quality, many businesses are compelled to incorporate Industry 4.0 technologies, such as IoT, AI, Simulation, FMS with autonomous Robots, AGVs, AS/RS, and Additive Manufacturing machines, since the industrial environment has grown more dynamic over the past ten years [1–3]. Additive Manufacturing (AM), often considered a key element of Industry 4.0, represents a revolutionary technology applied in various domains such as aerospace, medicine, food, construction, etc. [4].

Also known as 3D printing, this innovation enables the production of complex geometries that would be challenging with conventional manufacturing techniques, achieved through the layer-by-layer construction of materials [5, 6]. In addition to enhancing the flexibility of production processes, this technology plays a pivotal role in moving toward sustainability by adopting more effective material utilization and waste reduction [5, 7, 8].

Although the performance of AM is well-established across various domains, operational issues such as production losses due to process variability or inefficient machine utilization have become common concerns for companies adopting AM [9]. The Overall Equipment Effectiveness (*OEE*) metric is a widely used tool in traditional manufacturing to assess the efficient use of production capacity [10–14]. It encompasses availability, performance, and quality, making it a valuable tool for evaluating the overall performance of equipment and reducing production losses including machine downtimes, performance issues, and quality losses [10].

Moreover, *OEE* is particularly effective in contexts where the cost of performance loss is high, as emphasized by AM cost models [12]. When applied to 3D printing processes, as considered in this paper, quality losses are recognized as critical aspects that warrant precise attention [15–18]. In some industrial contexts, such as in the case of thin horizontal face of printed parts, a non-conforming part is not immediately rejected but can be reworked to be accepted again, reducing their environmental impact. In general, rework actions involve all activities transforming a non-conforming part into a compliant one. Parts may undergo rework either once or multiple times. When a part is subjected to rework, it can either be processed within the same manufacturing process or assumed to undergo rework in an external process [19–21].

In the literature, previous studies on the application of *OEE* to additive manufacturing have a limited understanding of how AM operations impact production losses [12]. Studies have looked at how production systems manage product inspection, separating acceptable and non-conforming products on the presumption that parts that don't meet standards must be discarded right away [12, 22–26]. However, to make sure that non-conforming components satisfy client needs, product inspection may involve extra steps, and rework activities, in addition to simply identifying parts as either excellent or rejected.

This paper aims to provide a comprehensive analysis of *OEE* in the context of a manufacturing cell incorporating a 3D printing machine susceptible to sporadic breakdowns and repairs, rework, and scrap. This comprehensive evaluation will serve as the basis for developing targeted strategies aimed at improving operational performance, thereby enhancing the role of additive manufacturing in the move toward more sustainable and efficient manufacturing practices. Simulation models were proposed to imitate the real dynamic and stochastic behaviors of the 3D printing machines under several reworking and scrapping strategies.

This is how the rest of the paper is structured: section two specifies additive production cell characteristics. In section three, we outline the analytical formulation of the Overall Equipment Efficiency (OEE) according to international standards and provide a formulation specific to 3D printing machines. In section four, we present comprehensive simulation models to evaluate the OEE of this production system. Additionally, it gives numerical findings and uses simulation results to validate analytically suggested models. Finally, conclusions are provided in section five.

2 Manufacturing System Description and Notations

The system being studied in this paper is a manufacturing cell composed of a 3D Additive Manufacturing printing machine dedicated to producing objects made of metal, plastics, or other materials (Fig. 1). Objects are manufactured layer by layer until they attain their final shapes. Every printed object is processed during a deterministic amount of time denoted as the cycle time (tc) and can be inspected for conformity.

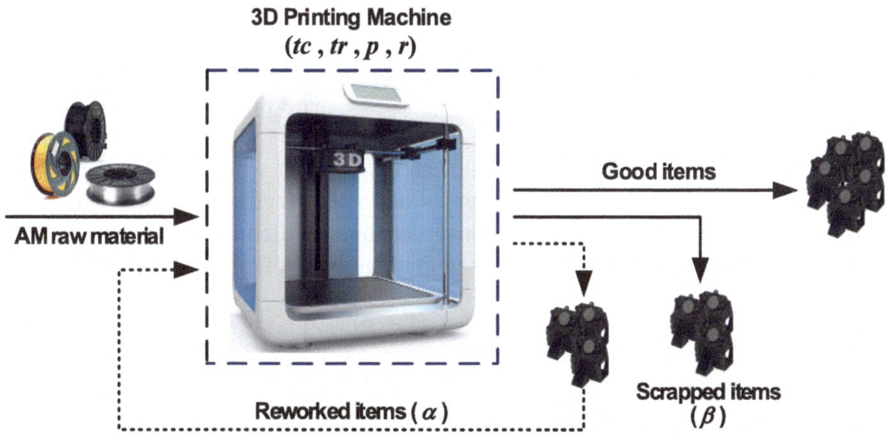

Fig. 1. 3D printer manufacturing cell

Upon inspection, the printed object can be immediately accepted if it is found to conform to specifications. In certain situations, the quality of the printed object may be susceptible to random defects, such as warping, over-extrusion, under-extrusion, burrs or strings between layers, sizing errors, etc. [13, 15–17]. In such cases, the printed object can be scrapped immediately with a probability β if its quality characteristic falls below the lower specification limit (L) (Fig. 2). Alternatively, if the printed object exhibits a thin top layer defect, it can be reworked on the same printing machine with a probability α if its measured quality characteristic falls above the upper specification limit (U). This rework process involves adding extra layers to rectify defects and occurs within a predetermined timeframe for each correction, independent of the part's complexity or the severity of the flaw. The rework operation is carried out immediately on the same printing machine only once, requiring tr time units. Upon completion of the rework operation, the printed object is deemed satisfactory and exits the 3D printer.

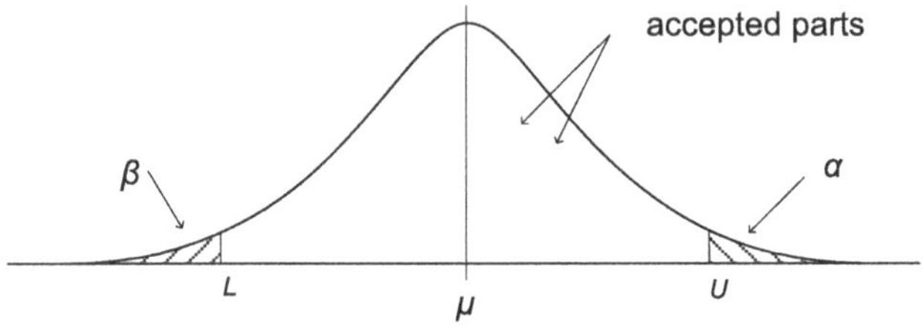

Fig. 2 Rework and scrap probabilities after the additive manufacturing process

Quality characteristics are assumed to follow a normal distribution with mean μ and standard deviation σ, as shown in Fig. 2. Therefore, the probabilities of rework and rejection can be evaluated by Eqs. (1) and (2), respectively.

$$\alpha = \int_{U}^{+\infty} \frac{1}{\sqrt{2\pi}\sigma} e^{-\frac{1}{2}\left(\frac{x-\mu}{\sigma}\right)} dx \tag{1}$$

$$\beta = \int_{-\infty}^{L} \frac{1}{\sqrt{2\pi}\sigma} e^{-\frac{1}{2}\left(\frac{x-\mu}{\sigma}\right)} dx \tag{2}$$

Additionally, sporadic repairs and breakdowns are possible with the 3D printer. With corresponding values p and r, the operational and maintenance times have geometric distributions; the distribution function for the equipment lifetime is given by Eq. (3), where n is the equipment lifetime expressed in the number of cycles.

$$F(n) = 1 - (1 - p)^n \tag{3}$$

The analytical methods for appraising OEE in various real-world scenarios, such as failures and repairs, scrap, and/or rework, are purported in the next section.

3 Overall Equipment Effectiveness of 3D Additive Manufacturing Printers

OEE is a critical performance indicator used to assess the operational efficiency of a machine or installation over the available time dedicated for production. It refers to the proportion of time during which the machine produces good parts involving the aspects of availability, performance, and quality (Eq. 4) [10, 27].

$$OEE = AR \times QR \times PR \tag{4}$$

where *AR* is the UpTime Ratio (*UTR*) corresponding to the availability rate, *PR* and *QR* are, respectively, the Performance and the Quality Ratios of the machine. *UTR* is defined as the ratio between operating time and planned loading time, where operating time includes planned operation time minus planned downtime. *PR* represents the net operating time divided by the operating time, with net operating time comprising operating time minus idling time. *QR* is defined as the ratio between the quantity of good parts and the total quantity produced, with the quantity of good parts being those that meet quality requirements. Therefore, *OEE* does not only measure the efficiency of equipment but also serves to identify specific areas for improvement by minimizing losses that reduce equipment effectiveness. There are six major losses involving aspects of availability, performance, and quality (Fig. 3) [10–14].

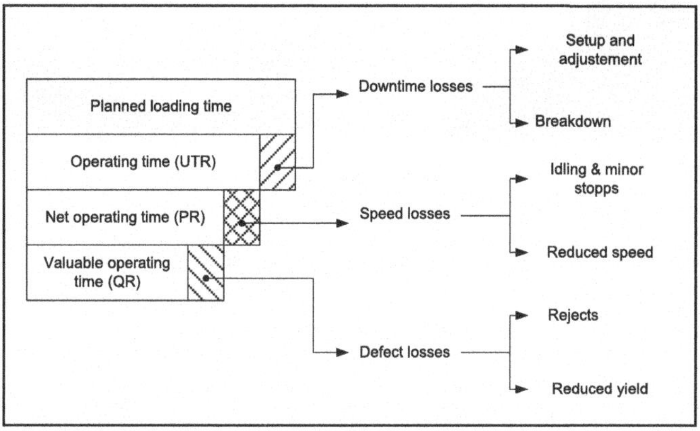

Fig. 3. Overall Equipment Effectiveness losses

In the context of automated 3D printers, as investigated in this paper, availability losses quantify breakdowns and setup/adjustments, which may be associated with warm-up, cool-down, or necessary cleaning between builds. Performance losses, related to minor stoppages and reduced speed, are negligible, so the performance ratio is considered 100%. Quality losses are associated with defects arising from inappropriate process parameters.

3.1 Manufacturing Cells Subject to Random Failures and Reparations

Breakdowns are a common fact with every piece of equipment, In the case where the 3D Additive Manufacturing printer is subject to random failures and reparations, the *OEE* can be reduced to its availability (Eq. 5) [28–31].

$$OEE = \frac{r}{p + r} \tag{5}$$

3.2 3D Printers with Scraped Parts in Light of Breakdowns

In certain industrial settings, 3D AM printers may produce defective parts when failures occur, as seen in the case of a breakdown in the extrusion system leading to an inability to extrude the filament properly (Fig. 4).

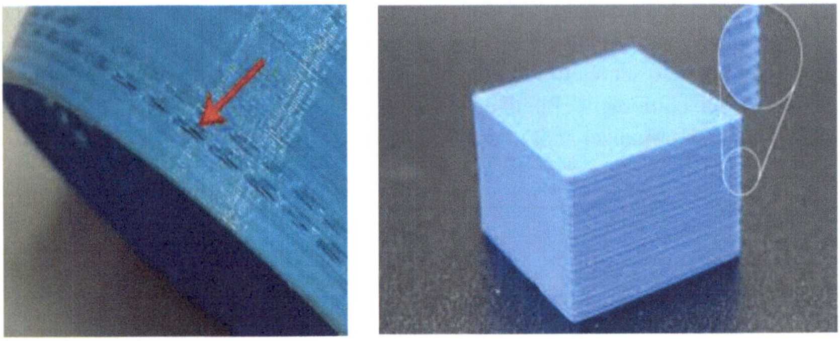

Fig. 4 Extrusion problems of 3D printing

Figure 5 illustrates a scenario representing the dynamics of the 3D printer susceptible to sporadic breakdowns and repairs with scrap due to failures, where *TTF* and *TTR* are the machine time to failure and time to repair, respectively.

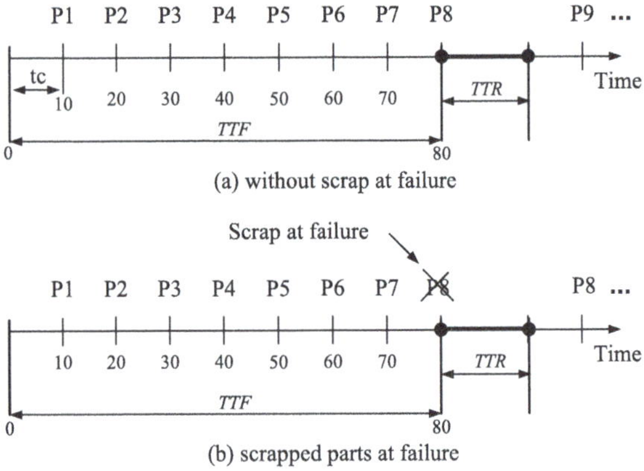

Fig. 5 3D Additive Manufacturing printer's dynamics (Scrap due to failure)

For example, at failure, item P8 is scrapped. Consequently, the *OEE* is determined by Eq. (6). Indeed, the quality ratio (*QR*) is equal to (1-*p*) (Eq. 7), where *NbTP*, *NbGP*, and *NbSP* are the respective number of total, good, and scrapped printed parts produced during a long horizon *T* [21, 27].

$$OEE = \frac{r}{p+r}(1-p) \tag{6}$$

$$QR = \frac{NbGP}{NbTP} = \frac{NbTP - NbSP}{NbTP} = 1 - \frac{NbSP}{NbTP} = 1 - p \tag{7}$$

3.3 3D Printers with Accidental Breakdowns and Scraps

Equation (8) determines the *OEE* if a non-compliant item is trashed with a probability β following inspection (Fig. 6). For example, parts P4 and P6 are inspected to be non-compliant and thus rejected. The quality ratio (*QR*) is defined by Eq. (9) [27].

$$OEE = \frac{r}{p + r}(1 - \beta) \tag{8}$$

$$QR = \frac{NbGP}{NbTP} = \frac{NbTP - NbSP}{NbTP} = 1 - \frac{NbSP}{NbTP} = 1 - \beta \tag{9}$$

This scenario can be illustrated by issues related to the first layer when it does not adhere properly (Fig. 7).

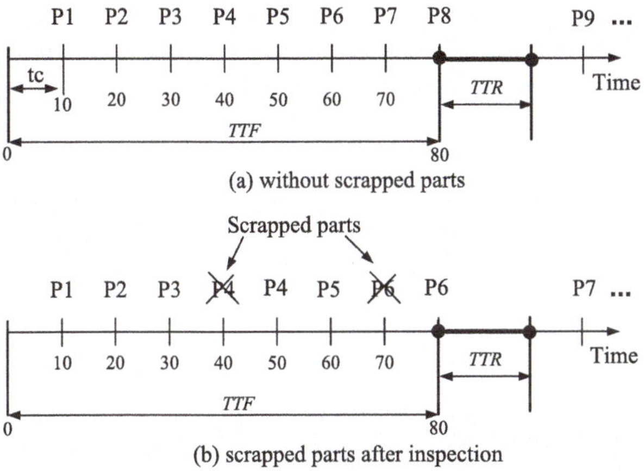

Fig. 6 3D Additive Manufacturing printer's dynamics (Scrap after inspection)

Fig. 7 First layer problems of 3D printing

3.4 3D Printers with Rework for Defective Parts

In specific circumstances, particularly when facing challenges such as the fragility of the upper surface (Fig. 8), non-conforming printed parts are retrieved to be rectified immediately on the same 3D printer transforming them into compliant components by adding an extra thin layer.

Fig. 8 Thin horizontal faces problem encountered in 3D printing

This additional process occurs during a defined rework time (tr), which may deviate from the standard cycle time. Figure 9 illustrates a scenario in which parts 3 and 6 undergo immediate rework after inspection. Rework actions are performed only once, and all reworked parts are processed at the same time (tr).

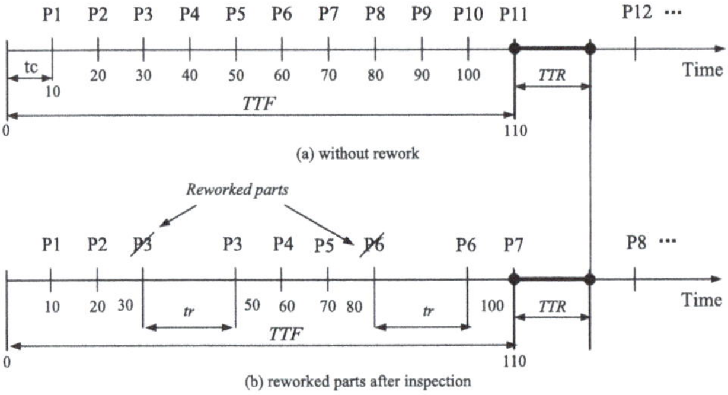

Fig. 9 3D Additive Manufacturing printer's dynamics (Reworked parts)

In this context, the number of good parts is determined using Eq. (10), where *NbRepCyc* and *NbRewCyc* are the respective numbers of reparation and rework cycles that occurred during a long horizon T.

$$NbGP = \frac{T - NbRepCyc \cdot tc - NbRewCyc \cdot tr}{tc} = \frac{T - T(1 - UTR) - \alpha \cdot a \cdot T \cdot UTR}{tc} = \frac{T}{tc} \cdot UTR(1 - a \cdot \alpha) \tag{10}$$

where $a = tr/tc$, (when $tr = tr$, $a = 1$).

The *OEE* is then computed using Eq. (11).

$$OEE = \frac{NbGP \cdot tc}{T} = UTR(1 - a \cdot \alpha) = \frac{r}{p + r}(1 - a \cdot \alpha) \tag{11}$$

3.5 3D Printers Susceptible to Sporadic Breakdowns, Repairs, Scraps, and Rework of Faulty Components

Think about the general scenario in which the 3D printer has sporadic breakdowns and repairs. Non-compliant printed items may be scrapped or reworked following inspection. If a failure does not impact the quality of the printed parts, the *OEE* of 3D printers is calculated using Eq. (12). However, if it leads to the destruction of parts, the *OEE* is determined by Eq. (13).

$$OEE = \frac{NbGP \cdot tc}{T} = UTR(1 - a \cdot \alpha) = \frac{r}{p + r}(1 - a \cdot \alpha)(1 - \beta) \tag{12}$$

$$OEE = \frac{r}{p + r}(1 - a \cdot \alpha)(1 - \beta)(1 - p) \tag{13}$$

4 Numerical Results

The accuracy of the analytical formulas for the Overall Equipment Effectiveness of 3D printing machines is evaluated in this section, which is exposed to sporadic breakdowns and repairs, as well as scrap and rework. This assessment is conducted in two scenarios: when a failure affects the quality of printed parts (Eq. 13) and when it does not (Eq. 12).

A simulation model was developed using the ARENA Software to imitate the stochastic and dynamic behavior of the studied 3D manufacturing system (Fig. 10) [32]. The flowchart consists of five main modules.

(a) The INITIALIZATION module sets system parameters like *tc*, *tr*, *UTR*, p, r, α, and β for each experiment.
(b) The ADDITIVE MANUF. The module manages part flow in the additive manufacturing cell.
(c) The INSPECTION module examines each finished item, sampling rework and scrap probabilities based on rework attempts and respective distributions.
(d) The BREAKDOWN module samples machine failure and repair times.
(e) The UPDATE PERFORMANCE module records the number of good parts, scrapped items, and rework actions during the simulation horizon, to update and then assess the 3D manufacturing printer *OEE* and throughput.

Numerous experiments were conducted, altering the 3D Additive Manufacturing printer's parameters such as p, r, α, β, and *tr*. Each simulation runs for 10,000,000 time units, with a warm-up period of 100,000 time units to ensure the steadiness of KPIs. Ten replications were carried out for each configuration.

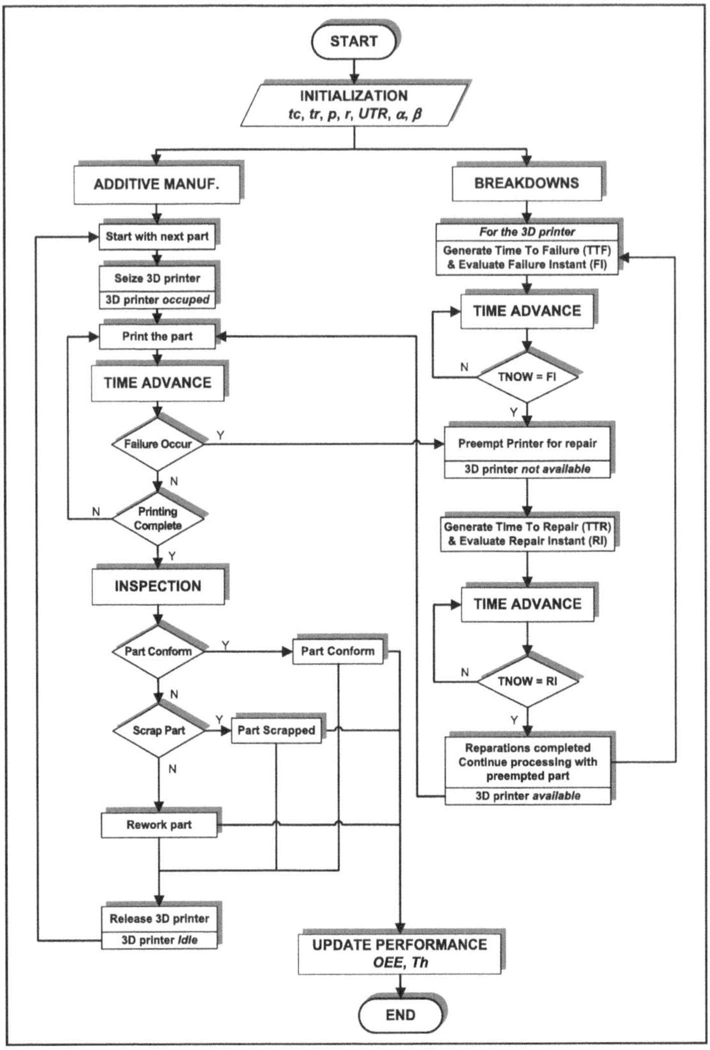

Fig. 10 Simulation flow chart for the 3D printing machine dynamics

Table 1 gives 13 experimented 3D printers with randomly generated reliability, maintainability, operational, and quality characteristics. In the base case, a 3D printer with a one-time unit for both operation and reworking is taken into account, along with respective breakdown and repair probabilities of 0.01 and 0.04. Once inspected, the probability of a printed part being rejected is 0.1 and the probability of being reworked is 0.05. Then, scenarios 1 to 12 were explored by varying the 3D printer base case parameters of p, r, α, β, and tr.

Table 1. Analytical versus simulation results

CASE	p	r	α	β	tr	OEE_{anal1}	OEE_{sim1}	ε_1 (%)	OEE_{anal2}	OEE_{sim2}	ε_2 (%)
BASE	0.01	0.04	0.05	0.1	1	0.6840	0.6853	0.1897	0.6772	0.6787	0.2210
1	0.01	0.04	0.05	0.1	1.5	0.6660	0.6732	1.0695	0.6593	0.6664	1.0654
2	0.01	0.04	0	0	1.3	0.8000	0.8000	0.0000	0.7920	0.7920	0.0000
3	0.01	0.04	0.05	0	0.7	0.7720	0.7707	-0.169	0.7643	0.7627	-0.210
4	0.01	0.04	0	0.1	1	0.7200	0.7198	-0.028	0.7128	0.7124	-0.056
5	0.01	0.04	0.02	0.1	0.7	0.7099	0.7096	-0.042	0.7028	0.7023	-0.071
6	0.01	0.04	0.1	0.1	1.1	0.6408	0.6475	1.0347	0.6344	0.6417	1.0759
7	0.01	0.04	0.05	0.05	1.8	0.6916	0.6996	1.1435	0.6847	0.6928	1.1692
8	0.01	0.04	0.1	0.2	0.3	0.6208	0.6141	-1.091	0.6146	0.6079	-1.102
9	0.005	0.02	0.05	0.1	0.7	0.6948	0.6943	-0.072	0.6913	0.6906	-0.101
10	0.1	0.4	0.02	0.1	1.2	0.7027	0.7038	0.1563	0.6324	0.6333	0.1421
11	0.01	0.023	0.05	0.1	0.9	0.5990	0.5991	0.0167	0.5931	0.5929	-0.034
12	0.01	0.09	0.05	0.1	1.6	0.7452	0.7520	0.9043	0.7377	0.7445	0.9134

The *OEE* produced by the simulation and analytical models are also included in Table 1. Equation (14), which compares the analytical and simulation findings, provides the relative errors.

$$\varepsilon_i(\%) = \frac{OEE_{sim\ i} - OEE_{anal\ i}}{OEE_{sim\ i}} \times 100 \quad ; \quad i = 1,\ 2 \tag{14}$$

where OEE_{sim1} and OEE_{anal1} (OEE_{sim2} and OEE_{anal2}) are, respectively, the 3D printer *OEE* provided by simulation and analytical model where breakdown doesn't influence the quality of printed parts (case where a failure results in the destruction of the produced part).

Table 1 shows that the mean relative errors produced by the analytical formulations are insignificant when compared to simulation findings, particularly when the 3D printer is merely susceptible to failure and/or scrap following inspection. In such cases, the analytical results align with the simulation outcomes. However, it can be observed that when the machine is concurrently susceptible to both rework and scrap, the mean relative error increases proportionally with the rework and scrap probabilities. This error is enlarged if *tr* value moves away from *tc* value.

5 Conclusion

This paper presents analytical formulations for the Overall Equipment Effectiveness of 3D printers susceptible to sporadic breakdowns and repairs, as well as scrap and rework. The proposed analytical models cover scenarios where the machine yields compliant items, scraps due to breakdowns, and scraps and reworks after inspection with specific proportions. Simulation models are developed to imitate the real dynamic and stochastic behavior of the 3D printers. Numerous experimentations are conducted by varying machine parameters (probabilities of breakdowns and repair, scrap and rework proportions, and rework time). A comparative study of analytical versus simulation results was carried out and shows that the proposed analytical models afford a worthy assessment of the *OEE*.

Nonetheless, the proposed model allows rework only once on the same printing machine. Future research could explore relaxing this constraint, enabling more occurrences or even infinite rework in other processes. Manufacturing multiple products should also be considered in the context of flexible manufacturing cells with various 3D printing machines. The evaluation of sustainability and environmental factors of manufacturing processes and maintenance activities is a relevant issue in the context of Industry 4.0. Additionally, the policy might support environmentally eco-friendly production processes by including sustainability criteria in the decision-making paradigm.

References

1. Abdullah, F.M., Saleh, M., Al-Ahmari, A.M., Anwar, S.: The impact of industry 4.0 technologies on manufacturing strategies: proposition of technology-integrated selection. IEEE Access. **10**, 21574–21583 (2022). https://doi.org/10.1109/ACCESS.2022.3151898
2. Wang, S., Wan, J., Li, D., Zhang, C.: Implementing smart factory of industry 4.0: an outlook. Int. J. Distrib. Sens. Netw. **12**, 1–10 (2016). https://doi.org/10.1155/2016/3159805
3. Dhouib, K., Ait-Kadi, D.: Expert system for AGV managing in bidirectional networks: KADS methodology based approach. Int. J. Prod. Econ. **33**(1), 31–43 (1994). https://doi.org/10.1016/0925-5273(94)90116-3
4. Jandyal, A., Chaturvedi, I., Wazir, I., Raina, A., Haq, M.I.U.: 3D printing–a review of processes, materials and applications in industry 4.0. Sustain. Oper. Comput. **3**, 33–42 (2022). https://doi.org/10.1016/j.susoc.2021.09.004
5. Gardan, J.: Additive Manufacturing Technologies. Additive Manufacturing Handbook, pp. 149–168. CRC Press, Boca Raton, FL (2017)
6. Saran, O.S., Reddy, A.P., Chaturya, L., Kumar, M.P.: 3D printing of composite materials: a short review. Mater. Today: Proc. **64**, 615–619 (2022). https://doi.org/10.1016/j.matpr.2022.05.144
7. Nadagouda, M.N., Ginn, M., Rastogi, V.: A review of 3D printing techniques for environmental applications. Curr. Opin. Chem. Eng. **28**, 173–178 (2020). https://doi.org/10.1016/j.coche.2020.08.002
8. Patti, A., Acierno, S., Cicala, G., Zarrelli, M., Acierno, D.: Recovery of waste material from biobags: 3D printing process and thermo-mechanical characteristics in comparison to virgin and composite matrices. Polymers. **14**, 1943 (2022). https://doi.org/10.3390/polym14101943
9. Mathlouthi, S., Gabsi, A.E.H.: Non-conformities and scrap costs reduction in a 3D printing workshop. In: Mabrouki, T., et al. (eds.) Advances in Additive Manufacturing: Materials, Processes and Applications. AIAM 2023 Lecture Notes in Mechanical Engineering, pp. 149–158. Springer, Cham (2023). https://doi.org/10.1007/978-3-031-47784-3_19
10. SEMI E79-2022: Specification for Definition and Measurement Of Equipment Productivity. Semiconductor Equipment & Materials Institute (SEMI), San Jose, CA (2022)
11. Bouti, A., Ait-Kadi, D., Dhouib, K.: Automated manufacturing systems failure analysis based on a functional reasoning. In: 10th ISPE/IFAC International Conference, pp. 423–429. OCRI Publications, Kanata, ON (1994)
12. Basak, S., Baumers, M., Holweg, M., Hague, R., Tuck, C.: Reducing production losses in additive manufacturing using overall equipment effectiveness. Addit. Manuf. **56**, 102904 (2022). https://doi.org/10.1016/j.addma.2022.102904

13. Hajji, A., Dhouib, K., Gharbi, A.: Evaluating the Effectiveness of Manufacturing Systems Under Random Failures and Quality Disruptions. In: 15th International Conference on Industrial Engineering and Operations Management. IEOM 2025, Singapore (2025). https://doi.org/10.46254/AN15.20250276
14. De Ron, A.J., Rooda, J.E.: Equipment effectiveness: OEE revisited. IEEE Trans. Semicond. Manuf. **18**(1), 190–196 (2005). https://doi.org/10.1109/TSM.2004.836657
15. Zhao, K., Su, Z., Ye, Z., Cao, W., Pang, J., Wang, X., Zhu, J.: Review of the types, formation mechanisms, effects, and elimination methods of binder jetting 3D-printing defects. J. Mater. Res. Technol. **27**, 5449–5469 (2023). https://doi.org/10.1016/j.jmrt.2023.11.045
16. Günaydın, K., Türkmen, H.S.: Common FDM 3D printing defects. In: International Congress on 3D Printing (Additive Manufacturing) Technologies and Digital Industry, Antalya, Turkey, pp. 671–677 (2018)
17. Haddad, T., Shaheen, B.W., Németh, I.: Improving overall equipment effectiveness (OEE) of extrusion machine using lean manufacturing approach. Manuf. Technol. **21**(1), 56–64 (2021). https://doi.org/10.21062/MFT.2021.006
18. Agnihotri, S.R., Kenett, R.S.: The impact of defects on a process with rework. Eur. J. Oper. Res. **80**(2), 308–327 (1995). https://doi.org/10.1016/0377-2217(94)00173-A
19. Flapper, S.D.P., Fransoo, J.C., Broekmeulen, R.A., Inderfurth, K.: Planning and control of rework in the process industries: a review. Prod. Plan. Control. **13**(1), 26–34 (2002). https://doi.org/10.1080/09537280110061548
20. Lin, Y.K., Huang, C.F., Chang, P.C.: System reliability evaluation of a touch panel manufacturing system with defect rate and reworking. Reliab. Eng. Syst. Safe. **118**, 51–60 (2013). https://doi.org/10.1016/j.ress.2013.04.007
21. Nobil, A.H., Nobil, E., Sarker, B.: R.: optimal decision-making for a single-stage manufacturing system with rework options. Int. J. Syst. Sci.: Oper. Logist. **7**(1), 90–104 (2018). https://doi.org/10.1080/23302674.2018.1514087
22. BouAbid, H., Dhouib, K., Gharbi, A.: Integrated production and maintenance policy for manufacturing systems prone to products' quality degradation. Adv. Prod. Eng. Manag. **19**, 512–526 (2024). https://doi.org/10.14743/apem2024.4.521
23. Shokri, A.: Reducing the scrap rate in manufacturing SMEs through Lean Six Sigma methodology: an action research. IEEE Eng. Manag. Rev. **47**, 104–117 (2019). https://doi.org/10.1109/EMR.2019.2931184
24. De Mattos Nascimento, D.L., Nepomuceno, M.R., Caiado, R.G., Maqueira, J.M., Moyano-Fuentes, J., Garza-Reyes, J.A.: A sustainable circular 3D printing model for recycling metal scrap in the automotive industry. J. Manuf. Technol. Manag. **33**, 876–892 (2022). https://doi.org/10.1108/JMTM-10-2021-0391
25. BouAbid, H., Dhouib, K., Gharbi, A.: Production/preventive maintenance comprehensive approach for manufacturing systems susceptible to quality degradation. Arab. J. Sci. Eng. (2024). https://doi.org/10.1007/s13369-024-09496-3
26. Ait El Cadi, A., Gharbi, A., Dhouib, K., Artiba, A.: Joint production, maintenance, and quality control in manufacturing systems with imperfect inspection. J. Manuf. Syst. **77**, 848–858 (2024). https://doi.org/10.1016/j.jmsy.2024.10.020
27. Hajji, A., Bouabid, H., Dhouib, K., Gharbi, A.: Throughput assessment of manufacturing cells subject to random disturbances. In: Bouraoui, T., et al. (eds.) Advances in Mechanical Engineering and Mechanics III. CoTuMe 2023 Lecture Notes in Mechanical Engineering, pp. 440–447. Springer, Cham (2024). https://doi.org/10.1007/978-3-031-70428-4_48
28. Dhouib, K., Gharbi, A., Ayed, S.: Availability and throughput of unreliable, unbuffered production lines with non-homogeneous deterministic processing times. Int. J. Prod. Res. **46**(20), 5651–5677 (2008). https://doi.org/10.1080/00207540701294635

29. Dhouib, K., Gharbi, A., Landolsi, N.: Throughput assessment of mixed-model flexible transfer lines with unreliable machines. Int. J. Prod. Econ. **122**, 619–627 (2009). https://doi.org/10.1016/j.ijpe.2009.06.016
30. Dhouib, K., Gharbi, A., Ayed, S.: Simulation-based throughput assessment of non-homogeneous transfer lines. Int. J. Simulat. Model. **8**, 5–15 (2009). https://doi.org/10.2507/IJSIMM08(1)1.111
31. Dhouib, K., Gharbi, A., Landolsi, N.: Availability modeling and analysis of multi-product flexible transfer lines subject to random failures. Int. J. Adv. Manuf. Technol. **50**, 329–341 (2010). https://doi.org/10.1007/s00170-009-2487-8
32. Kelton, W.D., Zupick, N.B., Ivey, N.J.: Simulation with ARENA, 7th edn. McGraw-Hill, New-York (2024)

Automation, Artificial Intelligence and Post-processing for AM Processes

Neural Network Optimization for 3D Printing Parameters in Biomedical Applications: An Anti-trichiral Honeycomb Structure Case Study

Houda Khaterchi[1](✉), Mondher Nasri[2], Nasri Mouhmed Toumi[2], Baderddine Larbi[2], Mondher Zidi[1], Ated Ben Khlifa[1], and Hédi Belhadjsalah[1]

[1] Mechanical Engineering Laboratory (LGM), LR99ES32, ENIM, University of Monastir, LGM, ENIM, Av. Ibn ElJazzar, 5019 Monastir, Tunisia
houda_enim@yahoo.fr

[2] LMPE, ENSIT, University of Tunis, 5 Avenue Hussein, BP, 56, 1008 Bab Menara, Tunisia

Abstract. The intersection of 3D printing technology and biomedical applications has opened promising avenues for developing personalized medical devices and anatomical models. However, achieving optimal print parameters for complex biomedical structures remains a significant challenge. This study introduces an optimization framework to determine 3D printing parameters tailored to biomedical applications, with a specific focus on the anti-trichiral honeycomb structure. A systematic parametric study and experimental characterization of this structure were conducted to generate a training dataset for the artificial neural network (ANN). The trained ANN then enables the replacement of experimental trials by providing rapid predictions of Young's modulus for the printed structure. Additionally, an ANOVA analysis was conducted to investigate parameter interactions, with results demonstrating high accuracy and reliability of the predictions. Our work contributes to advancing 3D printing in biomedicine, highlighting the potential of neural networks to accelerate the development of personalized biomedical solutions.

Keywords: 3D Printing · Tensile Strength · Neural Network · Honeycomb Structure

1 Introduction

In recent years, the alliance between 3D printing technology and biomedicine has led to groundbreaking advances in personalized healthcare solutions, offering unprecedented possibilities for the manufacture of custom medical devices, the production of anatomical models for surgical planning, and even the bioprinting of tissues and organs. This technology enables extensive customization of medical products, which can be adapted to the specific needs of each patient, thus paving the way for more efficient and tailored medicine. However, to fully exploit the potential of 3D printing in biomedicine, it is

M. Soula et al. (Eds.): AIAM 2024, LNME, pp. 165–180, 2025.
https://doi.org/10.1007/978-3-031-94889-3_13

essential to optimize the printing parameters, which is a major challenge. The optimization of these parameters, such as printing speed, temperature, material choice, and filling density, is paramount to ensure the quality, accuracy, and biocompatibility of biomedical structures.

As the complexity of the structures increases, it becomes increasingly difficult to find the right balance between these different parameters. This requires more innovative optimization methods to ensure optimal results and effectively guide research and development in this growing field.

3D printing, also known as additive manufacturing, is an advanced manufacturing process that constructs three-dimensional objects from digital 3D models. This groundbreaking technique enables the creation of intricate parts using a variety of materials, each possessing distinct mechanical and physical properties, typically layer by layer [1]. Initially utilized for prototyping to expedite part development and reduce associated costs, additive manufacturing technologies have evolved to a stage where they facilitate the production of final, functional components or tools. Beyond prototyping, 3D printers present immense potential for diverse applications across industrial design, aviation, automotive, and medicine,......

Within the realm of 3D printing methods, Fused Deposition Modeling (FDM) stands out as a significant approach, delivering parts crafted from thermoplastic filaments. This method involves stacking layers to form the final 3D part and employs various plastic materials, as well as composite materials comprising plastics and metal particles, carbon fibers, and more. FDM's popularity stems from its user-friendly nature and cost-effectiveness, making it widely employed for prototyping and manufacturing functional industrial products [1].

Despite its advantages, the default printing process parameters, sometimes predetermined by manufacturers, do not inherently guarantee the quality of the printed part. This is due to the significant number of process parameters involved, requiring careful consideration. Multiple studies have focused on optimizing 3D printing parameters [2–5].

To ensure the functionality of parts produced through 3D printing, meeting mechanical characteristics' requirements becomes imperative, depending on the chosen base material and technological process parameters.

The 3D printing technique involves numerous factors, including parameters related to the printing process, FDM printer parameters, orientation mode of the 3D model, material, and parameters related to the 3D model's environment. Scientific research has extensively explored the influence of these parameters on the quality and mechanical properties of 3D printed parts [6–9], particularly focusing on factors such as layer thickness, type and proportion of internal infill, the gap between layers, width, and angle of the raster lines, material density, orientation of the part along the x, y, and z axes, number of perimeter layers, nozzle temperature, etc. [10–15].

Mechanical properties of FDM printed samples are commonly evaluated through tensile strength testing, providing a comprehensive understanding of the material's mechanical characteristics. Tensile testing involves subjecting test samples (test pipes) to forces in the direction of the axis, aiming to extend and break them. This method, frequently employed in previous scientific research papers, involves the use of tensile testing machines, which are alternatively referred to as tension test machines.

This study addresses the challenge of optimization by proposing a new approach: the use of neural networks [5]. They are renowned for their ability to identify complex patterns, and relationships and represent a promising solution for optimizing 3D printing parameters in the context of biomedicine. By harnessing the power of neural networks, we aim to create a precise and effective optimization framework to improve the overall performance of 3D-printed biomedical structures. Our work focuses on the exploration of honeycomb meta-materials, adding dimension to the optimization challenge. The unique properties of these materials introduce extra complexity into the optimization process, further emphasizing the need for a flexible and intelligent approach. These structures found in nature have served as inspiration for creating lightweight yet rigid structures applicable across various domains [16, 17]. A honeycomb structure provides high bending strength relative to its low density and exhibits excellent energy absorption capabilities, particularly in the context of sandwich composite structures [18, 19].

While influential process parameters have been identified, this paper focuses on their optimization to enhance the quality and mechanical properties of 3D printed parts. This optimization is a significant research challenge, widely addressed by researchers. The complexity lies in the multitude of process parameters and their combinations, necessitating experimental design, considering the wide array of materials available, recognizing the unique specifications of each 3D printer, and introducing an artificial neural network system for the selection and optimization of process parameters and test specimen orientation.

2 Materials and Methods

2.1 Materials and Test Specimen

The anti-trichiral honeycomb structure obtained through 3D printing represents an innovative design inspired by trichiral geometry found in nature. This three-dimensional structure features interconnected hexagonal cells, providing exceptional robustness while minimizing overall weight. In the field of medicine, this structure holds promising applications. It offers a unique combination of lightweight and strength, making it ideal for manufacturing medical devices such as bone implants, lightweight prosthetics, and support structures for various biomedical applications. The anti-trichiral honeycomb geometry allows for even stress distribution while promoting excellent fluid circulation, which can be beneficial in contexts such as tissue regeneration.

The material used in this study is made from polylactic acid (PLA) reinforced with flax fibers. Figure 1 illustrates the geometric configuration of the anti-trichiral honeycomb structure along with its various geometric parameters. The specimens are designed with varying unit cell sizes (Table 1).

Each unit cell is positioned within an equilateral triangle. L represents the ligament's length, r indicates the cylindrical components' radius, and t denotes the thickness of the cell wall. The cell plane is determined by the (X, and Y) axes, while the thickness extends along the (Z) axis. H represents the specimen's width in the Y direction, and e signifies its thickness.

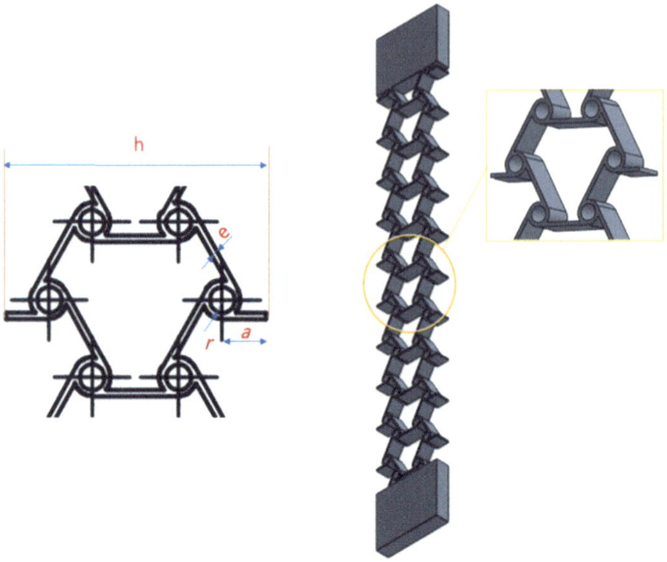

Fig. 1 Geometry of the anti-tri chiral honeycomb architecture and the unit cell

Table 1. Geometric parameters of materials [20]

Parameters	Values
r (mm)	1.2
$\rho/\rho s$ (%)	13.31
t (mm)	0.6
L (mm)	4.17
e (mm)	5
H (mm)	25

In this study, the RAISE3D Pro2 Plus 3D printer is employed. This printer is part of a new generation of professional and robust 3D printers, endowed with a range of capabilities that facilitate the production of precise and high-quality components.

2.2 Parameters and Conditions of the 3D Printing Process

The 3D printing process for creating the test specimens involved several key steps. Initially, a 3D model of the standard test specimen was designed, and an STL file was generated. This STL file was then imported into CreatWare software, where the G-code, the set of instructions for the 3D printer, was created. The G-code defined the printing path and parameters (Fig. 2).

Emphasizing the variation of three crucial parameters: print speed, printer head temperature, and print density; the experiment aimed to explore their influence on the

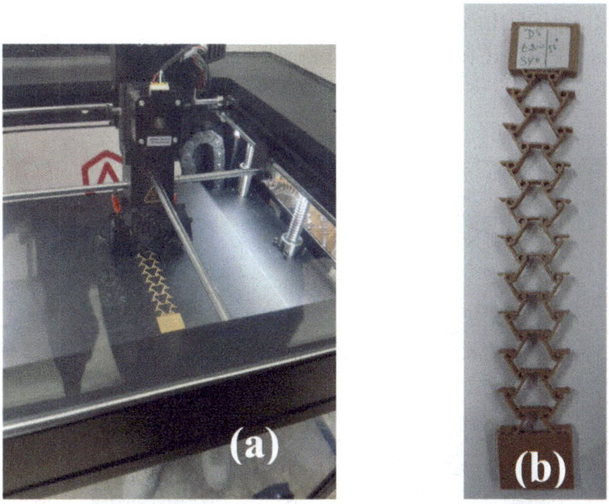

Fig. 2 Test specimen: (a) 3D printer manufactures the test specimen; (b) a sample produced by a 3D printer

characteristics of the printed specimens. Print speed was adjusted across a range from 40 to 80 mm/s, printer head temperature varied between 200 °C and 220 °C, and print density spanned from 5% to 25%. Using the Taguchi method, nine distinct samples were generated, representing a comprehensive set of combinations for the specified parameter variations. The samples were then sorted and labeled for subsequent analysis. PLA plastics in the form of 3 mm diameter wire were utilized as the fabrication material, utilizing a CreatBot DX2 3D printer equipped with FDM printing technology. The entire set of input data, encapsulating the process parameters, is represented in Table 2.

Table 2. Input data

Parameters	Values
Speed (mm/s)	40-60-80
Temperature (°)	200-210-220
Density (%)	5-15-25

2.3 Tensile Strength Testing and Design of Experiments

2.3.1 Tensile Strength Testing

The anti-trichiral honeycomb structure, produced through 3D printing, is examined for a critical parameter (tensile strength) particularly relevant for its use in biomedical implants within the medical field. Each specimen undergoes tensile strength testing using a universal testing machine. The specimens are securely clamped in the grips, and a load is gradually applied until failure (i.e., breaks) occurs (Fig. 3). Subsequently, the ultimate tensile strength is recorded for each sample. Figure 4 depicts the images of test parts after all tests.

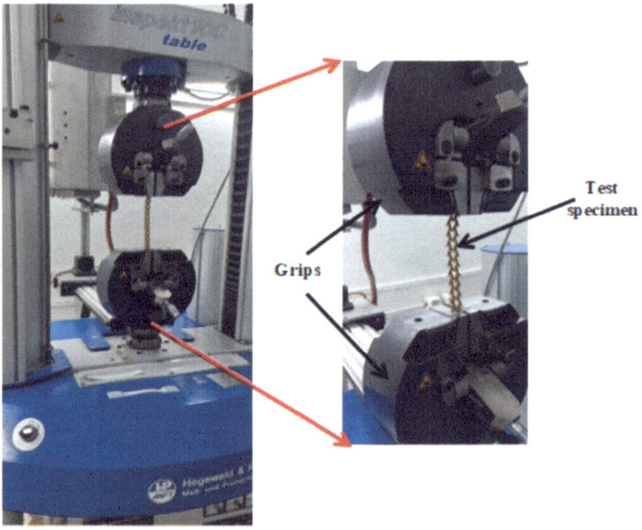

Fig. 3 Experimental tensile test equipment.

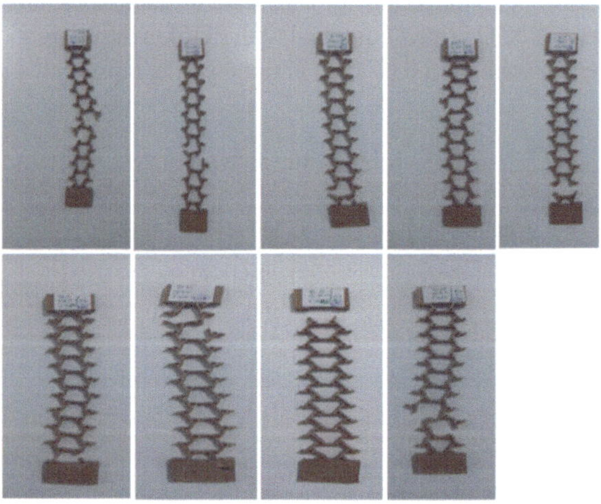

Fig. 4 Images of test parts after all tests

2.3.2 Design of Experiments

The Taguchi (L9) Experimental Design was employed to determine the optimal combination of process parameters. The results of the experiments conducted for various experimental runs are presented in the following Table 3.

Table 3 Experimental results

Parameters					Values				
Speed (mm/s)	40	40	40	60	60	60	80	80	80
Temperature (°)	200	210	220	200	210	220	200	210	220
Density (%)	5	15	25	15	25	5	25	5	15
Tensile modulus (MPa)	0.554	0.725	1.672	4.128	0.917	1.79	1.26	1.765	0.795

2.4 Artificial Neural Network Modeling

The use of predictive models based on soft computing techniques, such as Artificial Neural Networks (ANN), has become a major trend in the field of nonlinear modeling, particularly significant for 3D printing [1, 5, 21–24]. Artificial Neural Networks are specially designed to capture nonlinear effects and interaction, think to their ability to learn, store, and apply knowledge from experience, much like the way the human brain networks. In this context of 3D printing, using models based on ANNs offers a promising direction [5]. This is essential to predict mechanical properties such as tensile strength and yield strength of 3D printed parts in materials like PEEK [5]. The increasing adoption of this approach demonstrates the need to stay abreast of the latest technological advances to ensure more accurate and reliable modeling in the field of 3D printing.

In this work, we employed an in-house Artificial Neural Network (ANN) program developed in Fortran, to predict output data [25, 26]. Initially, a setup for the ANN was selected, typically consisting of one hidden layer with the number of hidden units set at half the sum of the input and output units. Through iterative processes, multiple calculations were performed for each configuration, and the best-performing network (based on verification error) was retained throughout the process.

The neural network architecture consists of three or more layers of neurons. The initial layer does not conduct any operations on the input data; instead, it directly transmits these values to the hidden neurons of the subsequent layer, which are influenced by weights represented as w_{im}. The individual neuron calculates a weighted sum using the inputs Xi, usually sourced from neurons' outputs in the preceding layer Vm. This summation involves the addition of a threshold value bi, resulting in the generation of an output:

$$v_m = f\left(\sum_{i=1}^{L} W_{im} x_i + b_i\right) \quad (1)$$

The cumulative input signals within the neuron block undergo activation through a function whether linear or nonlinear, resulting in a singular output denoted as Ym.

$$y_m = f(v_m) \tag{2}$$

In neural network applications, the sigmoid (logistic) function is commonly used among activation functions. Its mathematical expression is given by:

$$f(v_m) = \frac{1}{1 + e^{-\beta v_m}} \tag{3}$$

Where β represents a parameter that defines the slope of the function.

The mean square error (MSE) serves as the key parameter to evaluate the fitness of the ANN model during the training processes. This metric is calculated as:

$$MSE = \frac{1}{2}\sum_{j=1}^{p}\sum_{i=1}^{N}\left(t_{ij} - y_{ij}\right)^2 \tag{4}$$

Training neural networks involves presenting a set of input patterns along with their corresponding target outputs. The system adjusts the internal connection weights to minimize the discrepancies between the network's output and the expected output. The knowledge is encoded in the strengths of these connections between the network's components. Various algorithms can be used for training neural networks. In this study, the back-propagation (BP) algorithm was employed. This iterative gradient-based method is designed to optimize the connection weights, thereby reducing the mean square error between the actual output and the desired output. Initially, the weights are assigned randomly, and the training process involves iteratively comparing the expected and desired outputs with the calculated output using the current set of weights and thresholds [25].

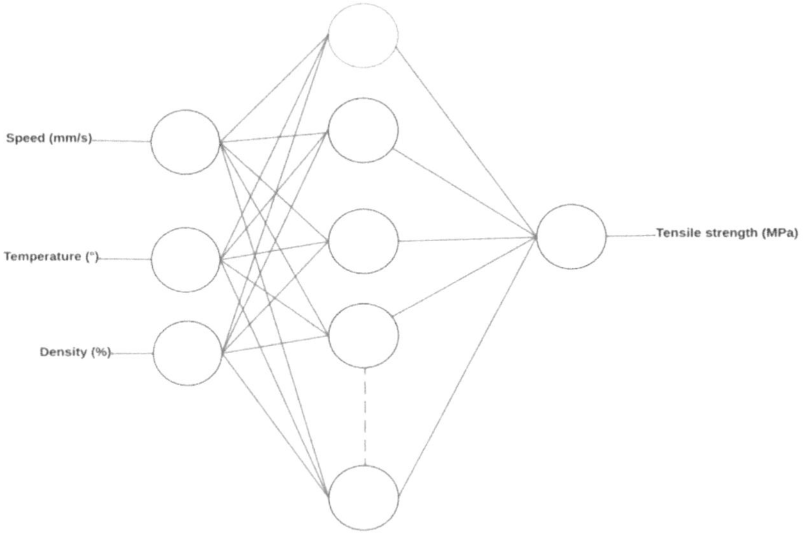

Fig. 5 ANN architecture

Table 3 was utilized as the database for training neural networks. Various neural network architectures were devised, each with differing numbers of neurons in the hidden layer, and their performance was assessed using mean squared error. The weighted connections are computed by the network to minimize the total mean squared error (Eq. 4) between its actual and desired output. The optimal architecture for the established ANN model consists of 3-12-2 neurons (Fig. 5).

The best prediction of tensile modulus by the ANN is illustrated in Fig. 6.

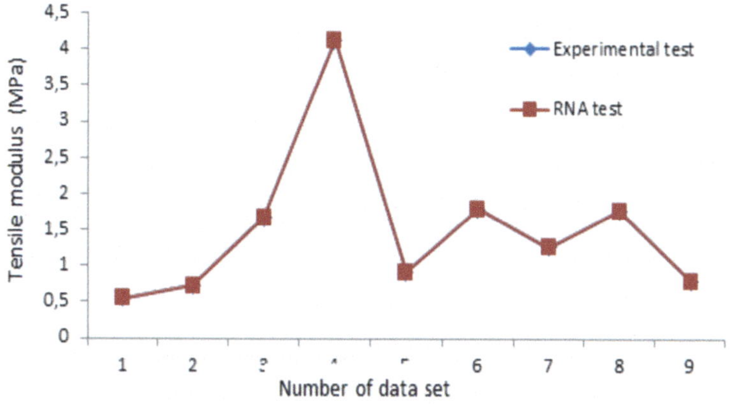

Fig. 6 Output data - Tensile modulus

To validate the developed NN model, Fig. 7 illustrates the progression of the MSE in relation to training iterations. With an increase in the number of training iterations, there is a notable decrease in the MSE. Upon reaching 100,000 epochs, the MSE reaches a value of 10^{-6}.

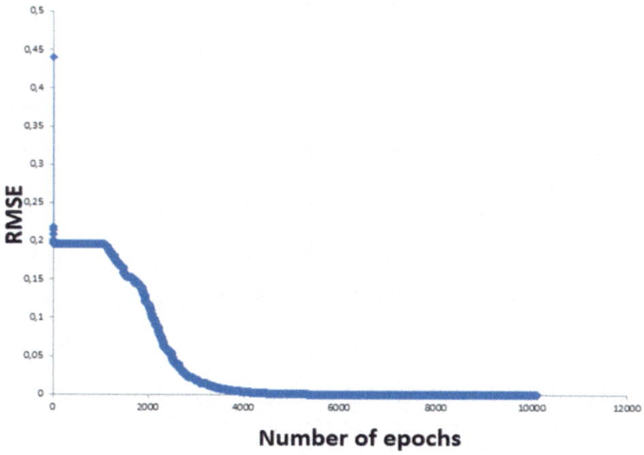

Fig. 7 Mean square error during the epochs

3 Results

In this section, our focus will be on exploring the significance of the main effects as well as potential interactions among the parameters, namely speed, density, and temperature. This will be accomplished through the utilization of ANOVA analysis, which allows for a quantitative assessment of the individual and combined impacts of these variables. Subsequently, we will delve into the examination of the predicted outcomes generated by the neural network, shedding further light on the relationship between the input variables and the corresponding responses in our study.

The interaction plots in Fig. 8 clearly illustrate that the combination of speed and density has a significant impact on the tensile modulus. Furthermore, it was noted that the tensile modulus tends to increase with lower temperatures. This observed trend suggests a material sensitivity to temperature, where lower temperatures promote an increase in material stiffness, as evidenced by the rise in tensile modulus. Such findings underscore the importance of simultaneously considering multiple variables in analyzing material mechanical properties, which may have significant implications for design and manufacturing processes.

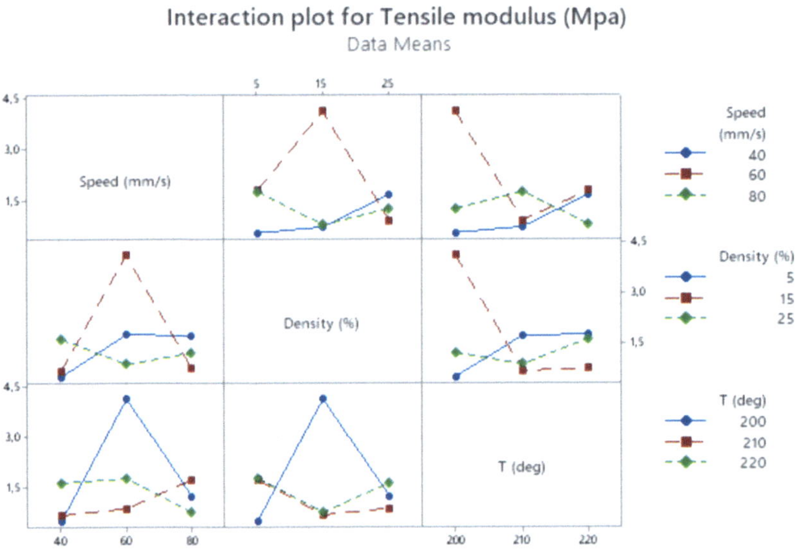

Fig. 8 Interaction plot for Tensile modulus

The 2D response surface is presented in Fig. 9. It represents a combination of an infinite number of potential solutions for the factors influencing the response. From Fig. 9, it was observed that the tensile modulus reached its maximum value with the lowest temperature.

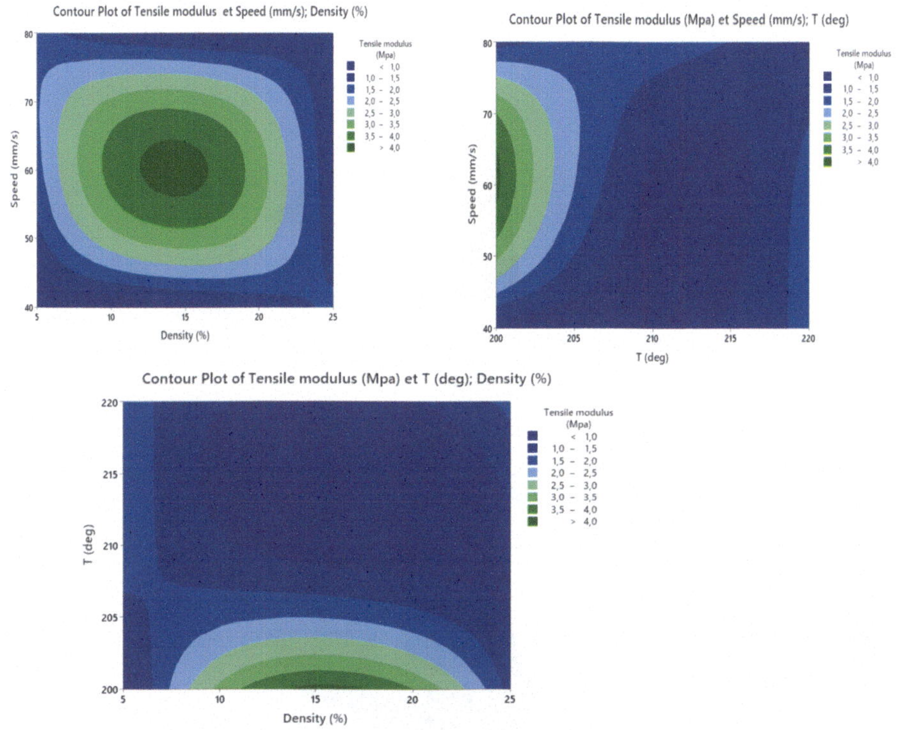

Fig. 9 Contour plot of Tensile modulus vs. speed, Temperature, and Density

Figure 10 illustrates a surface plot demonstrating how the tensile modulus varies with different process parameters (speed, temperature, and density). This plot demonstrates that the tensile modulus tends to increase as the temperature decreases. Additionally, it can be observed from this figure that the tensile modulus decreases for higher values of speed and density.

Figure 11 displays the main effects plot for tensile modulus, where curves that rise and then fall can indicate several important aspects regarding the relationship between the studied parameters and the tensile modulus:

- Optimal point: The ascending portion of the curve suggests that the tensile modulus increases with increasing levels of the parameter up to a certain point. However, beyond this point, the descending portion of the curve indicates that the tensile modulus begins to decrease with further increases in the parameter. This may signify that there exists an optimal level of the parameter that maximizes the tensile modulus.
- Nonlinear behavior: The presence of curves that rise and then fall reveals nonlinear behavior between the parameter and the tensile modulus. This implies that the effect of the parameter on the tensile modulus is not constant across the entire range of parameter levels.

In the context of our study on the tensile modulus derived from experimental tests, we conducted a normal probability plot (Fig. 12) to assess the distribution of the data.

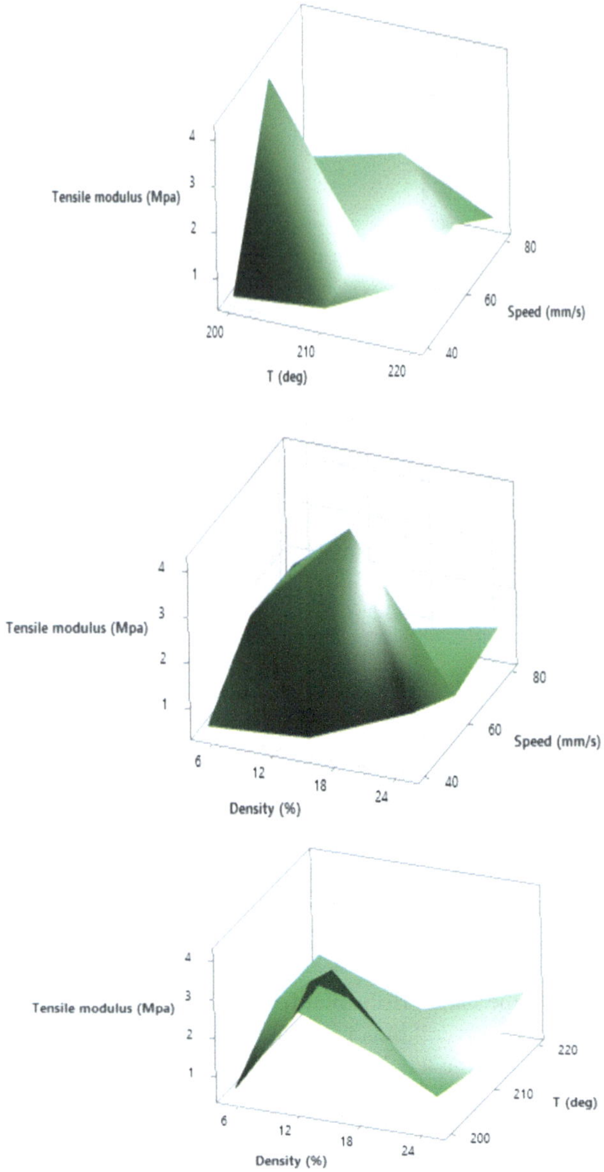

Fig. 10 Surface plot of Tensile modulus vs. speed, Temperature, and Density

Careful observation of the plot reveals that the data align closely along the straight line, suggesting a remarkable fit with a normal distribution. This alignment accuracy, combined with the distribution centered around the axis, demonstrates the balance and accuracy of our experimental model.

Moreover, the fact that the P-values associated with the normality test exceed 0.05 and that the observed residues follow a normal distribution reinforces the validity of our

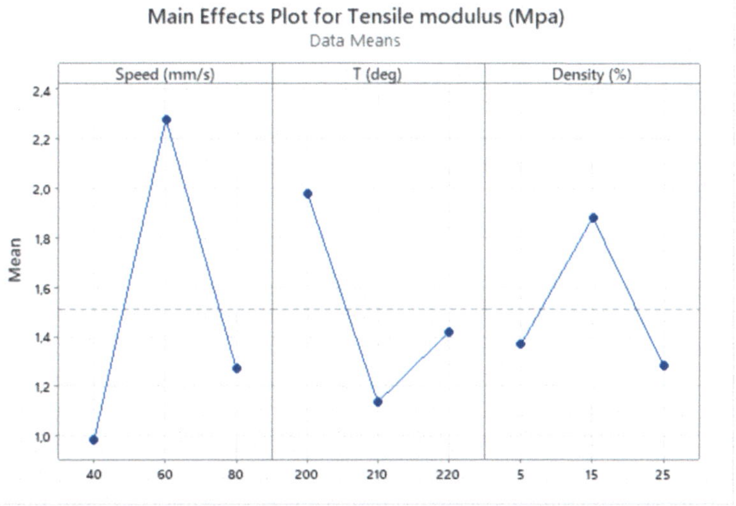

Fig. 11 Main Effects Plot for Tensile Modulus

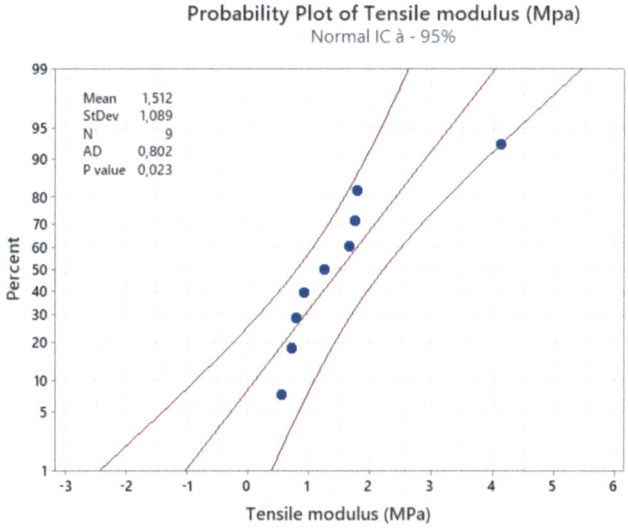

Fig. 12 Probability plot.

experimental results. This respect for normality is essential because it allows us to use various parametric statistical tests that require data following this distribution.

This study developed a neural network to replace experimental tests by allowing the prediction of the tensile module. Thanks to this neural model, it becomes possible to predict the Young module instantaneously, without requiring the manufacture of a specimen by 3D printing, nor the definition of parameters such as temperature, speed or density. This approach is a fast and cost-effective alternative to conventional materials characterization methods, reducing the time and costs associated with experimental testing. At this stage, the validation of this neural model is underway, aiming to confirm its effectiveness and accuracy in the prediction of the traction module. The results of this validation will be essential to demonstrate the reliability and practical applicability of this innovative approach in the field of materials characterization.

4 Discussion

This study presents an in-depth parametric analysis of the 3D printing process, focusing on its applications in biomedicine. Experimental tests were conducted to determine the tensile modulus, followed by the creation of a neural network model capable of instantaneously predicting this modulus. A statistical analysis was performed to evaluate the effects and interactions of various parameters on the tensile modulus, offering valuable insights for optimizing the 3D printing process specifically for biomedical applications. The results underscore the relevance of neural network modeling in simplifying material characterization procedures.

The use of artificial neural networks in this study highlights their significance in advancing 3D printing for biomedicine. By utilizing neural network modeling, the efficiency and accuracy of material characterization are improved, facilitating the development of tailored biomedical solutions. The simplicity and speed of neural predictions of mechanical properties further enhance this approach, enabling the quick acquisition of valuable data for medical device design.

5 Conclusion

The next step in this research will involve validating the neural network model with additional experimental tests to demonstrate its effectiveness. This validation aims to increase the model's accuracy and reliability, moving it closer to practical applications in real biomedical settings. Overall, this research significantly contributes to the understanding and optimization of the 3D printing process within the biomedical field.

References

1. Milićević, I., Popović, M., Dučić, N., Vujičić, V., Stepanić, P., Marinković, D., Ćojbašić, Ž.: Improving the mechanical characteristics of the 3d printing objects using hybrid machine learning approach. https://doi.org/10.22190/FUME220429036M
2. Taşcıoğlu, E., Kıtay, Ö., Keskin, A.Ö., et al.: Effect of printing parameters and post-process on surface roughness and dimensional deviation of PLA parts fabricated by extrusion-based 3D printing. J. Braz. Soc. Mech. Sci. Eng. **44**, 139. (2022) (2022). https://doi.org/10.1007/s40430-022-03429-7
3. Rosentritt, M., Preis, V., Behr, M., Sereno, N., Kolbeck, C.: The shear bond strength between veneering composite and PEEK after different surface modifications. Clin Oral Invest. **19**, 739–744 (2015)

4. Najeeb, S., Zafar, M.S., Khurshid, Z., Siddiqui, F.: Applications of polyether ether ketone (PEEK) in oral implantology and prosthodontics. J. Prosthodont. Res. **60**, 12–19 (2016)
5. Borah, J., Chandrasekaran, M.: Experimental investigation and development of Artificial neural network modeling of 3D printed PEEK bio-implants and its optimization; https://doi.org/10.21203/rs.3.rs-3204960/v1
6. Algarni, M., Ghazali, S.: 2021, comparative study of the sensitivity of PLA, ABS, PEEK, and PETG'S mechanical properties to FDM printing process parameters. Crystals. **11**(8), 995 (2021)
7. Nabipour, M., Akhoundi, B.: An experimental study of FDM parameters effects on tensile strength, density, and production time of ABS/Cu composites. J. Elastom. Plast. **53**(2), 146–164 (2021)
8. Wickramasinghe, S., Do, T., Tran, P.: FDM-based 3D printing of polymer and associated composite - a review on mechanical properties, defects and treatments. Polymers. **12**, 1529 (2020)
9. Garzon-Hernandez, S., Garcia-Gonzaleza, D., Jérusalem, A., Arias, A.: Design of FDM 3D printed polymers - an experimental-modelling methodology for the prediction of mechanical properties. Mater. Des. **188**, 108414 (2020)
10. Yadav, D., Chhabra, D., Gupta, R.K., Phogat, A., Ahlawat, A.: Modeling and analysis of significant process parameters of FDM 3D printer using ANFIS. Mater. Today Proc. **21**(3), 1592–1604 (2019)
11. Jatti, V.S., Jatti, S.V., Patel, A.P., Jatti, V.S.: A study on effect of fused deposition modeling process parameters on mechanical properties. Int. J. Sci. Technol. Res. **8**(11), 689–693 (2019)
12. Kamaal, M., Anas, M., Rastogi, H., Bhardwaj, N., Rahaman, A.: Effect of FDM process parameters on mechanical properties of 3D-printed carbon fibre–PLA composite. Progr. Addit. Manufact. **6**, 63–69 (2021)
13. Vicente, C.M.S., Martins, T.S., Leite, M., Ribeiro, A., Reis, L.: Influence of fused deposition modeling parameters on the mechanical properties of ABS parts. Polym. Adv. Technol. **31**, 501–507 (2020)
14. Prayitno, G., Imaduddin, F., Ubaidillah, Arifin, Z.: Recent progress of fused deposition modeling (FDM) 3D printing: constructions, parameters and processings. IOP Conf. Ser. Mater. Sci. Eng. **1096**, 012045 (2021)
15. Samykano, M., Selvamani, S.K., Kadirgama, K., Ngui, W.K., Kanagaraj, G., Sudhakar, K.: Mechanical property of FDM printed ABS: influence of printing parameters. Int. J. Adv. Manuf. Technol. **102**, 2779–2796 (2019)
16. Hönig, A., Stronge, W.J.: In-plane dynamic crushing of honeycomb. Part I: crush band initiation and wave trapping. Int. J. Mech. Sci. **44**(8), 1665–1696 (2002)
17. Cutler, J.: In: Liber, J. (ed.) Understanding Aircraft Structures. Blackwell Publishing (2005)
18. Jaehyung, J., Summers, J.D.: Compliant hexagonal periodic lattice structures having both high shear strength and high shear strain. Mater. Des. **32**(2), 512–524 (2011)
19. Ying, L., Zhang, X.C.: The influence of cell microtopology on the in-plane dynamic crushing of honeycombs. Int. J. Impact Eng. **36**(1), 98–109 (2009)
20. Hamrouni, A., Rebiere, J.-L., El Mahi, A., Beyaoui, M., Haddar, M.: Caractérisation Mécanique d'un Meta-Matériau de Forme Anti- Trichirale Fabriqué par la Technique d'Impression 3D avec un Composite Biosourcé et Biodégradable. In: 16ème Congrès Français d'Acoustique, CFA2022, Société Française d'Acoustique; Laboratoire de Mécanique et d'Acoustique,Marseille, France. ffhal-03848312 (Apr 2022)
21. Honigmann, P., Sharma, N., Okolo, B., Popp, U., Msallem, B., Thieringer, F.M.: Patient-specific surgical implants made of 3D printed PEEK: material, technology, and scope of surgical application. Biomed. Res. Int. **2018, 4520636**(1) (2018). https://doi.org/10.1155/2018/4520636

22. Józwik, J., Ostrowski, D., Milczarczyk, R., et al.: Analysis of the relation between the 3D printer laser beam power and the surface morphology properties in Ti-6Al-4V titanium alloy parts. J. Braz. Soc. Mech. Sci. Eng. **40**, 215 (2018). https://doi.org/10.1007/s40430-018-1144-2
23. Azouzi, R., Guillot, M.: On-line optimization of turning process using an inverse process neuron controller. J. Manuf. Sci. Eng. **120**, 101–108 (1998)
24. Risbood, K.A., Dixit, U.S., Sahasrabudhe, A.D.: Prediction of surface roughness and dimensional deviation by measuring cutting forces and vibrations in the turning process. J. Mater. Process. Technol. **13**, 203–214 (2003)
25. Hambli, R., Katerchi, H., Benhamou, C.-L.: Multiscale methodology for bone remodeling simulation using coupled finite element and neural network computation. Biomech. Model. Mechanobiol. **2011**(10), 133–145 (2011). https://doi.org/10.1007/s10237-010-0222-x
26. Khaterchi, H., Chamekh, A., BelHadjSalah, H.: Artificial neural network analysis for modeling fibril structure in bone. Int. J. Precision Eng. Manufact. **16**(3), 581–587 (2015). https://doi.org/10.1007/s12541-015-0078-1

New Energy Consumption Prediction Aid Model Applied to Additive Manufacturing Process

Kamel Bousnina(✉), Anis Hamza, and Noureddine Ben Yahia

Mechanical, Production and Energy Laboratory (LMPE), National School of Engineering of Tunis (ENSIT), University of Tunis, Avenue Taha Hussein, Montfleury, 1008 Tunis, Tunisia
kamel11111976@yahoo.fr

Abstract. In this article, the comparative analysis between regression methods and the hybrid PSO-ANN algorithm for predicting energy consumption in additive manufacturing processes is explored. The study found that the PSO-ANN algorithm performs better than traditional regression methods. The results demonstrate a higher correlation coefficient of 0.99 for the PSO-ANN algorithm compared to 0.95 for the regression method. The hybrid approach is believed to provide more effective and accurate predictions for energy consumption in the additive manufacturing process. The findings demonstrate the potential of incorporating sophisticated algorithms, including PSO-ANN, to enhance predictive modeling and improve energy efficiency in manufacturing processes. The use of artificial intelligence has significantly improved the prediction of energy consumption in the additive manufacturing process by using advanced algorithms to analyze process parameters. These developments contribute to more sustainable and efficient production in additive manufacturing.

Keywords: Energy consumption · PSO-ANN · Additive manufacturing · Artificial Intelligence

1 Introduction

Additive manufacturing (AM) is a technique that creates a physical object by adding layers of material based on a numerical model [1]. Additive manufacturing has been widely used due to its low cost and low energy requirements. The fastest growth is in the health, transportation, aerospace, and biotechnology sectors [2]. A patent was filed by Crump in 1988 concerning the method of fused filament fabrication (FFF) by three-dimensional printing (3DP) (Fig. 1).

Additive manufacturing (AM) has become a transformative process in modern manufacturing, offering unparalleled flexibility and customization. As industries increasingly adopt additive manufacturing processes, the focus is increasingly on optimizing various aspects, with an emphasis on the quality of the components manufactured. Many

M. Soula et al. (Eds.): AIAM 2024, LNME, pp. 181–185, 2025.
https://doi.org/10.1007/978-3-031-94889-3_14

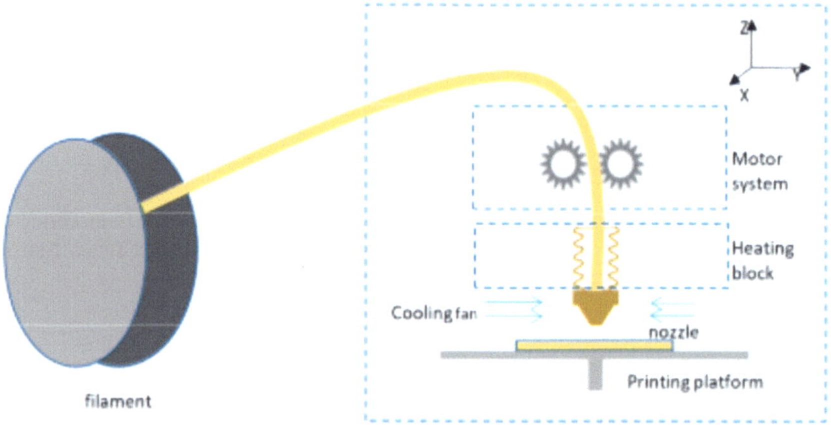

Fig. 1. Model of the FFF 3D printer [6]

studies have focused on the evaluation and improvement of parts quality in additive manufacturing [3], addressing factors such as material properties [4], layer adhesion, and geometric precision [5]. However, one critical aspect that has been underestimated is the energy consumption associated with additive manufacturing. In their research, Mushtaq et al. [6] aim to determine the optimal parameters for nylon-6 (PA6) 3D printing using the molten filament production process. Mathematical models of exit responses were identified using regression analysis. Mean surface roughness (Ra), quadratic mean surface roughness (Rq), printing time (PT), printing energy (PE), and tensile testing (T) were explored as response parameters to identify the impact of PA6 parameters (layer thickness, extrusion temperature, printing speed and filling density).

This research aims to develop a method to predict the energy consumption of 3D printers. This approach is based on the use of artificial intelligence by implementing a hybrid neural network model using a particle swarm algorithm. Next, we will study the impact of printing parameters on energy efficiency. In addition, it should be noted that 3D printing technologies are very sensitive to defects that may occur due to poor selection of printing parameters and properties of printed components.

2 Cases and Materials

In the context of sustainability and energy efficiency, it is essential to study energy consumption in the 3D printing process by filament fusion. In this perspective, Mushtak et al. [6] approached this problem using the regression method to predict energy consumption. Their approach provides a valuable understanding of the factors that influence energy consumption in this context. In general, our work is positioned as an extension and an innovative alternative to this established approach. We have implemented an alternative prediction method, using a hybrid algorithm combining particle swarm search (PSO) and artificial neural networks (ANN) [7] . This hybrid approach aims to improve prediction accuracy by taking into account the complex nuances of the 3D printing process through filament fusion. Table 1 shows the energy consumption results according to the printing parameters.

Table 1. Experimental design of the study.

Inputs				Output
LT (mm)	ET (°C)	S (mm/s)	IN (%)	PE (kWh)
0.12	240	40	10	0,338
0.12	255	55	50	0,344
0.12	270	70	90	0,356
0.20	240	55	90	0,263
0.20	255	70	10	0,169
0.20	270	40	50	0,286
0.30	240	70	50	0,152
0.30	255	40	90	0,228
0.30	277	55	10	0,152

3 Results and Discussion

The results of our study, based on the use of the PSO-ANN hybrid algorithm to predict the energy consumption of 3D printing by filament fusion, show a significant improvement over the previous approach proposed by Mushtaq et al. The latter obtained a correlation coefficient of 0.92 using the regression method. However, our method gave more remarkable results and obtained a remarkable correlation coefficient of 0.99. This significant difference highlights the effectiveness of our PSO-ANN hybrid approach to accurately model and predict energy consumption in the specific context of 3D printing. The strong correlation obtained indicates a better fit of our model to real data, reinforcing its reliability and potential for practical use in process optimization [8]. Figure 2 shows a comparison between predicted and experimental values for both approaches. PSO is used in the hybrid algorithm to improve the optimization of the weights and biases of the neural network, which allows a more improved adjustment of the results. By using this hybridization technique, the system will perform better in terms of prediction accuracy.

a)

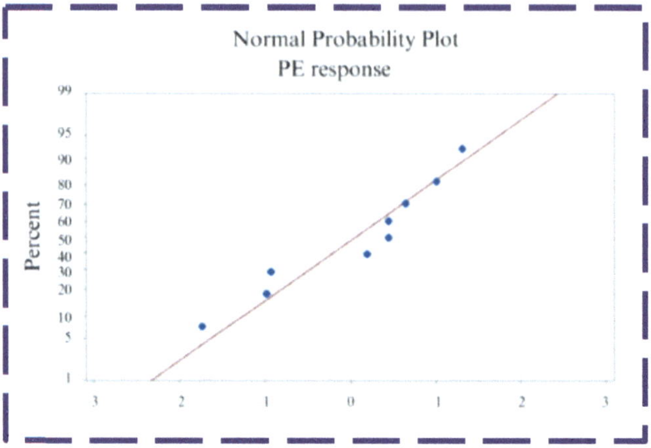

b)

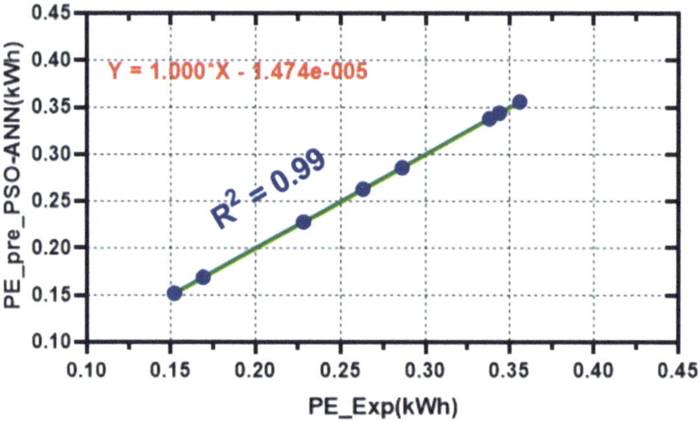

Fig. 2 Comparison between predicted and experimental values. (a) Regression method [6], (b) Hybrid algorithm (current study)

4 Conclusion

Our study highlighted the superior efficiency of the PSO-ANN hybrid algorithm compared to the regression method used by Mushtaq et al. to predict energy consumption in 3D printing by filament fusion. Our approach, which has a correlation coefficient of 0.99, significantly improves the modeling accuracy of this complex process. These results suggest that the integration of optimization techniques such as particle swarm optimization (PSO) with artificial neural networks (ANN) can significantly improve the prediction of energy parameters related to 3D printing. This promising advance offers tangible prospects for a more efficient use of energy resources in the field of 3D printing, paving the way for practical and sustainable applications. Our results encourage wider adoption of this hybrid approach in the modeling of complex processes and highlight the importance of algorithmic innovation to improve the performance of prediction methods.

In summary, our innovative approach that combines the PSO-ANN algorithm has significantly contributed to the scientific advance in the prediction of energy consumption in the field of additive manufacturing. Thanks to this hybrid methodology, the results obtained demonstrate its effectiveness and open new perspectives for more precise and sustainable energy management in this industrial context.

References

1. Kumbhar, N.N., Mulay, A.V.: Post processing methods used to improve surface finish of products which are manufactured by additive manufacturing technologies: a review. J. Inst. Eng. (India) Ser. C. **99**, 481–487 (2018)
2. Soares, J.B., Finamor, J., Silva, F.P., Roldo, L., Cândido, L.H.: Analysis of the influence of polylactic acid (PLA) colour on FDM 3D printing temperature and part finishing. Rap. Prototype J. **24**, 1305–1316 (2018)
3. Zhou, H., Yang, H., Yao, S., Jiang, L., Sun, N., Pang, H.: Synthesis of 3D printing materials and their electrochemical applications. Chin. Chem. Lett. **33**, 3681–3694 (2022)
4. Gautam, R., Idapalapati, S., Feih, S.: Printing and characterisation of Kagome lattice structures by fused deposition modelling. Mater. Des. **137**, 266–275 (2018)
5. Faveto, A., Lombardi, F., Chiabert, P., Segonds, F.: A circular approach to foster additive manufacturing early design stages sustainability: a methodological proposal. Int. J. Interact. Des. Manuf. (IJIDeM). **18**(2), 815–836 (2024)
6. Mushtaq, R.T., Alkahtani, M., Khan, A.M., Abidi, M.H.: Advancements in 3D-printed novel Nylon-6: a Taguchi method for surface quality sustainability and mechanical properties. Mach. Des. **11**(9), 885 (2023)
7. Bousnina, K., Hamza, A., Ben Yahia, N.: Predictive optimization of surface quality, cost, and energy consumption during milling alloy 2017A: an approach integrating GA-ANN and RSM models. Int. J. Interact. Des. Manuf. (IJIDeM). **18**(7), 5177–5196 (2024)
8. Bousnina, K., Hamza, A., Ben Yahia, N.: An energy survey to optimize the technological parameters during the milling of AISI 304L steel using the RSM, ANN, and genetic algorithm. Adv. Mater. Process. Technol. **10**(4), 3562–3580 (2024)

PLA Damage Properties in FDM Printing: A Hybrid Experimental and Machine Learning Study

Manel Dhouioui[1], Boutheina Ben Fraj[2], Hamdi Hentati[1,3](✉), Mounir Ben Amar[4], and Mohamed Haddar[1]

[1] Research Laboratory of Mechanics, Modeling and Manufacturing, National Engineering School of Sfax, University of Sfax, Sfax, Tunisia
hamdi.hentati@yahoo.fr

[2] Research and Technology Center of Energy, Technopole Borj Cedria, Tunisia

[3] High School of Sciences and Technology of Hammam Sousse, University of Sousse, Sousse, Tunisia

[4] Laboratory of Process and Materials Sciences, CNRS, Sorbonne University Paris Nord, Villetaneuse, France

Abstract. This paper investigates the damage properties of polylactic acid (PLA) polymers elaborated through fused deposition modeling (FDM) technology. Experimental tensile tests are conducted to investigate the influence of various FDM parameters on the mechanical performance of 3D-printed samples. Specifically, we focus on failure strain (εr) and the ultimate tensile strength (UTS) as key metrics. Our study involves the systematic variation of FDM parameters to assess their impact on quasi-static damage characterization. Additionally, leveraging insights from existing literature, we employ machine learning (ML) regression models to predict mechanical properties, particularly failure strain and the ultimate tensile strength. The findings offer valuable insights for optimizing FDM parameters to enhance the mechanical performance of PLA polymers in additive manufacturing applications. The random forest regression (RFR) and the XGBoost regression (XGB) models yield a lower value of root mean squared Error (RMSE), expressed as a percentage, compared to the linear regression (LR) model. That indicates that the RFR and XGB models are better at accurately predicting the ultimate tensile strength and failure strain of FDM-printed PLA parts.

Keywords: PLA polymer · FDM process · Damage properties · ML regression models

1 Introduction

The investigation into manufacturing parameters and their effects on different processes plays a crucial role in shaping the tool life, the thermomechanical characteristics, and the geometric features of manufactured components [1, 2]. While diverse studies in the literature have delved into the influence of manufacturing parameters on conventional

M. Soula et al. (Eds.): AIAM 2024, LNME, pp. 186–194, 2025.
https://doi.org/10.1007/978-3-031-94889-3_15

processes [3–5], advancements in additive manufacturing have been seen in recent years. This innovative process, also known as 3D printing has emerged as a transformative technology with profound implications across various industries. 3D printing offers significant cost advantages compared to traditional manufacturing methods. It offers capabilities in reducing prototyping costs and enabling the production of a diverse array of products [6, 7].

In addition, the main goal of manufacturing applications is to prevent structural failure. Experimental and numerical studies have increasingly been developed to predict the damage properties of various materials, whose responses are typically classified into brittle and ductile damage [8–10]. We are interested in the damage parameters of polymers. Research within the field of polymers is motivated to optimize materials by accurately selecting those with superior mechanical properties. Through a combination of numerical modeling techniques and experimental studies, scientists try to determine the main factors influencing the behavior of polymers across various scales, ranging from molecular interactions to macroscopic performance [11, 12]. The integration of 3D printing technology with thermoplastic polymers has gained significant traction in numerous prototyping. Among the array of printing thermoplastics, PLA has emerged as a prominent choice due to its exceptional properties. Its popularity is attributed to its biodegradability and ease of printing, and it is the preferred material in additive manufacturing processes [13–15]. Notably, PLA distinguishes itself with its high tensile strength and relatively lower flexural strength compared to other polymers. Recent studies have honed in on additive manufacturing applications within the aerospace, and automotive, with a specific focus on the FDM process. The primary objective is to delve into the intricate correlation between the mechanical properties of PLA and the various factors inherent in the FDM process [16–18]. Furthermore, despite the increasing prominence of 3D printing technology, a notable lack carries on in the development of optimal models for these additive manufacturing processes. Recent years have observed a significant development in optimizing 3D printing technologies, driven by notable advancements in ML tools. The integration of ML models has emerged as an effective tool for advancing innovations in additive manufacturing processes [19, 20]. Utilizing ML algorithms allows researchers to develop computational models that predict outcomes of manufacturing processes, particularly focusing on FDM, thereby enabling the prediction of performance outcomes across different printing conditions [21, 22].

This research aims to enhance our understanding of how FDM parameters influence the damage characteristics of PLA, paving the way for optimized manufacturing processes and improved material performance in diverse industrial applications. This study places a concentrated emphasis on the application of ML models tailored for regression tasks, with a specific goal of predicting the failure strain and the ultimate tensile strength for printed PLA components. Within this study, we delve into an exploration and comparative analysis of the effectiveness of three regression ML models: Random Forest Regression (RFR), XGBoost regression (XGB), and Linear Regression (LR). By analyzing the performance and predictive capabilities of these ML models, we aim to shed light on their efficacy in enhancing the accuracy and reliability of predictions within the context of optimizing 3D printing processes, particularly in the domain of FDM.

2 Materials and Experimental Dataset

In the FDM process, the PLA filament is guided into a nozzle equipped with an induction heater that maintains a precise temperature critical for the proper melting and flow of the material. Once heated to the required temperature, the molten filament is extruded through the nozzle, which has a specific diameter, and methodically deposited layer by layer onto the build platform, gradually forming the printed specimen. The representation of the FDM-printed PLA part is shown in Fig. 1, illustrating the sequential stages of PLA extrusion.

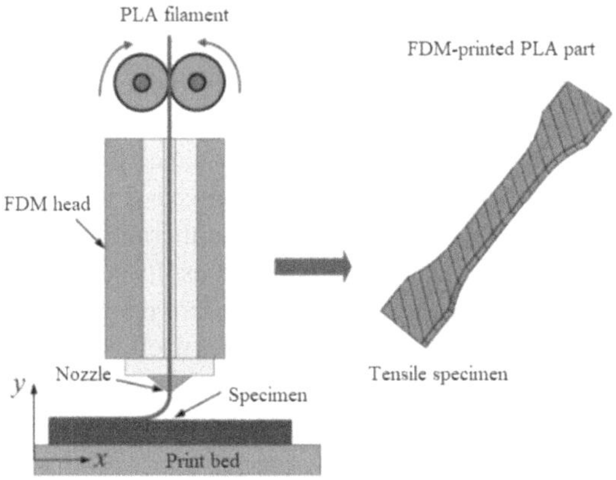

Fig. 1. Schematic of the FDM-printed PLA part

According to ASTM 638-02a, the standard specimen of PLA for tensile test is presented in Fig. 2.

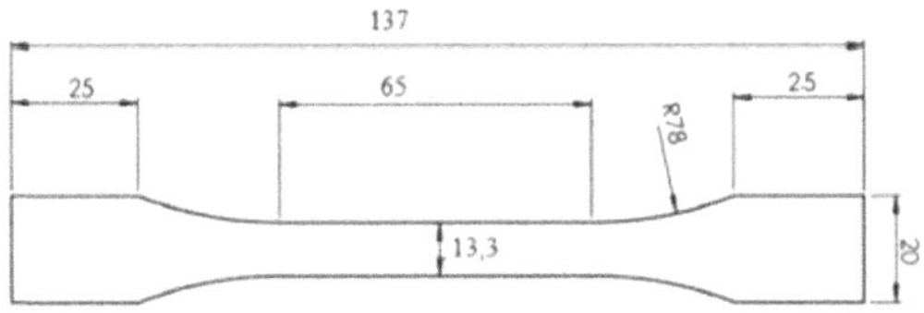

Fig. 2. Dimensions (mm) of a standard specimen of PLA for tensile test

By choosing different FDM parameters of the printed PLA parts, the damage properties are modified. This study is crucial for achieving desired outcomes in terms of the strength and damage of the final printed parts. The properties of FDM-printed PLA parts are exposed in Table 1.

Table 1. Properties of FDM-printed PLA parts [17]

Properties	values
Extrusion temperature (°C)	200
Diameter of filament (mm)	1.75
Printed speed (mm/sec)	20; 50
Printing angle (°)	0; 90°
Layer thickness (mm)	0.2; 0.3
Infill density (%)	80; 100

Uniaxial tensile tests are conducted on PLA samples until failure occurs. These tests are conducted at 25 °C. εr and UTS will be determined experimentally. UTS refers to the maximum stress a material can resist before it fails. εr is a measure of how much a material stretches before it breaks under tensile loading.

3 Experimental Results

Figure 3 presents an example of a tensile curve for an FDM-printed PLA part with the following FDM parameters: printing angle: θ = 90°, printed speed: V = 50 mm/sec, infill density: ρ = 100% and layer thickness: Th = 0.2 mm.

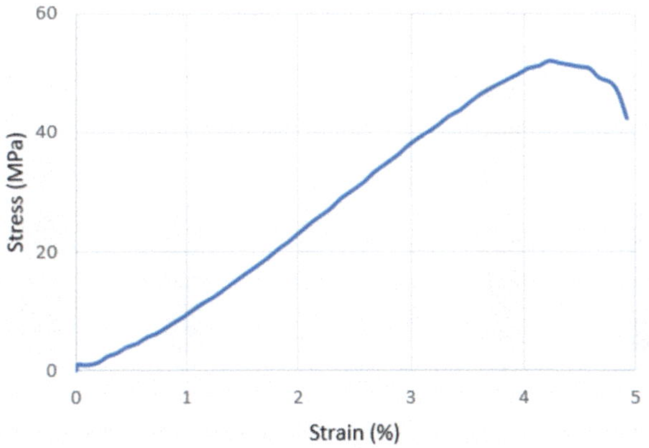

Fig. 3. Tensile curve for FDM-printed PLA part

In our study, a DOE plan was indeed carried out. Specifically, 16 tests were performed, each repeated twice to ensure the reliability and consistency of the results. We illustrate in Table 2 samples of the obtained results.

We show in detail, in Figs. 4 and 5, the influence of each FDM parameter on *UTS* and ε_r.

Table 2. Samples of experimental results of tensile test

Specimen N°	Printing angle: θ (°)	Printed speed V (mm/sec)	Infill density ρ (%)	Layer thickness Th (mm)	UTS (MPa)	ε_r (%)
1	90	50	100	0.3	44	1.95
2	0	20	100	0.3	46	2.28
3	90	50	100	0.2	52	4.9
4	0	50	80	0.3	37	1.3
5	0	50	100	0.3	58	2.25
6	0	50	80	0.2	48	4.27

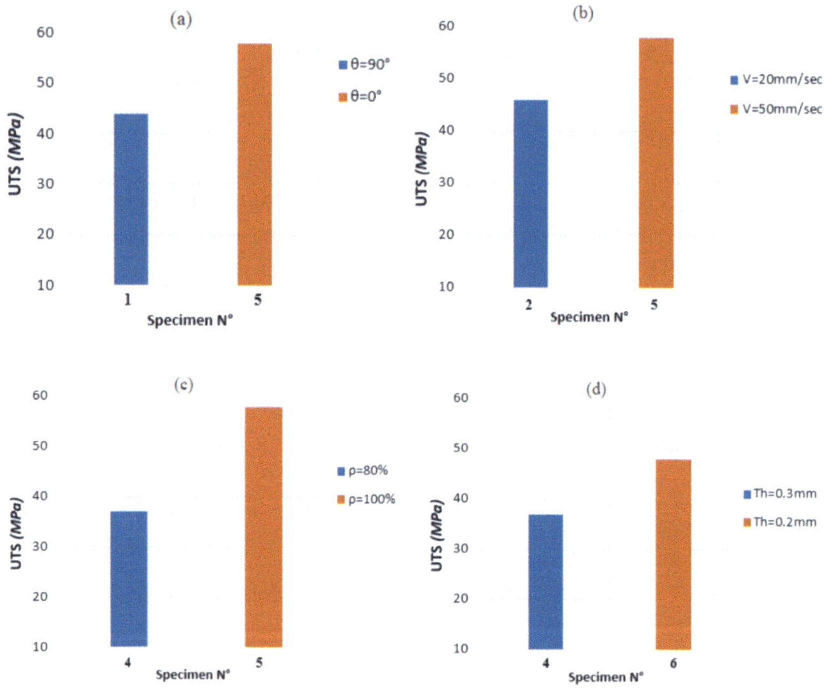

Fig. 4. Effect of FDM parameters on *UTS*: (a) Printing angle; (b) Printed speed; (c) Infill density; (d) Layer thickness

A smaller thickness and higher infill density enhance considerably the strength of the FDM-printed PLA part. However, a higher infill density value extends printing time and increases material consumption significantly. A lower infill density leads to weaker components. In addition, by reducing the printing speed, there is a decrease in UTS. However, with an increase of the printing speed, there is a moderate decrease of ε_r. In addition, with a raster angle $\theta = 0°$, the best UTS and ε_r are obtained.

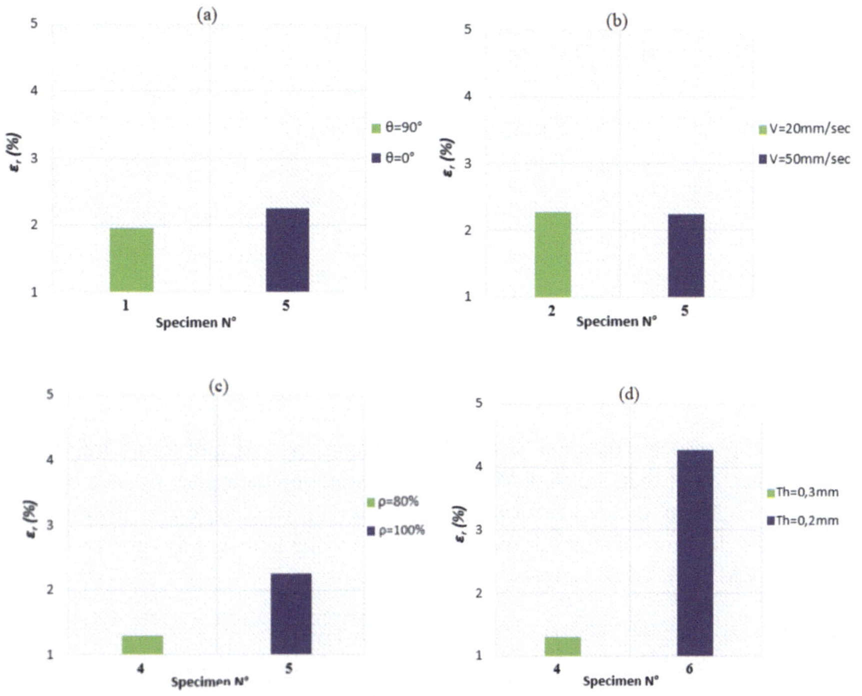

Fig. 5. Effect of FDM parameters on ε_r: (a) Printing angle; (b) Printed speed; (c) Infill density; (d) Layer thickness

4 Prediction with Machine-Learning Regression Models

In this section, ML models for predicting both *UTS* and ε_r of FDM-printed PLA parts using LR, RFR, and XGB are developed. The Linear Regression (LR) model is a straightforward approach to modeling the relationship between *UTS* or ε_r and FDM printing parameters. The model assumes a linear relationship between the input features and the output. However, the Random forest Regression (RFR) consists of a multitude of decision trees, where each tree is trained on a random subset of the training data and a random subset of the input features. The final prediction is then obtained by averaging of predictions from individual trees. The XGBoost regression (XGB) handles large datasets with complex relationships. It uses a sequence of decision trees, optimized through gradient boosting and regularization techniques. RFR and XGB are particularly useful for handling nonlinear relationships between FDM printing parameters and *UTS* or ε_r. To predict the output of FDM-printed PLA parts using ML models, we start by collecting a dataset comprising mechanical testing data and printing parameters. In fact, in our study, we initially used 16 experimental results from our tests. However, to enhance the model's robustness and address concerns about the dataset size, we also incorporated additional data from other tensile tests we performed, as well as results from FDM-printed PLA parts reported in the literature [23–26]. Printed speed, printing angle, layer thickness, Infill density, and extrusion temperature represent the input parameters of the

FDM-printed PLA part used in the prediction study. Table 3 illustrates the ranges of values for each input in the FDM-printed PLA part study.

Table 3. Input FDM parameters

Input parameters	values
Printed speed (mm/sec)	20–70
Printing angle (°)	0–90°
Layer thickness (mm)	0.06–0.6
Infill density (%)	60–100

Then, we preprocess the data, select relevant features, and split it into training and validation sets. After that, we train ML models, evaluating their performance by calculating the Root Mean Squared Error percent RMSE (%). We denote T as the number of tests, y_i as the definite value and $\hat{y}_i$ as the predicted value with a regression model. The formula for RMSE(%) is as follows.

$$RMSE(\%) = \left(\frac{\sqrt{\frac{1}{T}\sum (y_i - \hat{y}_i)^2}}{Range\ of\ the\ target\ variable} \right) * 100 \quad (1)$$

Finally, select the best-performing model and deploy it for predicting damage properties based on input parameters of FDM-printed PLA parts by calculating the RMSE(%) for LR and RFR models (Fig. 6).

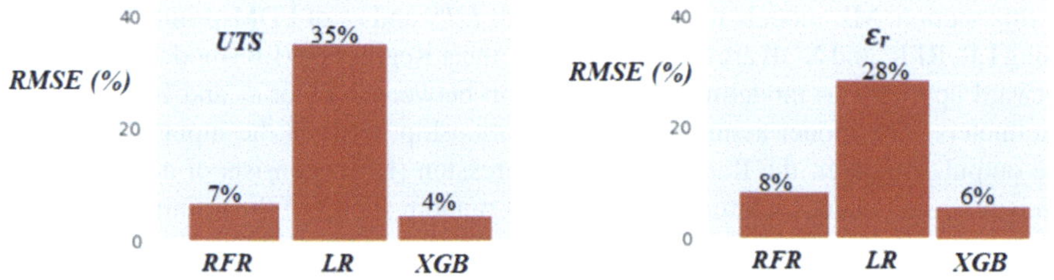

Fig. 6. RMSE(%) for damage properties prediction

The RFR and XGB models consistently yield a lower value of RMSE(%), expressed as a percentage, compared to the LR model. They provide more accurate predictions. The RMSE, as a percentage of the range of the target variable, gives a measure of the average deviation of predicted values of UTS and εr of FDM-printed PLA part from measured values. A lower RMSE(%) shows that the XGB model is better at capturing the complex relationships between FDM printing parameters and the damage properties of the printed parts, likely due to its ability to handle non-linearity and interactions among features more effectively than the LR model.

5 Conclusion

In this study, FDM was used to print PLA with different printing parameters. Four FDM parameters are chosen, which are the printing angle, the printed speed, the infill density, and the layer thickness. The main tensile test results showed that:

- Damage properties were influenced by these FDM parameters.
- Significant reductions in strength and strain at failure were observed with decreasing the layer thickness and infill density.

In addition, the RFR and mainly XGB models give a low value of RMSE(%), showing best predictive models than LR. This outcome underscores the superiority of the XGB model in accurately predicting the ultimate tensile strength and failure strain of FDM-printed PLA parts, making it the preferred choice for such predictive tasks. These results demonstrate that the linear assumption does not capture the complexity of the relationship as effectively as the nonlinear models.

References

1. Hanon, M., Dobos, J., Zsidai, L.: The influence of 3D printing process parameters on the mechanical performance of PLA polymer and its correlation with hardness. Procedia Manufact. **54**, 244–249 (2021)
2. Moakhar, S., Hentati, H., Barkallah, M., Louati, J., Haddar, M.: Parametric study of aluminum bar shearing using Johnson-cook material modeling. Proc. Inst. Mech. Eng. B J. Eng. Manuf. **235**(9), 1399–1411 (2021). https://doi.org/10.1177/0954405421995621
3. Ilyas, M., Hussain, G., Rashid, H., Alkahtani, M.: Influence of forming parameters on the mechanical behavior of a thin aluminum sheet processed through single point incremental forming. Metals. **10**, 1461 (2020). https://doi.org/10.3390/met10111461
4. Moakhar, S., Hentati, H., Barkallah, M., Louati, J., Haddar, M., Bonk, C., Behrens, B.A.: Modeling of the ductile damage—application for bar shearing. Mater. Werkst. **50**(11), 1353–1363 (2019). https://doi.org/10.1002/mawe.201800128
5. Ben Fraj, B., Kamoun, T., Hentati, H., Trabelsi, M., Ghazouani, N., Ahmed, M.: Optimization of forming force and Erichsen index using Taguchi design of experiments: mathematical models and experimental validation. Proc. Inst. Mech. Eng. B J. Eng. Manuf. **238**, 1316–1326 (2023). https://doi.org/10.1177/09544054231194107
6. Allouch, M., Ben Fraj, B., Dhouioui, M., Kesssentini, A., Hentati, H., Wali, M., Ferhi, M.: Mechanical, microstructural and numerical investigations of 3D printed carbon fiber reinforced PEEK. Proc. Inst. Mech. Eng. C J. Mech. Eng. Sci. **238**, 2131–2139 (2023). https://doi.org/10.1177/09544062231190534
7. Iftekar, S.F., Aabid, A., Amir, A., Baig, M.: Advancements and limitations in 3D printing materials and technologies: a critical review. Polymers. **15**(11), 2519 (2023)
8. Kriaa, Y., Hentati, H., Zouari, B.: Applying the phase-field approach for brittle fracture prediction: numerical implementation and experimental validation. Mech. Adv. Mater. Struct. **1**(12), 07 (2020). https://doi.org/10.1080/15376494.2020.1795957
9. Hentati, H., Dhahri, M., Dammak, F.: A phase-field model of quasistatic and dynamic brittle fracture using staggered algorithm. J. Mech. Mater. Struct. **11**(3), 309–327 (2016). https://doi.org/10.2140/jomms.2016.11.309
10. Ben Said, L., Allouch, M., Wali, M., Dammak, F.: Numerical formulation of anisotropic elastoplastic behavior coupled with damage model in forming processes. Mathematics. **11**, 204 (2023). https://doi.org/10.3390/math11010204

11. Raja, S.A., Muthukumaran, E., Jayakrishna, K.: A case study of 3D printed PLA and its mechanical properties. Mater. Today: Proc. **5**, 11219–11226 (2018). https://doi.org/10.1016/j.matpr.2018.01.146
12. Jamal, M.A., Shah, O.R., Ghafoor, U., Qureshi, Y., Bhutta, M.R.: Additive manufacturing of continuous fiber-reinforced polymer composites via fused deposition modelling: a comprehensive review. Polymers. **16**, 1622 (2024)
13. Salmi, M.: Additive manufacturing processes in medical applications. Materials. **14**, 191 (2021)
14. Sun, C., Wang, Y., McMurtrey, M.D., Jerred, N.D., Liou, F., Li, J.: Additive manufacturing for energy: a review. Appl. Energy. **282**, 116041 (2021)
15. Shady, F., Daniel, G.A., Robert, L.: Physical and mechanical properties of PLA, and their functions in widespread applications—a comprehensive review. Adv. Drug Deliv. Rev. **107**, 367–392 (2016). https://doi.org/10.1016/j.addr.2016.06.012
16. Hsueh, M.H., Lai, C.J., Wang, S.H.: Effect of printing parameters on the thermal and mechanical properties of 3D-printed PLA and PETG. Using Fused Depos. Model. Polym. **13**(11), 1758 (2021). https://doi.org/10.3390/polym13111758
17. Ammar, S., Ben Fraj, B., Hentati, H., Saouab, A., Ben Amar, M., Haddar, M.: Mechanical performances of printed carbon fiber-reinforced PLA and PETG composites. Proc. Inst. Mech. Eng. Part L J. Mater. Des. Appl. **238**(8), 1488–1499 (2024). https://doi.org/10.1177/14644207231225761
18. Rajan, K., Samykano, M., Kadirgama, K.: Fused deposition modeling: process, materials, parameters, properties, and applications. Int. J. Adv. Manuf. Technol. **120**, 1531–1570 (2022). https://doi.org/10.1007/s00170-022-08860-7
19. Goh, G.D., Sing, S.L., Yeong, W.Y.: A review on machine learning in 3D printing: applications, potential, and challenges. Artif. Intell. Rev. **54**(1), 63–94 (2021)
20. Dabbagh, S.R., Ozcan, O., Tasoglu, S.: Machine learning-enabled optimization of extrusion-based 3D printing. Methods. **206**, 27–40 (2022)
21. Vyavahare, S., Teraiya, S., Kumar, S.: FDM manufactured auxetic structures: an investigation of mechanical properties using machine learning techniques. Int. J. Solids Struct. **265**, 112126 (2023)
22. Sampedro, G.A.R., Rachmawati, S.M., Kim, D.S., Lee, J.M.: Exploring machine learning-based fault monitoring for polymer-based additive manufacturing: challenges and opportunities. Sensors. **22**(23), 9446 (2022)
23. Chacón, J.M., Caminero, M.A., García-Plaza, E., Núnez, P.J.: Additive manufacturing of PLA structures using fused deposition modelling: effect of process parameters on mechanical properties and their optimal selection. Mater. Des. **124**, 143–157 (2017)
24. Yao, T., Ye, J., Deng, Z., Zhang, K., Ma, Y., Ouyang, H.: Tensile failure strength and separation angle of FDM 3D printing PLA material: experimental and theoretical analyses. Compos. Part B. **188**, 107894 (2020)
25. Müller, M., Jirků, P., Šleger, V., Mishra, R.K., Hromasová, M., Novotný, J.: Effect of infill density in FDM 3D printing on low-cycle stress of bamboo-filled PLA-based material. Polymers. **14**(22), 4930 (2022)
26. Hsueh, M.H., Lai, C.J., Chung, C.F., Wang, S.H., Huang, W.C., Pan, C.Y., Hsieh, C.H.: Effect of printing parameters on the tensile properties of 3D-printed polylactic acid (PLA) based on fused deposition modeling. Polymers. **13**(14), 2387 (2021)

Effect of Post-treatment on Additively Manufactured Medical Implant: Numerical Study

Oumayma Bougadouha[1,5](✉), Mounir Frija[2,5], Raouf Fathallah[3,5], and Malika Khodja[4]

[1] National Engineering School of Sousse (ENISo), University of Sousse, Sousse, Tunisia
oumayma.bougadouha@eniso.u-sousse.tn

[2] Higher Institute of Applied Sciences and Technology (ISSATSo), University of Sousse, Sousse, Tunisia
mounir.frija@issatso.u-sousse.tn

[3] National Engineering School of Sousse (ENISo), University of Sousse, Sousse, Tunisia
raouf.fathallah@eniso.u-sousse.tn

[4] Faculty of Engineering, Built Environment, and Information Technology (EBIT), Department of Materials Science and Metallurgical Engineering, University of Pretoria, Pretoria, South Africa
malika.khodja@up.ac.za

[5] Unit of Mechanical and Materials Production Engineering (UGPMM), National Engineering School of Sfax (ENIS), University of Sfax, Rte de la Soukra km 4, 3038 Sfax, Tunisia

Abstract. Additive Manufacturing (AM), commonly known as 3D printing, has transformed the production of medical implants by offering advantages such as material optimization, design flexibility, and customization. Selective Laser Melting (SLM) stands out as the most widely used metallic AM process in the medical sector. Titanium alloy Ti6Al4V is recognized as a biomaterial due to its lightweight, corrosion resistance, fatigue resistance, and high strength. SLM has successfully produced various functional medical implants using Ti6Al4V. Achieving functional medical implants requires post-treatments to enhance their mechanical properties. Heat treatments are typically employed to alleviate residual stresses generated during the additive manufacturing process. In this study, we explored the effects of heat treatment and Hot Isostatic Pressing (HIP) treatment on a cervical fusion cage manufacturing SLM using Ti6Al4V titanium alloy. The results indicate that both heat treatment and HIP treatment significantly reduce the internal residual stresses by over 90%, in metallic medical implants produced by SLM. This reduction positively impacts the material strength of the additively manufactured metallic device.

Keywords: Additive manufacturing · post-treatment · medical implant

M. Soula et al. (Eds.): AIAM 2024, LNME, pp. 195–202, 2025.
https://doi.org/10.1007/978-3-031-94889-3_16

1 Introduction

Additive manufacturing is a widely used technique in several sectors such as medical, aerospace, construction, and automotive [1]. SLM is one of the most used powder bed fusion manufacturing processes to produce metallic medical implants. This technique has the advantages of producing complex geometries and reducing material waste compared to traditional manufacturing processes [2, 3]. Ti6Al4V titanium alloy is a lightweight material, with high strength, good corrosion resistance, and fatigue resistance. It is considered a biomaterial and it is widely used for the manufacturing of different medical implants such as dental, hip, knee, and spinal implants.

However, when manufacturing metal parts by additive manufacturing processes, it is still necessary to carry out appropriate heat treatments with the main goal of reducing residual stresses often generated during the manufacturing process. It is important to study the effect of different treatments on the printed parts to choose the most suitable one that allows the desired mechanical properties. Also, it is interesting to optimize the parameters of the post-treatment to obtain the best results.

Silvia Gaiani et al. [4] investigated the impact of heat treatments on Ti6Al4V specimens. A microstructure analysis was carried out on the additively manufactured samples, before and after, heat treatment. The results showed that heat treatment changes the microstructure of the material. Residual stresses were also evidenced by the XRD technique. The findings indicate that once the material microstructure was modified, the internal residual stresses were reduced. Chenyu Su et al. [5] examined the impact of heat and HIP treatment on the tensile and fatigue properties of Ti6Al4V specimens produced by SLM. The tensile strength was increased from 895 MPa to 1085 MPa and 1010 MPa, respectively after a heat treatment at 850 °C followed by water quenching and after HIP treatment. The fatigue cycle number was enhanced from $10^{4\sim5}$ to $10^{5\sim6}$ at a stress amplitude of 500 MPa, after HIP treatment. Gerrit M. Ter Haar [6] found that heat treatment may increase tensile properties.

In this study, the process of manufacturing a cervical fusion cage by SLM in Ti6Al4V was developed numerically. "Simufact additive" which is a numerical software dedicated to simulating metal additive manufacturing processes was used in this study. The effect of heat treatment and HIP treatment on the studied cage was investigated numerically. The rest of this paper is composed of 3 parts. A methodology section contains the manufacturing process and post-treatment parameters and stages. A results and discussion section contains the findings of this study. A conclusion and perspectives section.

2 Methodology

In this study, three simulations were carried out. The first simulation focused on the SLM process itself, modeling the layer-by-layer fabrication of the cage. The second simulation examined the effects of heat treatment applied to the additively manufactured cage while it was still attached to the supports and base plate. The third simulation investigated the impact of HIP treatment performed on the SLM-ed cage after it had been removed from the supports and base plate.

2.1 Material

The material used for the spinal cages is the EOS Titanium Grade 23 which is a Ti-6Al-4V alloy. In contrast to standard Ti-64 alloy, it contains a reduced quantity of oxygen and iron. Table 1 contains the chemical composition of EOS Titanium Grade 23. This material has low weight combined with high strength, excellent corrosion resistance, and high fatigue resistance compared to other lightweight alloys.

Table 1. Chemical composition of EOS Titanium Grade 23 [7]

Element	Al	V	O	N	C	H	Fe	γ	Ti
Powder chemical composition (Wt. %)	5.5–6.5	3.5–4.5	0.13	0.05	0.08	0.012	0.25	0.005	Balance

In this study, the material behavior is considered thermal-mechanical to effectively simulate the SLM and the heat treatment processes. The mechanical and thermal properties of the material needed for simulation are given in Table 2. The stress-strain curves at different temperatures of the Ti6Al4V are shown in Fig. 1.

Table 2. Mechanical and thermal properties of Ti6Al4V needed for simulation (software material database)

Property	Value at 20 °C	Value at 850 °C	Value at 1600 °C
Elasticity modulus (GPa)	115.61	71.16	43.24
Density (mg/mm^3)	4.48 e^{-9}	4.35 e^{-9}	4.21 e^{-9}
Poisson's ratio	0.26	0.26	0.26
Thermal expansion ($°C^{-1}$)	8.84 e^{-6}	1.19 e^{-5}	1.24 e^{-5}
Thermal conductivity (N/(s.k))	6	20	34.3
Specific heat capacity ($mm^2/(s^2.°C)$)	5.34 e^8	9.6 e^8	7.97 e^8

2.2 SLM Process

Simufact additive software was used to simulate the SLM manufacturing process of the cervical fusion device. The powder particle size of Ti6Al4V alloy is 50 μm. The SLM machine has a 280 * 280 * 365 mm build space and a 10 mm base plate thickness. The SLM process parameters are illustrated in Table 3. The implant design under study was created by Southern Medical, a South African company specializing in the manufacturing of spinal implants (Fig. 2).

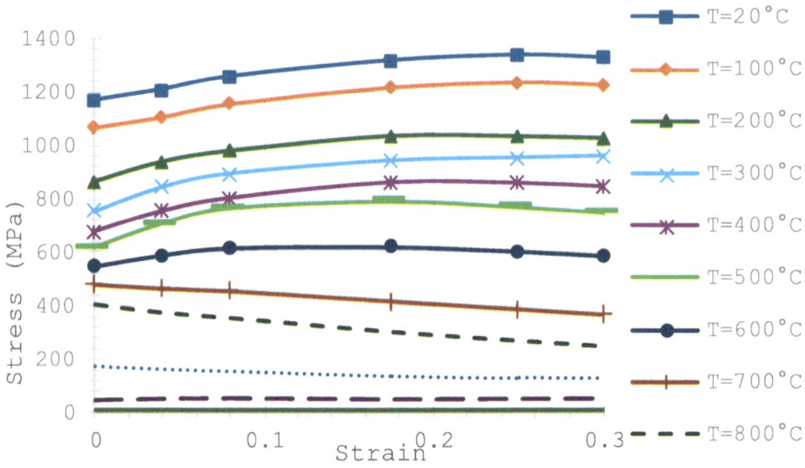

Fig. 1. Temperature-dependent stress-strain curves of Ti6Al4V

Table 3. SLM process parameters

Parameter	Value
Laser power	200 W
Scanning speed	1000 mm/s
Beam width	150 μm
Layer thickness	30 μm

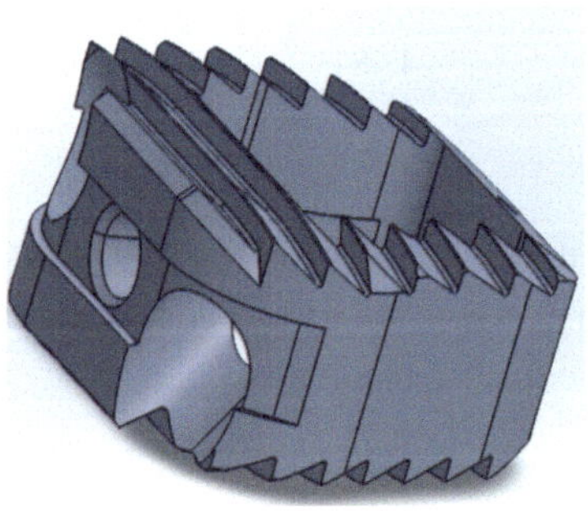

Fig. 2. Studied cervical fusion cage

2.3 Post Processing

According to the ASTM F2924–14 (2021), the HIP treatment of Ti6Al4V additively manufactured components should be done in an inert atmosphere to at least 100 MPa in the range 895 °C to 955 °C, held at the selected temperature at ±15 °C for 180 ± 60 min, then cooled in an inert atmosphere to a temperature below 425 °C [8]. A hot isostatic pressing treatment was applied to the implant after being manufactured and removed from the base plate and the supports. As illustrated in Fig. 3(a), the HIP treatment was conducted as follows; heating to 910 °C at a rate of 5 °C/min, holding for 2 hours under a pressure of 130 MPa, and finally cooling at a rate of 3 °C/min.

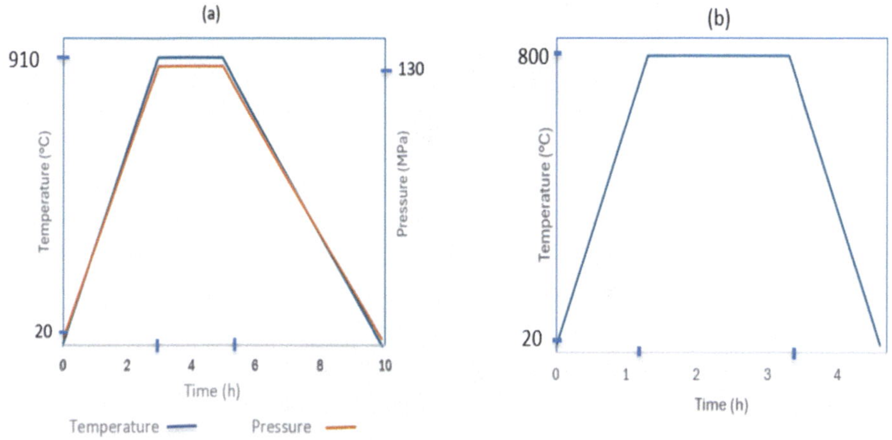

Fig. 3. The cycle of applied post-processing: (a) HIP treatment 910 °C-130 MPa / 2 h, (b) heat treatment 800 °C / 2 h

A heat treatment was applied to the build cage, before removing the base plate and the supports, as indicated in Fig. 3(b), as follows; heating to 800 °C at a rate of 10 °C/min, holding for 2 hours, and then cooling at a rate of 10 °C/min.

A voxel mesh size of 0.25 mm was used for all simulations. The effective stresses (ES) and the total distortion (TD) before and after treatments were simulated.

3 Results and Discussion

The ES and DT distributions are illustrated respectively in Fig. 4 and Fig. 5, before treatment (a), after HIP treatment (b), and after heat treatment (c). Results showed that the application of post-processing significantly reduced the ES. After HIP treatment, maximum and minimum ES were decreased respectively from 1174 MPa to 23 MPa and from 15 MPa to 7 MPa. After heat treatment, the maximum and minimum ES values were respectively equal to 97 MPa and 2 MPa. The findings demonstrate that both heat and HIP treatments could eliminate over 90% of residual stresses of Ti6Al4V components. These results are in line with those found by Hanjun Gao et al. [9], where the effect of heat treatment on Ti6Al4V was studied and it was found that it has the capability of reducing internal residual stresses by about 90%. S. Zhang et al. [10] studied the influence of heat treatment on Ti6Al4V manufactured by laser solid forming. The internal residual stresses were decreased by over 70%.

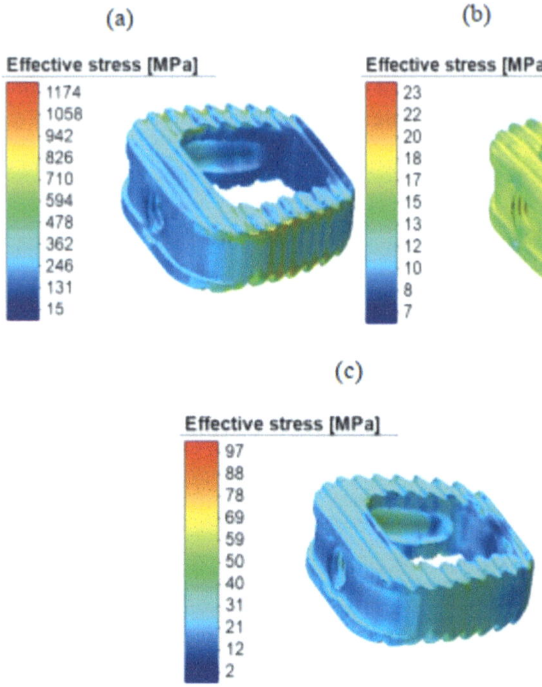

Fig. 4. Effective stress distribution; (a): as-built cage, (b): after HIP treatment, and (c): and after heat treatment

Heat treatment does not influence DT. For the as-built cage and the heat-treated cage, the DT measures 48 μm. While there is an insignificant effect on DT after HIP treatment, it decreases to 46 μm.

The reduction in residual stresses in Ti6Al4V after heat treatment and HIP treatment involves several mechanisms, including phase transformation, which plays an important role. The martensitic microstructure transformed into a lamellar $\alpha + \beta$ microstructure, where the α phase is in the form of fine needles located among the β ridges, with the presence of α prior β grain boundary [4, 11, 12]. This microstructure transformation improves the tensile properties and fatigue performance of the treated additively manufactured part [5, 6].

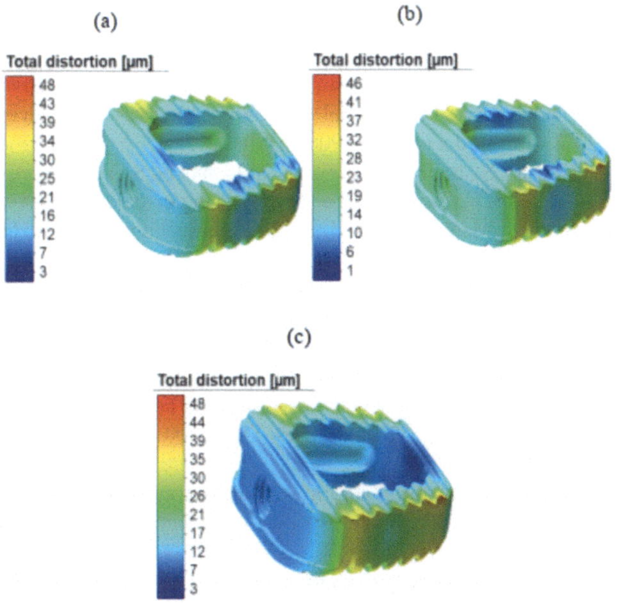

Fig. 5. Total distortion distribution; (a): as-built cage, (b): after HIP treatment, and (c): and after heat treatment

4 Conclusion and Perspectives

In this study, the manufacturing of a cervical fusion device by selective laser melting in Ti6Al4V titanium alloy was simulated. A heat treatment at a temperature of 800 °C for 2 h and a HIP treatment at a temperature of 910 °C and a pressure of 130 MPa for 2 h were applied on the medical implant to investigate its influence on the internal residual stresses. The results showed that the residual stresses decreased from 1174 MPa to 23 MPa after HIP treatment and to 97 MPa after heat treatment. The application of combined heat and HIP treatments will be studied in the future to determine its effect on the internal residual stresses. The parameters of heat treatment and HIP processes should be optimized to achieve the desired mechanical properties of the medical implant. It is also interesting to study the effects of other mechanical and chemical post-processing techniques, as they have the potential to positively impact the mechanical and chemical properties of medical implants.

Acknowledgement. The authors are thankful to the southern medical company located in Centurion, South Africa for providing the cervical fusion cage design.

Disclosure of Interest. The authors have no competing interests to declare that are relevant to the content of this article.

References

1. Vafadar, A., Guzzomi, F., Rassau, A., Hayward, K.: Advances in metal additive manufacturing: a review of common processes, industrial applications, and current challenges. Appl. Sci. **11**(3), 1213 (2021). https://doi.org/10.3390/app11031213
2. Gupta, S.K., Shahidsha, N., Bahl, S., Kedaria, D., Singamneni, S., Yarlagadda, P.K.D.V., et al.: Enhanced biomechanical performance of additively manufactured Ti-6Al-4V bone plates. J. Mech. Behav. Biomed. Mater. **119**, 104552 (2021). https://doi.org/10.1016/j.jmbbm.2021.104552
3. Balasubramanian, G., Kumar, Y., Kumar, R.L., Ramalingam, V.V., Priyadharshini, G.S., Kumar, K.S., et al.: Additive manufacturing of Ti-6Al-4V alloy for biomedical applications. J. Bio Tribo Corros. **8**(4) (2022). https://doi.org/10.1007/s40735-022-00700-1
4. Gaiani, S., Ferrari, E., Gozzi, M., Di Giovanni, M.T., Lassinantti Gualtieri, M., Colombini, E., et al.: Impact of post-process heat treatments performed on Ti6Al4V titanium alloy specimens obtained using LPBF technology. Technologies. **11**(4), 100 (2023). https://doi.org/10.3390/technologies11040100
5. Su, C., Yu, H., Wang, Z., Yang, J., Zeng, X.: Controlling the tensile and fatigue properties of selective laser melted Ti–6Al–4V alloy by post-treatment. J. Alloys Compd. **857**, 157552 (2021). https://doi.org/10.1016/j.jallcom.2020.157552
6. Ter Haar, G., Becker, T.: Selective laser melting produced Ti-6Al-4V: post-process heat treatments to achieve superior tensile properties. Materials. **11**, 146 (2018). https://doi.org/10.3390/ma11010146
7. EOS: EOS Titanium Ti64 Grade 23 Material Data Sheet. (2022)
8. ASTM F2924-14: Standard Specification for Additive Manufacturing Titanium-6 Aluminium-4 Vanadium with Powder Bed Fusion. ASTM, West Conshohocken, PA (2021)
9. Gao, H., Li, X., Li, B., Wu, Q., Ma, Y., Jian, X., et al.: Residual stress and microstructure of Ti6Al4V treated by thermal-vibratory stress relief process. J. Mater. Res. Technol. **18**, 5161–5181 (2022). https://doi.org/10.1016/j.jmrt.2022.04.137
10. Zhang, S., Lin, X., Chen, J., Huang, W.: Heat-treated microstructure and mechanical properties of laser solid forming Ti-6Al-4V alloy. Rare Metals. **28**(6), 537–544 (2009). https://doi.org/10.1007/s12598-009-0104-5
11. Eshawish, N., Malinov, S., Sha, W., Walls, P.: Microstructure and mechanical properties of Ti-6Al-4V manufactured by selective laser melting after stress relieving, hot isostatic pressing treatment, and post-heat treatment. J. Mater. Eng. Perform. **30**(7), 5290–5296 (2021). https://doi.org/10.1007/s11665-021-05753-w
12. Zhang, Y., Feng, L., Zhang, T., Xu, H., Li, J.: Heat treatment of additively manufactured Ti-6Al-4V alloy: microstructure and electrochemical properties. J. Alloys Compd. **888**, 161602 (2021). https://doi.org/10.1016/j.jallcom.2021.161602

Comparative Analysis of Simulation Models for LENS and LWAM Processes in Ti-6Al-4V Additive Manufacturing

Larbi Yousfi[1,2(✉)], Amir Guizani[1], Moncef Hammadi[2], Slim Bouaziz[1], and Mohamed Haddar[1]

[1] LA2MP, National Engineering School of Sfax, University of Sfax, Sfax, Tunisia
larbi.yousfi@enis.tn
[2] QUARTZ Laboratory EA7393, ISAE-SUPMECA, Saint-Ouen, France

Abstract. This study undertakes a comprehensive comparison of two prominent metal additive manufacturing processes from the Directed Energy Deposition (DED) family, Laser Engineered Net Shaping (LENS) and Laser Wire Additive Manufacturing (LWAM), specifically focusing on their application in printing the Ti-6Al-4V titanium alloy. Beyond evaluating these processes in practical terms, our research delves into the comparative analysis of advanced simulation models developed to accurately predict the thermo-microstructural evolution within Ti-6Al-4V parts fabricated by these methods. Through integrating transient finite element thermal models with density-based microstructural models to predict the α-phase fractions, β-phase fraction, and alpha lath widths, we aim to elucidate the nuanced differences in microstructure and phase transformations in the titanium alloy printed by each process. The study identifies four stages of phase transformations, including the transformation from β phase to α phase, and the elimination of α' phase during the LWAM process. In addition, this study compares these two processes, from the point of view of their efficiency and their potential for industrial applications. Results show that LWAM exhibits higher deposition rates and flexibility, while LENS offers precise control over microstructural characteristics. Our findings provide valuable insights for optimizing process parameters to achieve desired mechanical properties, enhancing the potential for industrial applications.

Keywords: Additive manufacturing · Titanium alloy · Thermo-microstructural simulation

1 Introduction

Additive manufacturing (AM), often referred to as 3D printing, is a promising way of producing metal parts with complex geometries and near-net shapes and is currently one of the most widely studied processes. Additive manufacturing offers many advantages over machining processes, such as reduced manufacturing times and costs thanks to low material loss [1]. Therefore, metal additive manufacturing can be categorized into

M. Soula et al. (Eds.): AIAM 2024, LNME, pp. 203–216, 2025.
https://doi.org/10.1007/978-3-031-94889-3_17

two main groups: Direct Energy Deposition (DED) processes [2, 3], involving the use of thermal energy to melt the material and form the desired part [4], and Powder Bed Fusion (PBF) processes [5, 6], which selectively fuse powder using thermal energy. Within the DED classification, five primary processes exist: Direct Metal Deposition (DMD), Laser Wire-Feed Additive Manufacturing (LWAM), Laser Engineered Net Shaping (LENS), Laser Metal Deposition (LMD), and Wire Arc Additive Manufacturing (WAAM). In the PBF classification, there exist five primary techniques: Selective Laser Sintering (SLS), Selective Laser Melting (SLM), Laser Metal Fusion (LMF), Direct Metal Laser Sintering (DMLS), and Electron Beam Melting (EBM). In this study, we concentrate on comparing thermos-microstructural simulation models for the titanium alloy Ti-6Al-4V, fabricated using the LENS/LWAM additive manufacturing processes from the DED family. These two methods are employed to determine the microstructural characteristics of various materials, especially for Ti-6Al-4V. This material constitutes a heat-treatable two-phase alloy, consisting of approximately 90% α—phase at room temperature and 100% stable β-phase beyond the beta-transit temperature (T_{β}^{trans} Approximately 1000 °C) [7]. Several research studies on microstructural modeling have shown that the microstructural properties of Ti-6Al-4V titanium alloy parts depend strongly on the temperature evolution during the DED process, which generally involves high and non-uniform thermal gradients [8, 9]. The use of finite element method (FE)—based macroscale modeling is suitable for representing thermal phenomena during DED processing for industrial parts [10]. Most finite element models used in DED process modeling draw inspiration from the numerical simulation of physical phenomena, such as heat transfer and fluid flow. Heat input models are employed to explain the laser heat source [11]. The density-based approach for microstructure modeling [12] allows predicting the evolution of the microstructure during treatments such as laser welding and forming, thanks to the direct coupling with transient finite element thermal models, and requires reduced computational cost. Kelly and Kampe [13, 14] achieved a numerical and experimental study of the microstructural evolution of the Ti-6Al-4V titanium alloy along the LENS process using a two-dimensional finite-difference formulation that takes into account the Widmanstätten phases α in the colonies. Suárez et al. [15] formulated a transient thermo-metallurgical finite element (FE) model to simulate the deposition process of Ti-6Al-4V. Charles and Järvstrat [16] developed numerical thermo-microstructural models to study the microstructural evolution of Ti-6Al-4V deposited by welding processes using the same kinetic model as described in references [13, 14]. However, they incorporated the martensitic transformation and the calculation of the α-lath width for their model. The microstructural model was applied based on the microstructural models of Kelly and Charles [13, 16], which are used to calculate the volume fractions of the α-phase, β-phase, and alpha lath width in the LENS process. Studies of the phase transformations of titanium alloy Ti-6Al-4V printed by additive manufacturing using LENS technology are being further extended into other additive manufacturing technologies, including laser-wire additive manufacturing (LWAM). The LWAM process [17] is proving attractive because of its higher deposition rate and flexibility. The laser induces the formation of a molten pool on the substrate material, where the metal wire is introduced and melted, creating a metallurgical bond with the substrate [18]. Through the movement of the laser processing head and wire feeder or by shifting the substrate, a bead is shaped during the solidification process. This relative motion between the welding tool and the substrate can be accomplished using a robotic arm or a computer-numerically controlled worktable. Many researchers are dedicated to exploring the correlation between the microstructure of additively manufactured (AM) products and their mechanical properties. The two main aims of optimizing the microstructure of the Ti-6Al-4V titanium alloy are to refine the β-grain [17, 19] and to eliminate the undesirable phase [20–23]. The Density-Based Constituent Phase Simulation (DCPM) method is an alternative approach for predicting the phase transformation of Ti-6Al-4V. The discrete Johnson-Mehl-Avrami-Kolmogorov (JMAK) model is applicable for simulating the diffusion phase transformation, while the discrete Koistinen-Marburger (KM) model is suitable for simulating the diffusionless phase transformation involved in $\beta \rightarrow \alpha$ and $\beta \rightarrow \alpha'$ transformations. This research aims to study the phase transformation in the Ti-6Al-4V material during the WLAM and LENS processes using the DCPM algorithm and the microstructural model. The paper is structured as follows. Following the introduction, Sect. 2 presents the experimentation and simulation methods already developed in the literature for these two processes, with a particular focus on the study of five deposition layers. In Sect. 3, we will provide a comparative study of the two processes, LENS and LWAM, based on metallurgical structure and in terms of cost efficiency, material properties optimization, and potential industrial applications.

2 Experimental Setup and Thermo-Microstructural Modeling

2.1 Experimental Setup for LENS

The experimental setup for LENS, as described by Baykasoglu et al. [24], involves a rectangular-shaped Ti-6Al-4V specimen composed of five layers manufactured using the Optomec LENS system, a Ti-6Al-4V square substrate, and an aluminum fixture. The laser is operated at a power of 300 W, featuring a beam intensity of 18,679 MW/m^2 and a minimum spot size of 143 μm at the focal point. The travel speed is set to 2 mm/s, accompanied by a powder feed rate of 8.6 g/min. Two Omega SA1XL-K-72 thermocouples are affixed to both the upper and lower surfaces of the substrate to record temperature data throughout the heating and cooling phases.

2.2 Experimental Setup for LWAM

The experimental setup for Laser Wire Additive Manufacturing (LWAM), as detailed by Sun et al. [25], involves an LWAM machine, inert gas protection, and a controlled wire feeding system. In the LWAM process, the feedstock wire is melted using a laser, and the molten material is deposited onto the additive part or substrate. The laser scans, or the workbench moves, with a path determined by computer numerical control machine tools (CNC). The product is additively manufactured layer by layer. Stringent requirements for wire feeding are necessary to maintain stable mass transfer from the wire to the additive part. The end of the wire should be positioned ahead of the laser center to maintain a liquid bridge between the wire and the additive part. The additive material is deposited onto a substrate with dimensions of 100 mm × 100 mm × 10 mm. Five layers were fabricated, and the size of a single layer is approximately 90 mm in length, 5 mm in width, and 0.3 mm in thickness. The experimental processing parameters include a laser power set at 3000 W and a scanning velocity of 10 mm/s. Each layer deposition takes 9 seconds, with a 3-second interval for the laser device to return to the starting position before proceeding to the next layer. Transversal planes are examined using an Optical Microscope (OM) and Back Scattered Electron (BSE).

2.3 Thermo-Microstructural Modeling

Heat Source Modeling In thermal models, the laser-material interaction is simulated by a moving surface or a volumetric heat source. The numerical heat source model for laser energy depends on the laser intensity distribution. Researchers have used various types of heat sources, detailed in the next section and shown in Table 1 [26].

Table 1. Different types of numerical heat sources are used in the literature for Laser DED [26].

Name / Type	Mathematical expression
Gaussian: Surface	$Q = \frac{AP}{\pi r_0^2} \exp\left(\frac{-2r^2}{r_0^2}\right)$ where A, P, r_0 and r are the laser Absorptivity, laser power, laser radius, and the distance between the laser power and the node
Top-Hat: Surface	$Q = \begin{cases} \frac{AP}{\pi r_0^2}, & -r_0 \leq r \leq r_0 \\ 0, & r > r_0 \end{cases}$
Conico-Gaussian: Volumetric	$Q = \frac{2AP}{\pi r_0^2 H}\left(1 - \frac{z}{H}\right) \exp\left(1 - \frac{r^2}{r_0^2}\right)$ where H and z are the characteristic length of the conical-gaussian beam and the distance along the beam's axis
Cylindrical: Volumetric	$Q = \begin{cases} \frac{AP}{\pi r_0^2 H}, & -r_0 \leq r \leq r_0 \\ 0, & r > r_0 \end{cases}$
Goldak: Volumetric	$Q = \frac{6\sqrt{3}APf_{f,r}}{a_{f,r}bc\pi\sqrt{\pi}} \exp\left(-\frac{3x^2}{a_{f,r}^2} - \frac{3y^2}{b^2} - \frac{3z^2}{c^2}\right)$ where Q, A, P, $f_{f,\,r}$ $a_{f,\,r}$, b, c and x, y, z are the volumetric heat source, Absorptivity, laser power, the scaling factor, the transverse, depth, and longitudinal semi-axes of the ellipsoid and the local coordinates

Thermo-Microstructural Model of LENS To predict the thermal stability of the melt pool and calculate the phase fractions in Ti-6Al-4V, Baykasoglu et al. [24] utilized the density-based simulation method. The overall simulation procedure, illustrating the prediction of temperature history and volumetric fractions for each phase, is presented in

Fig. 1. The model is implemented and simulations are conducted using the finite element software ABAQUS. To specify the heat source and calculate the microstructural characteristics, two FORTRAN-written subroutines, namely DFLUX and USDFLD, are incorporated into the model. As shown in Fig. 1, at each time step, initial computations involve the determination of thermal outputs at integration points, followed by the calculation and storage of microstructural outputs through a user subroutine.

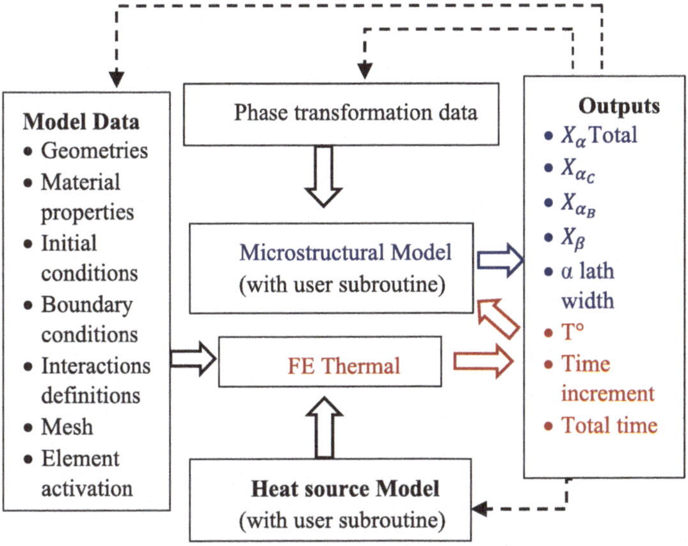

Fig. 1. The comprehensive method for predicting thermal and microstructural outputs [24]

For the Thermal Model, the governing equation for energy balance in heat transfer can be formulated as:

$$\rho C_P \frac{dT(x,t)}{dt} = -\nabla . q(x,t) + Q(x,t) \tag{1}$$

Within any evolving subvolume V of the AM part. In this context, T represents temperature, ρ signifies material density, C_P is the temperature-dependent specific heat, t denotes time, x represents spatial coordinates, q denotes the heat flux vector, and Q stands for the body heat source.

Employing Goldak's double ellipsoid heat source model [27] to characterize the laser heat source term in the LENS process, it is defined as:

$$q\left(x', y', z'\right) = \frac{6\sqrt{3}P\eta f}{abc\pi^{3/2}} e^{\left(-3\frac{x'^2}{a^2} - 3\frac{y'^2}{b^2} - 3\frac{z'^2}{c^2}\right)} \tag{2}$$

Here, P represents laser power; η stands for absorption efficiency; f is the scaling factor; x', y', and z' denote the local coordinates with the origin centered where the moving heat source, with a scan speed v, reaches its maximum value; and a, b, and c are the transverse, depth, and longitudinal semi-axes of the ellipsoid, respectively (refer to Fig. 2).

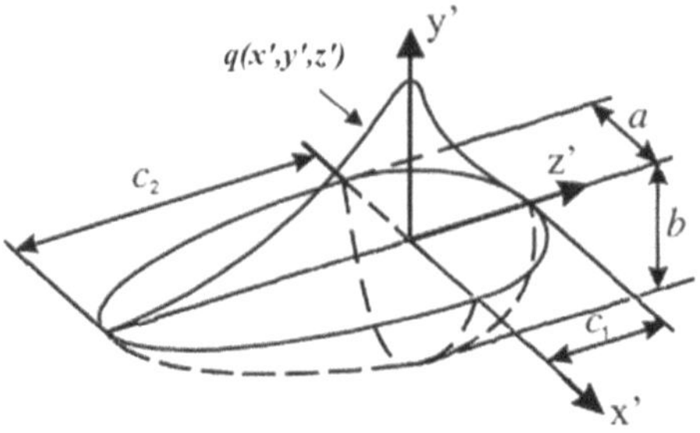

Fig. 2. Goldak's double ellipsoid heat source model [27]

Based on the microstructural model, Baykasoglu et al. [24] predicted the evolution of the Ti-6Al-4V titanium alloy in its solid state using volumetric phase fractions [18]. This microstructural modeling specifically includes fractions of the Widmanstätten colony and basketweave α-phase during the controlled phase transformation from β, governed by diffusion. The calculation of the transformed α-phase volume fraction during cooling is determined by the JMAK equation:

$$X_\alpha = \left(1 - e^{\left(-k_{\beta\alpha}\,\tau^{N_{\beta\alpha}}\right)}\right)X_\alpha^{eq} \tag{3}$$

Where, X_α represents the computed fraction of α-phases, while $k_{\beta\alpha}$ and $N_{\beta\alpha}$ are material parameters obtainable from the time-temperature–transformation diagrams. τ denotes a relative time measured from the initiation of the transformation, and X_α^{eq} signifies the equilibrium fraction of the α-phase. While heating, the choice between the colony or basketweave morphology as the source of the dissolved α-phase is determined by the intragranular nucleation temperature T_{IG}.

Thermo-Microstructural Model of LWAM In the case of the LWAM process, the prediction of the microstructural evolution of Ti-6Al-4V is based on the DCPM algorithm mentioned by Sun et al. [25], as illustrated in Fig. 3. The computational complexity of DCPM aligns with that of thermal transfer simulations and can be integrated as a function within it. The inputs for the model are $X_{\alpha_B}^n$, $X_{\alpha_c}^n$, $X_{\alpha_{gb}}^n$, $X_{\alpha'}^n$, X_β^n, α_{Lath}^n, T^n, T^{n+1} and dt^n. Where $X_{\alpha_i}^n$ is the $\alpha_i(i = B, C, gb', \beta)$-phase fraction at the nth time step, α_{Lath}^n represents the thickness of the α lath at time step n, T^n and T^{n+1} denote the temperatures at time steps n and n + 1, respectively, while dt^n represents the time increment at time step n. The results include $X_{\alpha_B}^{n+1}$, $X_{\alpha_c}^{n+1}$, $X_{\alpha_{gb}}^{n+1}$, $X_{\alpha'}^{n+1}$, X_β^{n+1}, α_{Lath}^n. As illustrated in Fig. 3(b), these are obtained through four stages of phase transformations.

The thermal transfer model utilizes the Gaussian surface heat source [27], characterized by:

$$S = \frac{Pd}{\pi r_b^2}\eta \exp^{\left(-d\frac{r^2}{r_b^2}\right)} \tag{4}$$

Here, P represents the laser power, d is equal to 3, r_b denotes the laser radius, η is the energy distribution factor, and r is the distance between the laser power and the node.

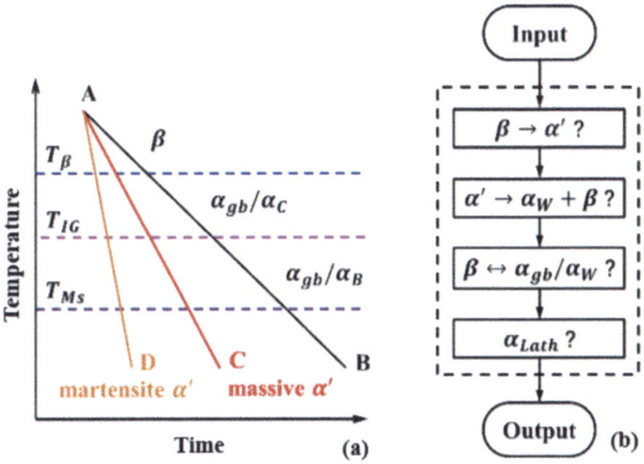

Fig. 3. Illustrations of (a) Ti-6Al-4V phase transformations during the cooling process and (b) the DCPM algorithm [25].

3 Comparative Study of LENS and LWAM Processes

3.1 Metallurgical Study of Ti-6Al-4V

The results of phase transformations in Ti-6Al-4V during deposition by the additive manufacturing process are listed in Table 2. Our results for the LWAM process demonstrate that when the temperature is below the β transus $T_\beta = 1269\ K$ and the cooling rate is gradual (approximately 20 K/s), the diffusion-controlled transformation $\beta \to \alpha_{gb}/\alpha_w$ occurs. Here, α_w represents the Widmanstätten α-phase, comprising the colony α_C and basket-weave α_B. If the temperature T exceeds $T_{IG} = 1100$ K, the newly formed α_W is α_C. Otherwise, it is α_B. If the cooling rate is rapid ($\geq$ 410 K/s) and the temperature T is below the start temperature of martensite transformation. T_{MS} = 923 K, a full martensite transformation takes place, transitioning from β to martensite α'. If the cooling rate is moderate (≥20 K/s and <410 K/s) and T is below T_{MS}, a partial martensite transformation occurs, resulting in $\beta \to$ massive α'.

For experimental validation, as reported by Baykasoglu et al. [24], the measurement of α-lath width is conducted using scanning electron microscope (SEM) micrographs of each layer deposited by the LENS process. This is achieved using the IMAGEJ software, and the arithmetic mean is calculated. However, in the LWAM process, experimental measurements are performed using Optical Microscope (OM) and Backscattered Electron (BSE) techniques. In the case of the LENS process, building upon the findings of Baykasoglu et al. [24], we compare the simulated results from the microstructural model with SEM measurements, which reveals excellent agreement between the microstructural model results and experimental data, as depicted in Fig. 4(a). This accurate prediction of α-lath width facilitates the estimation of mechanical properties.

For the LWAM process, based on the results of Sun et al. [25], we find that the predicted values are within the range of the experimental error bars (as shown in Fig. 4(b)). Specifically, the predicted α-lath is larger than the average experimental α-lath. Table 3 presents the results obtained from the calculation of the width of the alpha laths in Ti-6Al-4V. For LWAM, the experimental results indicate a maximum α-lath thickness of

Table 2. Overview of phase transformations in the titanium alloy Ti-6Al-4V.

LENS (DED-Powder)	LWAM (DED Wire)
• Room T°: (90% α-phase) • T° > Ttransus: (100% β-phase) • During heating, the α → β transformation until T°transus • Fast cooling rate: β → α' massive martensitic phase. • Low cooling rates: β ↔ α_{gb}/α_{w} transformation • Widmanstätten α can be divided into colony and basketweave morphologies based on an intragranular nucleation temperature	• Four stages for β → α/α' phase transformation • The β → α' transformation • The α' → β + α_{W} transformation • The β → α_{gb}/α_{W} transformation • The α_{Lath} sub-function

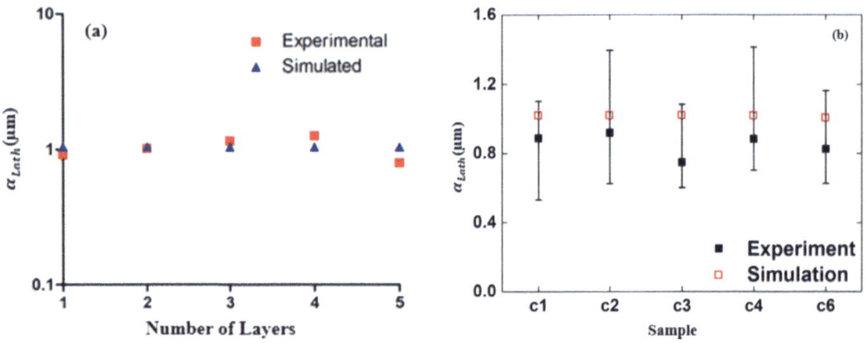

Fig. 4. Experimental and simulated mean α-lath widths as a function of layer number: (a) for LENS, and (b) for LWAM process [24, 25].

1.102 μm and a minimum of 0.529 μm. Simulated values closely align, with a maximum of 1.124 μm and a minimum of 0.818 μm. In the case of LENS, experimental findings reveal a maximum α-lath thickness of 1.198 μm and a minimum of 0.628 μm. Simulated values for LENS exhibit a maximum thickness of 1.101 μm and a minimum of 0.815 μm. Overall, the comparison between experimental and simulated results provides insights into the consistency and accuracy of the α-lath widths in the context of these two additive manufacturing processes.

To study the temperature profile and stability of the melt pool during the deposition of five layers by the LENS process, two thermocouples were utilized. The locations of the thermocouples are as follows: P1 (point 1) is positioned on the top surface of the base, while P2 (point 2) is located on the bottom surface of the base plate. Numerical temperature data were also obtained at the same thermocouple locations. Figure 5 illustrates the temperature comparison between experimental measurements and the results of finite element analysis (FEA).

The observation reveals higher temperature values at P2 compared to P1, particularly during heating cycles, as P2 is in closer proximity to the heat-affected area. It is

Table 3. Calculations of α-lath widths

Additive Manufacturing Process	Results	α_{Lath} Thickness (μm)	
		Max	Min
LWAM	Experimental	1.102	0.529
	Simulated	1.124	0.818
LENS	Experimental	1.198	0.628
	Simulated	1.101	0.815

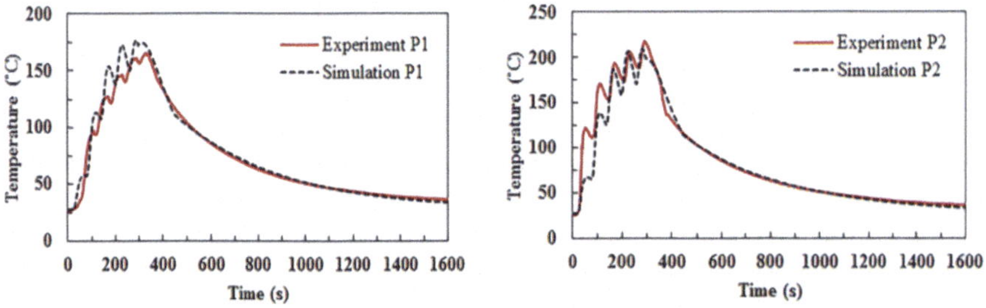

Fig. 5. Comparative analysis of temperature histories between simulation and experimental measurements at points P1 and P2 for LENS process [24].

noteworthy that the finite element (FE) field temperature results at both P1 and P2 align well with experimental thermocouple data. This alignment underscores the validity of the proposed FE model throughout the thickness direction. In our study of the LWAM process, the temperature evolution in the five deposited layers is shown in Fig. 6(a). The comparison of results is presented in Table 4. It can be observed that the simulation time for the LENS process is multiplied by 10 compared to the LWAM process. However, for the temperature field, the values in LWAM are significantly higher compared to the temperature in the LENS process. It can be noted that LWAM requires more energy than the LENS process. The phase fraction of the α'-phase and α_B phase is illustrated in Fig. 6(b).

Our results from Fig. 6 are summarized in Table 5. It is concluded that due to the poor heat dissipation of Ti-6Al-4V and the accumulation of excess heat inside the added part, the minimum temperature for each layer increases. The volumetric fraction of the α' phase decreases as the layer height increases, disappearing entirely by the fifth layer of deposition, highlighting the elimination of the undesirable α' phase during the LWAM process. The fraction of the desirable α_B phase increases with the increasing deposition height.

Figure 7 displays the 3D distribution of α_B and β in the five layers deposited by LWAM using a 3000 W laser power. It can be concluded that the transformed phase is predicted to be in the form of α_B In the Five Layers. In the center, the fraction of α_B is around 0.8, and the fraction of β is around 0.2. However, in the end, the fraction of α_B is about 0.5, and the fraction of β is about 0.5.

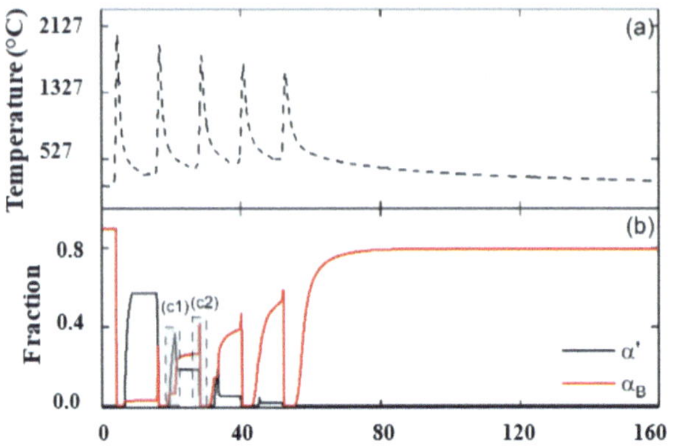

Fig. 6. Visual representations of (a) temperature and (b) $X_{\alpha'}$ and X_{α_B} over time for LWAM [25].

Table 4. Comparison of temperature history and simulation time between LENS and LWAM processes

AM Process	Laser power (W)	Results		
		T°min (°C)	T°Max (°C)	Simulation time (s)
LENS	300	25	180	1600
		25	220	
LWAM	3000	25	2105	160

Table 5. Thermal and microstructural results for five layers deposited by the LWAM process with 3000 W.

Laser power (3000 W)	T°min (°C)	T°Max (°C)	$X_{\alpha'}$	X_{α_B}
Layer 1	371	2105	0.57	0.03
Layer 2	463	1959	0.18	0.26
Layer 3	524	1832	0.05	0.36
Layer 4	567	1715	0.02	0.45
Layer 5	598	1610	0	0.80

3.2 Implications of Phase Transformations on Mechanical Properties

The different phases resulting from the phase transformations of Ti-6Al-4V can influence the strength, ductility, toughness, and other mechanical properties of the material. In the case of LENS, the thermo-microstructural modeling approach has shown that lath width is closely related to hardness, with thicker laths generally correlated with lower hardness [28]. The coupling of the microstructural model with thermal-mechanical models enables the development of a more precise framework, facilitating the prediction of

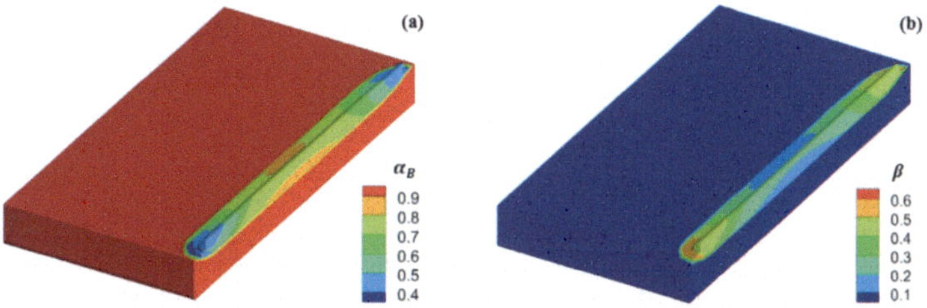

Fig. 7. 3D distribution of α_B and β phases in the Ti-6Al-4V Five Layer deposited by the LWAM process with a 3000 W laser power [25].

material responses such as distortions and stresses arising from both microstructure and temperature [11]. comparative with the case study of LWAM process [25], the transition from the β phase to the α phase in Ti-6Al-4V can significantly impact its mechanical properties. The α phase is generally softer and more ductile, while the β phase is harder and stronger. The presence of the α phase can enhance the material's toughness and ductility, making it more resistant to deformation and fracture. On the other hand, the β phase contributes to higher strength and hardness, providing improved load-bearing capabilities. By controlling the phase transformation process, such as through heat treatments and cooling rates, it is possible to tailor the microstructure and achieve desired mechanical properties in Ti-6Al-4V parts.

3.3 Cost Efficiency, Optimization of Material Properties, and Potential Industrial Applications

To compare LENS with the LWAM process, it is essential to consider factors such as process capabilities, material compatibility, build speed, surface finish, part complexity, cost efficiency, and overall suitability for specific applications. LENS technology offers flexibility by allowing the use of non-proprietary materials in the construction process, opening up many possibilities for existing and future combinations of metal alloys. It can be used to manufacture components for a variety of industries, including critical sectors such as medicine and aerospace (Guo and Leu, [29]; Dutta and Froes, [30]). Laser-wire additive manufacturing exhibits potential by facilitating the production of sizable components with significantly higher deposition rates (ten times greater than powder bed additive manufacturing). It also provides moderate forming flexibility and entails lower investment costs. Employing these technologies, particularly additive manufacturing (AM) by wire, enables achieving a substantial deposition rate [31]. Consequently, AM techniques are relatively pertinent and crucial for developers aiming to reduce the production costs of components in the Ti-6Al-4V titanium alloy. This alloy stands out as one of the most expensive but widely used in the aerospace industry due to its low density, excellent mechanical properties, and corrosion resistance. Owers current comparative study focuses on the prediction of the evolution of the microstructure of Ti-6Al-4V titanium alloy during the powder laser process (LENS) and the wire laser process (LWAM) and does not address the wider applicability of the model studied to other materials or additive manufacturing processes. Furthermore, the mechanical modeling of Ti-6Al-4V parts manufactured by LENS and LWAM to predict residual stress and distortion to guarantee the quality of the parts produced was not concluded in this study. This aspect remains a central point for our future research efforts.

4 Conclusion

The comparative study shows that the FE-based thermo-microstructural modeling approach is capable of predicting the evolution of the microstructure of Ti-6Al-4V titanium alloy during the LENS process and helping to select the optimal process parameters to achieve the desired microstructure and mechanical properties for additive manufacturing (AM) parts. The effects of processing parameters on microstructural behavior were not investigated in this study, but future research will systematically explore these effects and improve model validation using different characterization techniques. Simulation of the phase transformation in Ti-6Al-4V material during laser wire additive manufacturing has significant potential for predicting phase changes, revealing the underlying mechanisms, and optimizing processing parameters. Experimental validation confirms the accuracy of the density-based constitutive phase simulation method. The study concludes that increasing the preheating temperature and decreasing the cooling rate can contribute to a reduction in the fraction of the α' phase, or even its elimination. The optimum laser power for the process is determined to be 3000 W, which results in increased transformation of the α_B phase without increasing the thickness of α-laths and improves forming stability for multilayer deposition.

References

1. Frazier, W.E.: Metal additive manufacturing: a review. J. Mater. Eng. Perform. **23**(6), 1917–1928 (2014)
2. Dezaki, M.L., Serjouei, A., Zolfagharian, A., Fotouhi, M., Moradi, M., Ariffin, M., Bodaghi, M.: A review on additive/subtractive hybrid manufacturing of directed energy deposition (DED) process. Adv. Powder Mater. **1**, 100054 (2022)
3. Svetlizky, D., Zheng, B., Vyatskikh, A., Das, M., Bose, S., Bandyopadhyay, A., Schoenung, J.M., Lavernia, E.J., Eliaz, N.: Laser-based directed energy deposition (DED-LB) of advanced materials. Mater. Sci. Eng. A. **840**, 142967 (2022)
4. Majeed, A., Ahmed, A., Lv, J., Peng, T., Muzamil, M.: A state-of-the-art review on energy consumption and quality characteristics in metal additive manufacturing processes. J. Braz. Soc. Mech. Sci. Eng. **42**, 249 (2020)
5. Wang, D., Liu, L., Deng, G., Deng, C., Bai, Y., Yang, Y., Wu, W., Chen, J., Liu, Y., Wang, Y., et al.: Recent progress on additive manufacturing of multi-material structures with laser powder bed fusion. Virt. Phys. Prototype. **17**, 329–365 (2022)
6. Kotadia, H., Gibbons, G., Das, A., Howes, P.: A review of laser powder bed fusion additive manufacturing of aluminum alloys: microstructure and properties. Addit. Manuf. **46**, 102155 (2021)
7. Froes, F.H., Dutta, B.: The additive manufacturing (AM) of titanium alloys. Adv. Mater. Res. **1019**, 19–25 (2014)
8. Bian, L., Thompson, S.M., Shamsaei, N.: Mechanical properties and microstructural features of direct laser-deposited Ti-6Al-4V. JOM. **67**(3), 629–638 (2015)

9. Marshall, G.J., Young II, W.J., Thompson, S.M., Shamsaei, N., Daniewicz, S.R., Shao, S.: Understanding the microstructure formation of Ti6Al-4V during direct laser deposition via in-situ thermal monitoring. JOM. **68**(3), 778–790 (2016)
10. Thompson, S.M., Bian, L., Shamsaei, N., Yadollahi, A.: An overview of direct laser deposition for additive manufacturing—part I: transport phenomena, modeling and diagnostics. Addit. Manuf. **8**, 36–62 (2015)
11. Goldak, J.A., Akhlaghi, M.: Computational Welding Mechanics. Springer, New York, NY (2005)
12. Grong, Ø., Shercliff, H.R.: Microstructural modelling in metals processing. Prog. Mater. Sci. **47**(2), 163–282 (2002)
13. Kelly, S.M., Kampe, S.L.: Microstructural evolution in laser deposited multilayer Ti-6Al-4V build—part I: microstructural characterization. Metall. Mater. Trans. A. **35**(6), 1861–1867 (2004)
14. Kelly, S.M., Kampe, S.L.: Microstructural evolution in laser deposited multilayer Ti-6Al-4V build—part II: thermal modeling. Metall. Mater. Trans. A. **35**(6), 1869–1879 (2004)
15. Suárez, A., Tobar, M.J., Yánez, A., Pérez, I., Sampedro, J., Amigó, V., Candel, J.J.: Modeling of phase transformations of Ti6Al4V during laser metal deposition. Phys. Procedia. **12**(Part A), 666–673 (2011)
16. Charles, C., Järvstrat, N.: Modelling Ti–6Al–4V microstructure by evolution Laws implemented as finite element subroutines: application to TIG metal deposition. In: Eighth International Conference on Trends in Welding Research (TWR), Pine Mountain, GA, June 1–6, pp. 477–485 (2009)
17. Donoghue, J., Antonysamy, A.A., Martina, F., Colegrove, P.A., Williams, S.W., Prangnell, P.B.: The effectiveness of combining rolling deformation with wire-arc additive manufacture on b-grain refinement and texture modification in Ti6Al–4V. Mater. Charact. **114**, 103–114 (2016)
18. Heralic, A.: Monitoring, and control of robotized laser metal wire deposition. Doctoral thesis, Chalmers University of Technology (2012)
19. Todaro, C.J., Easton, M.A., Qiu, D., Zhang, D., Bermingham, M.J., Lui, E.W., Brandt, M., StJohn, D.H., Qian, M.: Grain structure control during metal 3D printing by high-intensity ultrasound. Nat. Commun. **11**(1), 142 (2020)
20. Simonelli, M., Tse, Y.Y., Tuck, C.: The formation of $\alpha + \beta$ microstructure in as fabricated selective laser melting of Ti–6Al–4V. J. Mater. Res. **29**(17), 2028–2035 (2014)
21. Xu, W., Brandt, M., Sun, S., Elambasseril, J., Liu, Q., Latham, K., Xia, K., Qian, M.: Additive manufacturing of strong and ductile Ti–6Al–4V by selective laser melting via in situ martensite decomposition. Acta Mater. **85**, 74–84 (2015)
22. Xu, W., Sun, S., Elambasseril, J., Liu, Q., Brandt, M., Qian, M.: Ti-6Al-4V additively manufactured by selective laser melting with superior mechanical properties. JOM. **67**(3), 668–673 (2015)
23. Xu, W., Lui, E.W., Pateras, A., Qian, M., Brandt, M.: In situ tailoring microstructure in additively manufactured Ti-6Al-4V for superior mechanical performance. Acta Mater. **125**, 390–400 (2017)
24. Baykasoglu, C., Akyildiz, O., Candemir, D., Yang, Q., A. C. To: Predicting microstructure evolution during directed energy deposition additive manufacturing of Ti-6Al-4V. J. Manuf. Sci. E. T. ASME. **140**(5), 1–11 (2018)
25. Sun, W., Shan, F., Zong, N., Dong, H., Jing, T.: A simulation and experiment study on phase transformations of Ti-6Al-4V in wire laser additive manufacturing. Mater. Des. **207**, 109843 (2021)
26. Nain, V.: Efficient thermomechanical modeling of large parts fabricated by directed energy deposition additive manufacturing processes. In: Mechanics of Materials [Physics.Class-Ph]. Université de Bretagne Sud (2022)

27. Goldak, J., Chakravarti, A., Bibby, M.: A new finite element model for welding heat sources. Metall. Trans. B. **15**(2), 299–305 (1984)
28. Irwin, J., Reutzel, E.T., Michaelis, P., Keist, J., Nassar, A.R.: Predicting microstructure from thermal history during additive manufacturing for Ti-6Al-4V. ASME J. Manuf. Sci. Eng. **138**(11), 111007 (2016)
29. Guo, N., Leu, M.C.: Additive manufacturing: technology, applications and research needs. Front. Mech. Eng. **8**, 215–243 (2013)
30. Dutta, B., Froes, F.H.: The additive manufacturing (AM) of titanium alloys. Metal Powder Rep. **72**(2), 96–106 (2017)
31. Ding, D., Pan, Z., Cuiuri, D., Li, H.: Wire-feed additive manufacturing of metal components: technologies, developments, and future interest. Int. J. Adv. Manuf. Technol. **81**(1–4), 465–481 (2015)

Computer-Aided Design for AM Processes

Explanation of the Slowing Infill in 3D Printing and the Influence of CAM Patterns

El Bechir Msaddek[1(✉)], Zeinab Hmila[2], Zoubeir Bouaziz[1], and Maher Baili[3]

[1] Laboratory of Applied Fluid Mechanics, Process Engineering and Environment, ENIS, University of Sfax, Sfax, Tunisia
elbechir.msaddek@gmail.com

[2] Department of Mechanical Engineering, Higher Institute of Industrial Systems ISSIG, University of Gabes, Gabes, Tunisia

[3] The Production Engineering Laboratory, ENIT, UTTOP, University of Toulouse, Tarbes, France

Abstract. Unlike subtractive and formatted manufacturing, additive manufacturing, especially rapid prototyping brings together manufacturing processes by adding material to parts from model data, generally layer by layer. There is a lot of slicing software that generates diverse infill tool paths or patterns. Part productivity is reduced by slowdowns that occur during filling, caused by the toolpath strategy containing discontinuities in the tool path. In this study, the dynamic behavior of a printing machine is modeled and a simulation tool, for filling trajectories and actual feed rates, is proposed to optimize the manufacturing process for complex-shaped parts. To do this, analytical models expressing the speed, acceleration, and jerk of a printer are developed. As a result, the diversification of CAM patterns in AM has affected important criteria for the printing operation of complex shapes such as the filling speed with other printing defaults. Certain patterns have penalized the dynamics of the machine, contrary to others.

Keywords: Slowing down · modeling · Simulation · Dynamics · strategies

1 Introduction

Nowadays, Additive Manufacturing (AM) is progressively converted to an important process of manufacturing in the modern industry. It is now able to bring solutions and opportunities in many fields such as medicine, automotive, aeronautics and energy production [1]. As 3D printers become available with affordable prices and online services, there is an accelerated market growth [2]. They work with a concept based on three-dimensional CAD models after slicing from a digital file. Several materials can be used: thermoplastics, ceramics, and metals [3]. Numerous ranges of CAD/CAM systems are applied. Among the software are Esprit, hyperMILL, Fusion 360, Slicer, Magics, Cura, and others. They offer different methods of interpolation and filing strategies or patterns. To produce a 3D functional part, there are various processes with different specificities [4, 5]. Because of the supposed high potential of these technologies, industrial engineers

M. Soula et al. (Eds.): AIAM 2024, LNME, pp. 219–230, 2025.
https://doi.org/10.1007/978-3-031-94889-3_18

conduct massive investigations in this field to anticipate the quality prospects offered by this process. One of the most influenced parameters that degrades the product quality is the CAM cutting method of the tool path [6].

Several research works have detailed some parameters influencing the 3D printing process. Gu et al. [7] have focused in their study on surface textures obtained in 3D printing. They have printed a thermoplastic part with geodesic paths to investigate the influence of the planning path on the surface condition. Zhang et al. [8] have proposed a new program of printing, which applies to Fermat's spiral tool path to realize a continuous filling. Tuli [9] has developed a new approach that includes the tool path parameter and the controller NC code to determine the 3D printing alignment. This operation is influenced by other factors such as the total time, the support material volume, the support contact area, and mechanical properties. Fleming et al. [10] have established an algorithm for post-processor tool path instructions that uses a problem solver to reduce the distance traveled between curves and fill layers, reducing unnecessary extrusion for simple object models on an open-source 3D printer. Jiang et al. [11] have provided a study to improve the strength of 3D-printed concrete by path optimization. Paths include crosswise, vertical, arched, and diagonal patterns. They have shown that arched-path concrete has 30% more flexural strength than others because it makes better use of filament-wise strength. Adame [12] has optimized the infill of 3D parts by improving the path planning by reducing sharp angles and having a continuous path. New algorithms that optimize the infill of an object are created. This was done by examining fill patterns with a continuous path, less sharp turns, and having consecutive layers. However, after reviews, we found that in all the articles presented, clarification of the actual cause of slow printing in sharp corners and other patterns is missing.

Accordingly, the research work carried out in this paper focuses on the modeling of the dynamic behavior of the 3D printer and CAM tool paths of filling (patterns) provided by the CAD/CAM software "Fusion 360©". This study aims to investigate the influence of CAM patterns on the dynamic behavior of the 3D printing machine, to explain the main cause of the excessive reduction in cycle times and the increase in filling errors generated during this manufacturing process.

2 Materials and Methods

Additive manufacturing is a set of processes that makes it possible to manufacture, layer after layer, by adding material, a physical object from a digital object. To achieve this objective, we must follow 4 steps: creation of digital data, preparation of files, manufacturing, and finishing of main parts.

The tool path in 3D printing is defined by the trajectory of the nozzle moved to deposit material. Since the filling quality is influenced by the tool path form, many research works have studied the optimization of CAM patterns. Jin et al. [13–16] have confirmed that the dynamic problem in AM is like the High-Speed Milling (HSM). The part quality can be damaged by the presence of enormous numbers of small segments in the tool path. Also, slowing downs can occur during crossing sharp corners (colored points in Fig. 1). Thus, the average feed rate of the nozzle decreases because of the acceleration and the deceleration. To show that, they have proposed a test deposited path (See Fig. 1), and they have presented a corresponding feed rate profile.

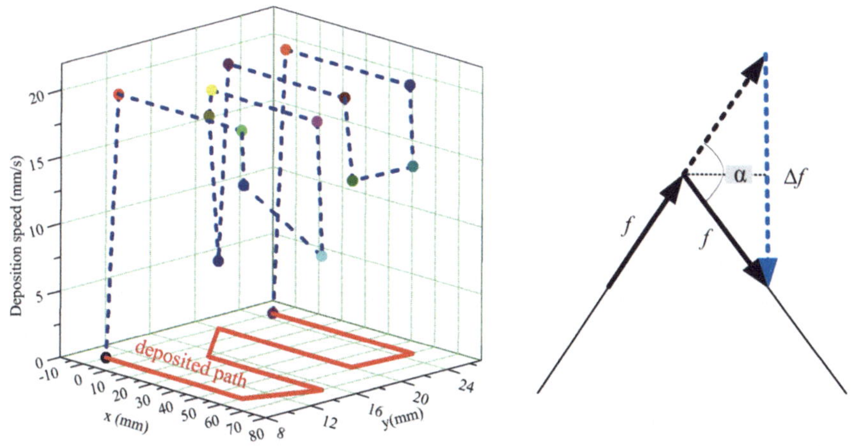

Fig. 1. "Zig zag" tool path and its associate feed rate variation [13, 15].

They have defined the real feed rate as: $\Delta f = 2f * \sin(\alpha/2)$, where f is the feed rate at the corner and α is the angle between the current line segment and the extension of the previous segment, as shown in Fig. 1. However, this equation is very simple and does not well describe the real dynamic of the machine of printing, because it neglects very important parameters like the maximal jerk of the axis. In the following, the dynamic modeling of 3D printing machines with more necessary parameters will be detailed.

2.1 Dynamic Modeling of the 3D Printing Machine

The dynamic modeling of the 3D printing machine consists of the modeling of the interaction between the controller and the axis. After design, the curve generated contains complex forms, sharp angles, and small segments. In CNC filling, quick variations impose decelerations of the axis for the overpass of discontinuities. The NC director incorporates linking arcs to smooth tool paths. The 3D printing processor includes the static look ahead that estimates the feed rate in costume time for the executing block. These limitations have been escaped by the integration of the dynamic look ahead with the anticipation of the next feed rate to avoid the overtaking caused by severe movements. Dynamic look ahead computes the following feed rate with its two deceleration/acceleration stages on the running block. It allows us to calculate kinematic parameters in advance. The modeling of the dynamic behavior of the 3D printing machine is presented in Fig. 2.

The additive manufacturing machine is made up of numerous mechanisms (axis, controller…) having their restrictions. This makes the machine modeling reliant on a large number of factors (smoothing arcs, segment time mini., Jerk maxi…). After modeling, a simulation of actual feed rates and cycle times becomes possible.

The simulation methodology is presented in the following diagram (Fig. 3):

After designing and slicing a workpiece on "Fusion 360©", the NC file is generated containing the coordinates of the points of the tool path of the cutting method (zigzag,

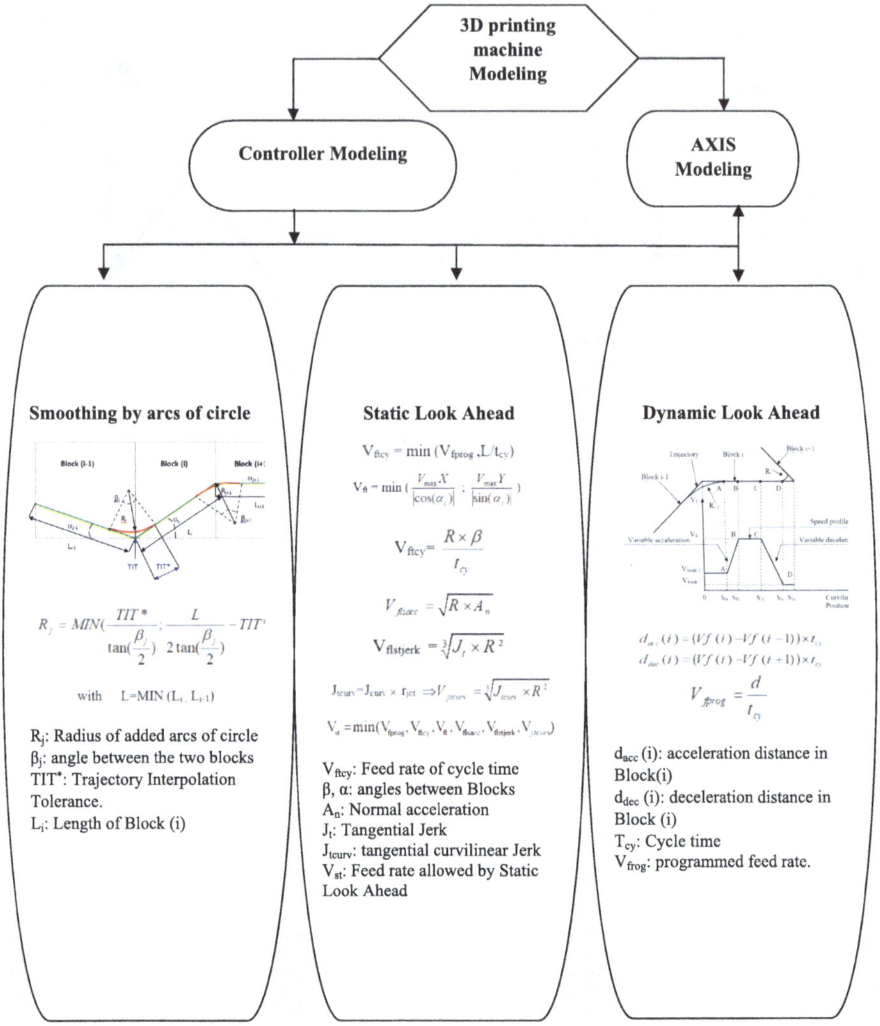

Fig. 2. Dynamic modeling of the 3D printer [17–22].

star, etc.). It is, therefore, necessary to determine the real points of the CAM trajectory as calculated by the controller, so that it can be processed on Matlab©. Thus, the programmed feed speed is not constant during the filling operation. So, modeling the dynamic behavior of the 3D printing machine is essential to determine the actual feed rate allowed by the axis limits and anticipate each trajectory block (look ahead). The influence of pattern types on actual feed rates and cycle times is provided. Finally, the simulation results are interpreted.

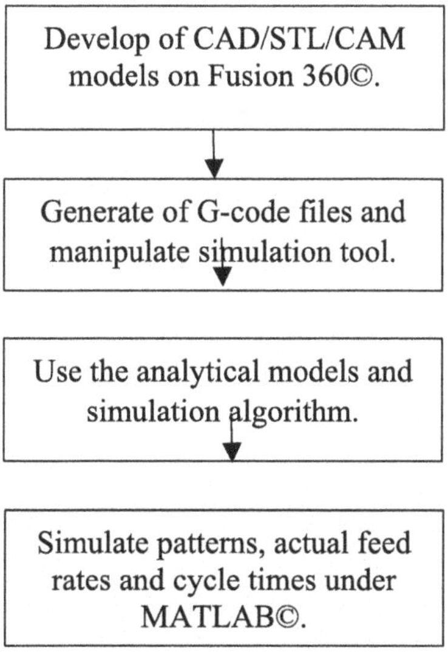

Fig. 3. Simulation steps.

2.2 Influence of Infill CAM Strategies on the Manufacture of Free Forms in 3D Printing

Some filling strategies are favorable for making complex shapes while others are not. Linear interpolation is widely used in CAM strategies provided by Fusion 360© software for the manufacture of several shapes, but it generates remarkable limitations in the manufacture of complex shapes with high precision. Thus, recent research works show that there is an increase in the number of generated CAM discretization nodes. Consequently, the generation of tangency discontinuities causes sudden decelerations [20], vibrations, shocks, and marks on the part. Indeed, there is not enough research carried out on this subject, we have started to study the influence of the type of CAM filling strategy on the cycle time of 3D printing of complex shapes.

To fulfill the stated objectives, a typical part collecting shape variations in the advanced direction of the tool is used. Then, the part printed according to different types of CAM filling strategies provided by Fusion 360© is simulated. To conclude, the influence of each strategy on the CAM cycle time is studied.

Definition of the Workpiece

The designed workpiece (see Fig. 4) is a compound of linear paths with single and combined axes. It contains shapes that can be met in an industrial part. Therefore, interpolations are diversified, which is interesting to accurately evaluate manufacturing errors when filling a complex shape in 3D printing.

Splitting the STL File into Layers and G-code File

After reading the data contained in the STL file, the CAD/CAM software Fusion 360© splits the part into layers of fixed thickness and generates nozzle paths to manufacture

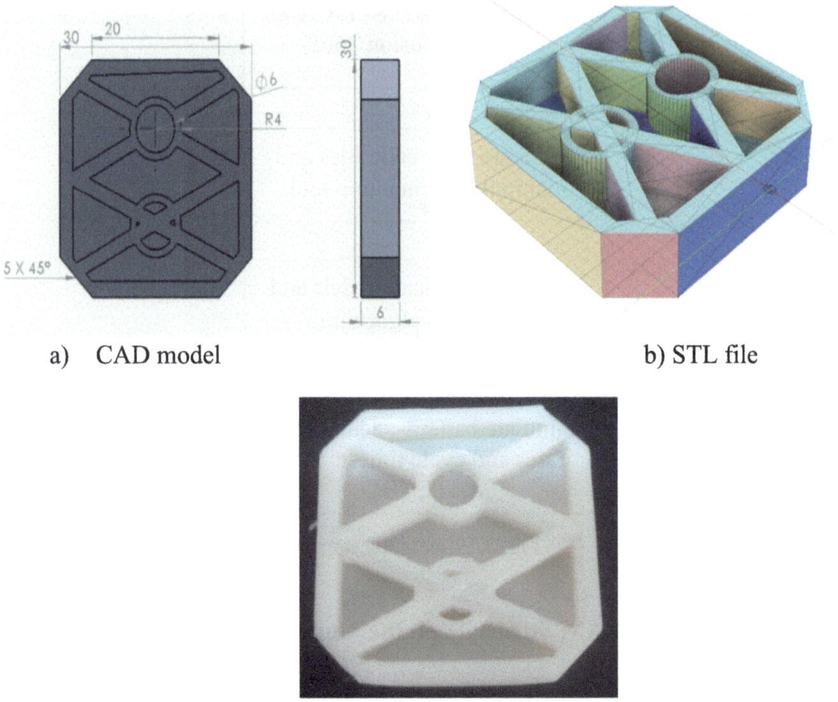

Fig. 4. CAD (a), STL (b), and realized workpiece (c).

the part. Thus, the programmer defines manufacturing parameters, such as the feed rate of the manufacturing head, the height of each layer, the layer-filling strategy, stepover,

3D Printing Parameters

The simulation of the test part consists in printing of the CAD model through the G-code file generated by the Fusion 360© software on the 3D printer Artillery Sidewinder X2. The following Table 1 gives the kinematic characteristics of the Artillery Sidewinder X2 printer.

The objective of simulation tests is to know which filling strategy is the most penalizing the dynamic behavior of the 3D printing machine since there is a risk of having vibrations and overruns during the filling operation. Oscillations can lengthen the printing time as they affect the precision and the quality of the final part. Thus, each strategy generates a different path from the other according to the mathematical formula that describes it. In the following paragraph, simulation results will be presented and studied.

Table 1. Kinematic parameters of the 3D printing machine.

Parameters	Values
Programmed feed rate	60 mm/s
Filament material	PLA (Polylactic acid)
Printing precision	±0.1 mm
Maximum print feed rate	80 mm/s
Maximum print acceleration	3000 mm/s^2
Maximum print jerk	10 mm/s^3
Minimum segment time	20 ms
Curvilinear jerk rate	0.6
Slice thickness	0.1 mm–0.32 mm

3 Results

Autodesk Fusion 360© software offers 13 material deposition strategies for 3D printing machines. We distinguish patterns: Rectilinear, Gyroid, Star, Honeycomb, Squares, Spherical, Rhombic Dodecahedron, Elongate Dodecahedron, Truncated Octahedron, Flower, Rhombus 3D, Rhombille, and Auxetic. To avoid the extension of the paper, we will only present the simulation of 2 types of strategies: Rectilinear and Star, because they are the closest to machining.

Figure 5 presents the simulation and the experiment test of the "Rectilinear" strategy for the material deposition in layer 7 (choice). This is a layer where the process is stabilized, and we are a little far from the plateau.

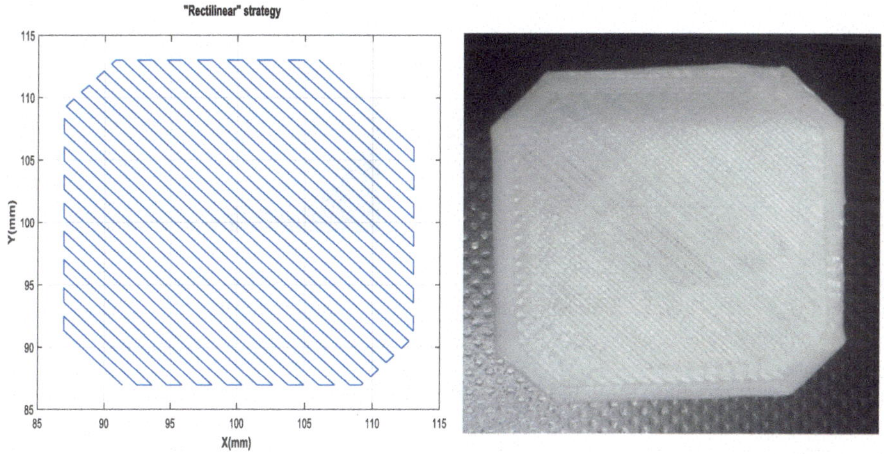

Fig. 5. Simulated and realized "Rectilinear" infill tool path.

Figure 6 shows the simulated feed rate and the cycle time evolution in the "rectilinear" strategy.

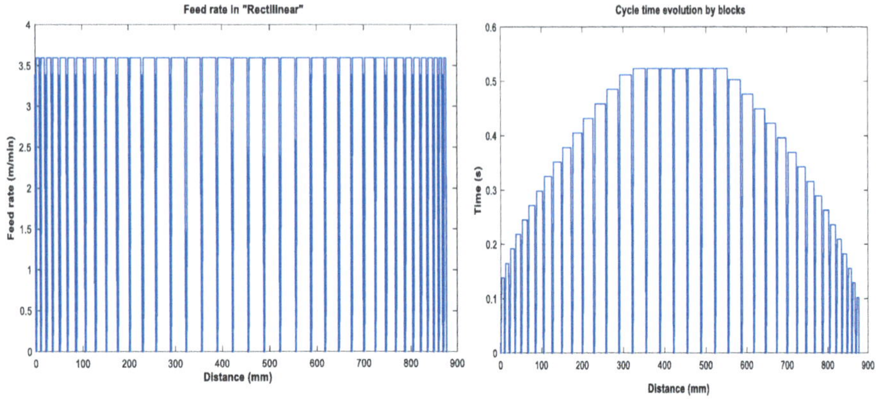

Fig. 6. Simulated feed rate and cycle time evolution in the "Rectilinear" strategy.

Figure 7 presents the simulation and the experiment test of the "Star" strategy for the material deposition in layer 7.

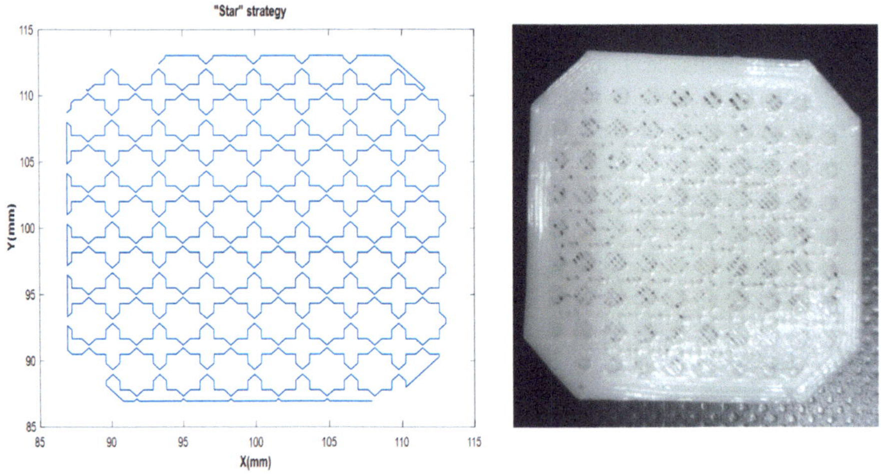

Fig. 7. Simulated and realized "Star" infill tool path.

Figure 8 shows the simulated actual feed rate and the cycle time evolution in the "star" strategy.

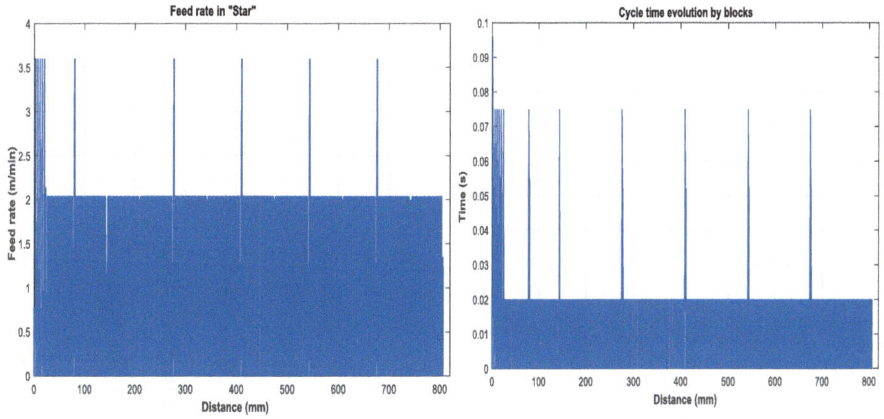

Fig. 8. Simulated feed rate and cycle time evolution by blocks in the "Star" strategy.

4 Discussion

The dynamic look ahead permits us to predict the kinematic parameters of 3D printers. The controller calculates the feed rate profile with its acceleration and deceleration stages on each block, according to the feed rate for the following block. Indeed, with the anticipation, the surpassing produced by the intense slowing downs can be avoided.

Results demonstrate the influence of the dynamic behavior of the 3D printing machine on the feed rate profile and the filling cycle time. By integrating the look ahead, decelerations are identified in advance and are supported on the running block. Feed rate profiles of patterns simulated are very close to the profile presented in Fig. 1. In tangential discontinuities, the feed rate is very near zero. Geometric constraints are due to the crossing of the tool by a small linking radius. More exactly, the feed rate is restricted by the 3D printer's maximal jerk J_{max}. In curvature discontinuities, feed rate restrictions on circular slices are due to the curvilinear tangential Jerk J_{tcurv}. For other portions of tool paths, we have seen that the feed rate continues to diminish if we have small segments or sharp angles. The main causes of the variation are the axis capacities, mainly the maximal jerk and the minimum segment time. Therefore, the controller enforces these limitations by the Look Ahead system. This explains alterations in feed rate profiles and the slowing down phenomena of the 3D printing process.

Thus, the actual cycle time of complex shapes manufacturing in 3D printing will be evaluated. So, we have noticed variations of cycle time profiles by blocks. Afterward, cycle times in Layer 7 (choice) for all patterns are provided by Fusion 360© (See Fig. 9).

The most favorable strategy towards the dynamics of the 3D printing machine Artillery Sidewinder X2 is the "Rectilinear" pattern with tcy = 17.33 s, because of the simplicity of its linear filling paths. The bad strategy is that of the "Rhombus3d" strategy with tcy = 218.59 s because this strategy has a lot of small circular paths which penalizes the dynamics of the machine and slows down the manufacturing operation.

In experimental tests, we have measured the cycle time of only a few critical examples of patterns, because it requires a lot of time on the machine. The measured and simulated cycle time results for layer 7 are shown in Table 2.

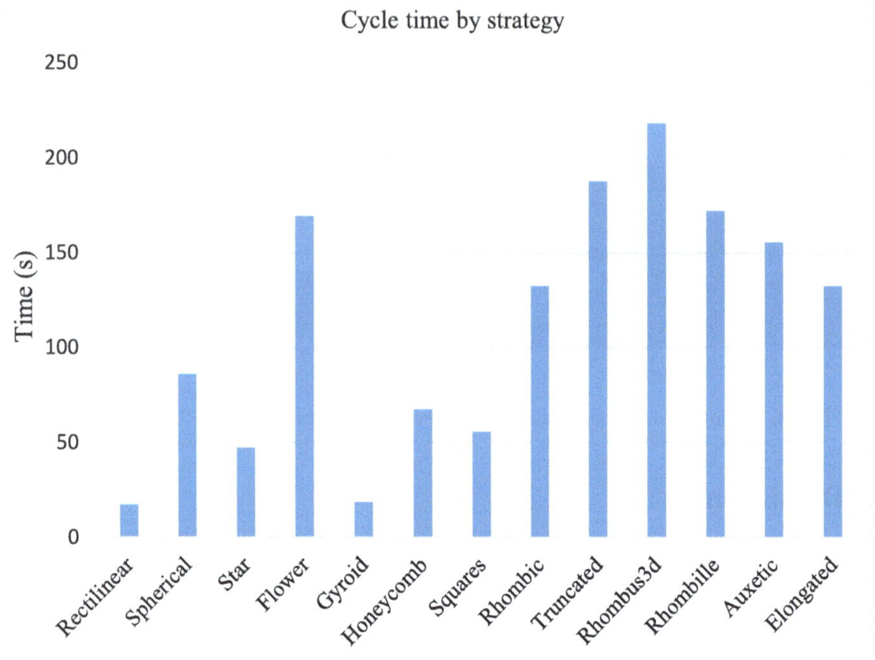

Fig. 9. Cycle time for each type of material deposition strategy.

Table 2. Measured and simulated cycle times of patterns.

	Measured	Simulated
Rectilinear	16.87	17.3324
Flower	61.67	169.3537
Spherical	23.3	86.0299
Star	30.84	47.486

Experimental results have confirmed that the most favorable strategy towards the dynamics of the 3D printing machine is the "Rectilinear" pattern with $t_{cy} = 16.87$ s. Although the strategy Flower makes more time with $t_{cy} = 61.67$ s because this strategy has a lot of short paths and direction changes that hamper the machine dynamics and slow down the manufacturing operation. Thus, the simulation results are close to the experimental measurements with significant gaps for certain CAM strategies, which proves the validity of our analytical model with requirements for deep revisions.

5 Conclusion

The right choice of the pattern type of the 3D printing machine during the material deposition serves to benefit from good manufacturing conditions regarding the dynamics of the machine and the quality of the product. The innovation of this paper is the comprehension of the important slowing down of the 3D printing machine during the filling operation of complex parts. To do this, dynamic modeling is developed considering the CAM tool path planning. First, dynamic modeling is described. After that, some CAM filling strategies provided by Fusion 360© software and their corresponding feed rates and cycle times are simulated and measured. As a result, the "Rectilinear" pattern has provoked a faster filling. Contrary, the "Rhombus3d" strategy has penalized the printing operation and has made it very slow because of large direction changes altering the dynamics of the machine, especially the jerk maximal, the curvilinear tangential Jerk, and the minimum segment time.

Indeed, the great diversification of CAM patterns in AM also affects other important criteria in the printing of complex shapes like the filling orientation and other printing defaults. As perspectives, it remains to ensure the same mechanical resistance to the different parts with a minimum cycle time. So, we must focus on optimizing the part quality produced by this modern process. There are other criteria to consider: topological optimization, surface condition, mechanical characteristics etc.

References

1. Gibson, I., Rosen, D., Stucker, B.: Additive manufacturing technologies: 3D printing. In: Rapid Prototyping, and Direct Digital Manufacturing, 2nd edn. Springer, New York, NY (2015)
2. Rayna, T., Striukova, L.: From rapid prototyping to home fabrication: how 3D printing is changing business model innovation. Technol. Forecast. Soc. Chang. **102**, 214–224 (2016)
3. Dubois, P., Aoussat, A., Duchamp, R.: Prototypage rapide, généralités. Techniques de l'Ingénieur, traité Génie mécanique, pp. 1–10 (2000)
4. Williams, S.W., Martina, F., Addison, A.C., Ding, J., Pardal, G., Colegrove, P.: Wire + arc additive manufacturing. Mater. Sci. Technol. **32**, 641–647 (2015)
5. Ponche, R., Kerbrat, O., Mognol, P., Hascoët, J.-Y.: A novel methodology of design for additive manufacturing applied to additive laser manufacturing process. Robot. Comput. Integr. Manufact. **30**(4), 389–398 (2014)
6. Nuñez, P.J., Rivas, A., García-Plaza, E., Beamud, E., Sanz-Lobera, A.: Dimensional and surface texture characterization in fused deposition modelling (FDM) with ABS plus. MUSIC 2015 Proc. Eng. **132**, 856–863 (2015)
7. Jianzhe, G., Breen, D.E., Jenny, H., Zhu, L., Tao, Y., Van de Zande, T., Wang, G., Zhang, Y.J., Yao, L.: Geodesy, Self-Rising 2.5D Tiles by Printing along 2D Geodesic Closed Path. CHI, Glasgow, Scotland (2019)
8. Zhang, Y., Li, H., Wang, T., Liu, B., Wang, G.: A hybrid tool-path with no pause generation algorithm for 3D printing. J. Phys. Conf. Ser. **1754**, 012222 (2021)
9. Tuli, T.B.: Path planning and simulation of 3D printer machine tool for development of prototypes. In: 3r[d] Annual Research Conference on the theme "Science Technologies and Innovations for Transforming Economies", Addis Ababa, Ethiopia (2017)
10. Fleming, C., Walker, S., Branyan, C., Nicolai, A., Hollinger, G.A., Mengüç, Y.: Toolpath Planning for Continuous Extrusion Additive Manufacturing. Oregon State University (2017)

11. Jiang, X., Li, Y., Yang, Z., Li, Y., Xiong, B.: Harnessing path optimization to enhance the strength of three-dimensional (3D) printed concrete. Buildings. **14**, 455 (2024)
12. Adame, C.D.B.: Improving the path planning and the printing time for an optimized infill of 3D objects by reducing sharp angles and having a continuous path. Master Dissertation, School of Engineering and Sciences, Tlalpan, Ciudad de Mexico (2021)
13. Jin, Y.-a., He, Y., Jian-zhong, F., Gan, W.-f., Lin, Z.-w.: Optimization of tool-path generation for material extrusion-based additive manufacturing technology. Addit. Manuf. **1–4**, 32–47 (2014)
14. Jin, Y., He, Y., Fu, J.Z.: A look-ahead and adaptive speed control algorithm for parametric interpolation. Int. J. Adv. Manuf. Technol. **69**(9–12), 613–2620 (2013)
15. Jin, Y., Jianke, D., Ma, Z., Liu, A., He, Y.: An optimization approach for path planning of high-quality and uniform additive manufacturing. Int. J. Adv. Manuf. Technol. **92**, 651–662 (2017)
16. Jin, Y.-a., He, Y., Jian-Zhong, F., Lin, Z.-w., Gan, W.-f.: A fine-interpolation-based parametric interpolation method with a novel real-time look-ahead algorithm. CAD Comput. Aided Des. **55**, 37–48 (2014)
17. Msaddek, E.B., Bouaziz, Z., Baili, M., Dessein, G.: Influence of interpolation type in high-speed machining (HSM). Int. J. Adv. Manuf. Technol. **72**(1–4), 289–302 (2014)
18. Msaddek, E.B., Baili, M., Bouaziz, Z., Dessein, G.: Influence of the compensation method of machining errors of B-spline and Cspline. Int. J. Comput. Integr. Manuf. **34**(3), 282–292 (2021)
19. Vincent, P., Emmanuel, D., Pascal, R.: Corner optimization for pocket machining. Int J Mach Tool Manu. **44**, 1343–1353 (2004)
20. Msaddek, E.B., Bouaziz, Z., Baili, M., Dessein, G.: Modeling and simulation of high-speed milling centers dynamics. Int. J. Adv. Manuf. Technol. **53**(9–12), 877–888 (2011)
21. Tapie, L., Mawussi, K.B., Anselmetti, B.: Circular tests for HSM machine tools: bore machining application. Int J Mach Tool Manu. **47**, 805–819 (2006)
22. Msaddek, E.B., Bouaziz, Z., Baili, M., Dessein, G.: Optimization of pocket machining strategy in HSM. Int. J. Adv. Manuf. Technol. **62**(1-4), 69–81 (2012)

Design for Additive Manufacturing (DfAM) in the Automobile Industry

Anis Hamza(✉), Kamel Bousnina, Issam Dridi, and Noureddine Ben Yahia

Mechanical, Production and Energy Laboratory (LMPE), National School of Engineering of Tunis (ENSIT), University of Tunis, Avenue Taha Hussein, Montfleury 1008, Tunis, Tunisia
anis7amza@gmail.com

Abstract. Additive design encompasses two primary approaches: adapting existing conventional designs for 3D printing and developing designs specifically for Design for Additive Manufacturing (DfAM). The latter prioritizes two key objectives: assembly consolidation and lightweighting. These designs are particularly suited for biomimicry and complex geometries that utilize negative space optimization, often achieved through advanced techniques such as generative design and topology optimization. The initial phase of the design process considers critical factors including printing processes, material selection, support structures, and subsequent operations like depowing and post-processing. DfAM offers unparalleled capabilities in complexity, customization, and optimization, surpassing traditional subtractive manufacturing methods. Its advantages have been recognized across various industries, notably aerospace, medical devices, and automotive sectors. Despite existing challenges within the field, they are not insurmountable. Continued innovation, dedicated research efforts, and technological advancements will expand the potential of additive manufacturing. This research thoroughly examines and validates DfAM as a transformative manufacturing process within the automotive industry, illustrating its potential to fundamentally redefine vehicle design, production, and perception in today's manufacturing landscape.

Keywords: Additive Manufacturing · Artificial Intelligence · Electric Vehicle · Lightweight Parts · Design for Additive Manufacturing (DfAM) · Automobile

1 Introduction

Additive manufacturing (AM), commonly known as 3D printing represents a transformative processing technique with the potential to revolutionize the production of intricate objects with complex geometries and diverse functionalities. Its application in the automobile sector has garnered significant attention in recent years, marked by considerable progress and implementation [1]. Interestingly, this surge in interest coincides with the automotive industry's pivotal shift toward embracing e-mobility [2]. However, despite the remarkable potential, the field faces constraints due to the lack of advancements in manufacturing technology [3]. The conventional manufacturing technologies currently in use pose limitations on creativity, as they lack abundant automation or machinery.

M. Soula et al. (Eds.): AIAM 2024, LNME, pp. 231–243, 2025.
https://doi.org/10.1007/978-3-031-94889-3_19

While human employees may brainstorm and adapt during problem-solving, machinery operates based on optimized instructions, albeit with inherent complications [4]. Additionally, conventional manufacturing presents challenges such as inferior surface finish, complexities in machining intricate parts, increased tool wear, poor dimensional accuracy, and noisy operations [5]. Automakers are increasingly attracted to advanced fabrication technology because they expect it to reduce manufacturing costs [6]. Consequently, the incorporation of Additive Manufacturing (AM) emerges as a promising avenue for the development of diverse automobile components, offering advantages that extend to the production of specific parts or entire vehicles, ensuring robust and adequate physical qualities [7].

Over the past decade, the automotive industry has witnessed a transformative shift from the meticulous subtraction of clay in traditional vehicle modeling to the rapid adoption of computer-aided design (CAD) data transformed into 3D prototypes [8]. Driven by the volatile economic environment and heightened consumer demands, automakers are actively exploring new prospects and materials to stay abreast of developments in other industries, highlighting the industry's inherent ability to turn necessity into creativity [9]. The advantages provided by 3D printing solutions in the automobile sector are immediately quantifiable in terms of performance parameters. With the capability to supplant expensive and time-consuming traditional CNC manufacturing, the realm of 3D printing empowers the generation of meticulously crafted yet gracefully light components, presenting an abundance of artistic liberty [10]. This culminates in reduced expenses during production, particularly in the realm of designing intricate bodily structures. For instance, 3D-printed plastic parts are not only less expensive but also require less time for in-house production. The on-site manufacturing characteristic of in-house 3D prototyping contributes to the regulation of Intellectual Property (IP) infringements and information leaks [11], offering a solution to minimize turnaround times across all manufacturing phases and enhancing overall agility. In addition to cost-effectiveness, 3D printing in automobile design facilitates decreased material consumption and waste, benefiting all phases of manufacturing [12, 13]. Furthermore, the rising demands for customization underscore the need for automotive manufacturers to embrace additive manufacturing, particularly for low-volume production scenarios [14].

While additive manufacturing (AM) is gaining traction in the automotive sector, existing literature mainly focuses on the fabrication of individual components rather than a comprehensive understanding of AM's overall impact. There is a significant gap in studies that explore how AM integrates into broader manufacturing processes, including design philosophies, supply chain logistics, and business models. The research highlights the need for a detailed analysis of design considerations specific to AM, material innovations, and strategies to overcome integration challenges. This study aims to address these gaps by examining how 3D printing is transforming automotive manufacturing now and in the future, with subsections dedicated to design considerations, materials, and digital transformation strategies for automakers.

2 Design for Additive Manufacturing (DfAM)

Additive manufacturing (AM), or 3D printing, is a transformative technology that creates three-dimensional objects from digital files, gaining traction across various industries due to declining costs and increased capabilities. Its applications range from rapid prototyping to on-demand production, but the full potential of AM is realized only when designs are specifically tailored for its processes, adhering to the principles of Design for Additive Manufacturing (DfAM). This approach allows for significant cost reductions, particularly in low- to medium-volume production, by eliminating the need for expensive tooling and enabling the creation of complex geometries. Industries such as aerospace, consumer products, and medical devices benefit greatly from AM's capabilities for mass customization and efficient spare part manufacturing. As AM technologies continue to evolve, they are reshaping manufacturing practices and paving the way for more efficient and customizable production methods.

2.1 Principles of Design for Additive Manufacturing (DfAM)

Design for Additive Manufacturing (DfAM) involves customizing product designs specifically for 3D printing to maximize the advantages of additive manufacturing technologies. The design phase is crucial, as it requires careful consideration of how parts will be manufactured, leading to stronger, lighter components with improved performance. One of DfAM's key benefits is the ability to consolidate complex assemblies into single parts, simplifying production and reducing costs. DfAM also optimizes support structures and build directions, minimizing material usage and print times. Intelligent design features, such as conformal cooling ducts, enhance functionality and process reliability. To succeed in AM, designers should avoid features like thin walls and intricate channels that complicate printing. Instead, using thicker cross-sections and managing internal cavities can prevent issues during production. Overall, DfAM represents a strategic approach that integrates thoughtful design considerations from the outset to improve the efficiency and quality of additive manufacturing processes. To enhance additive manufacturing (AM) efficiency, it is important to minimize overhanging geometries that require support structures, as these complicate post-processing. Optimizing designs by using rounded edges instead of sharp corners can improve print quality (see Fig. 1). Following minimum feature size guidelines ensures components are within the AM system's capabilities, contributing to better overall quality. Additionally, simplifying complex surface finishes is essential since AM has limitations in replicating intricate details compared to traditional methods. By adhering to these design principles, the effectiveness and quality of the additive manufacturing process can be significantly improved, leading to superior final products.

Topology optimization is a design approach that uses simulation software to enhance material distribution within a specified space, aiming to create lighter structures while maintaining optimal mechanical performance. This process systematically removes unnecessary material, preserving essential load paths to minimize mass and improve strength (see Fig. 2). It is particularly beneficial for additive manufacturing (AM), as

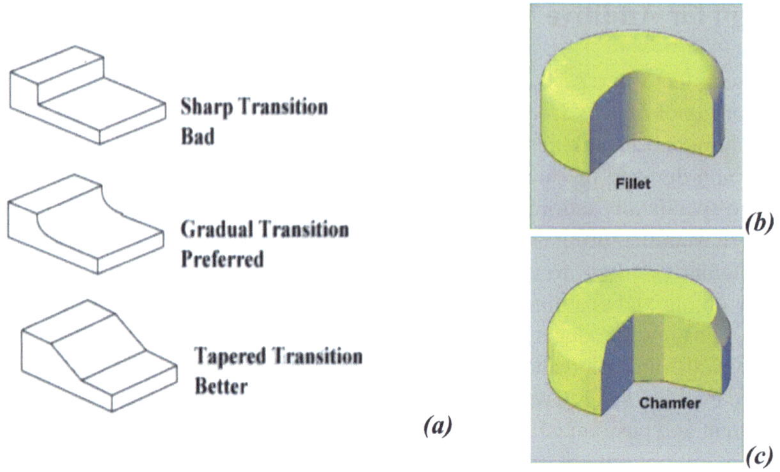

Fig. 1. Illustrations of Design Simplicity (**a** Using DfAM Transitions should be tapered rather than sharp, (**b**) Fillets should have rounded corners for load bearing, and (**c**) Chamfers should have slanted edges facilitating assembly.

it allows for the creation of complex geometries that are difficult to achieve with traditional methods. By leveraging the flexibility of AM, designers can produce efficient load-bearing shapes, maximizing performance while minimizing material usage.

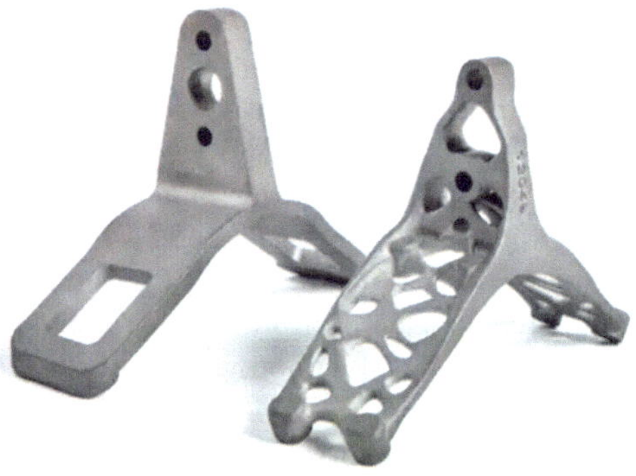

Fig. 2. Examples of Design Topology Optimization: The final ENDY additive bracket part (left) weighs about 40% less than the original (right), weighing in at 152 g instead of 250 g [15].

Accomplishing Portion Union through Solid Plan includes engineers recognizing components that would customarily exist as partitioned pieces and consistently joining them inside the show. This all-encompassing approach permits for the consolidation of complex inner structures and pathways that would be unreasonable or inconceivable to make utilizing routine strategies (see Fig. 3). Within the domain of moving congregations, the required person pieces to be joined are forestalled, as the complete component can be coordinated into a single printed substance. Interfacing that was once dependent on welds or clasps presently consistently changes into morphological moves inside the bound-together body geometry (see Fig. 4). A striking advantage of this approach lies within the exceptional decrease of back structures, accomplished by killing overhanging geometries through portion solidification.

Fig. 3. Examples of Conformal Cooling Channel [16].

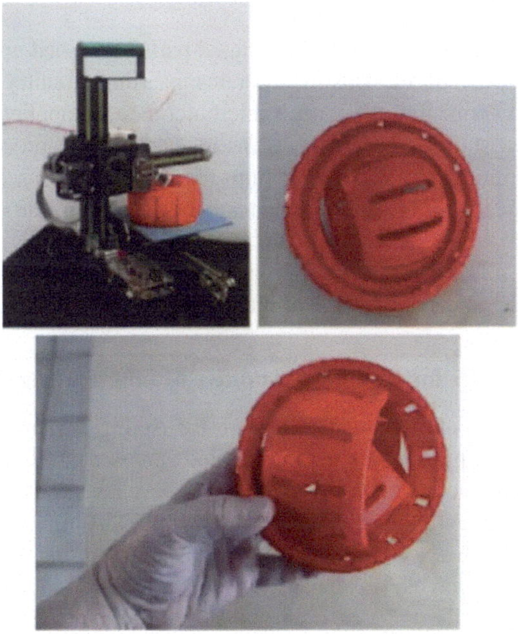

Fig. 4. Examples of Print moveable parts as a single piece [17].

Post-processing efforts are minimized with monolithic designs, as they eliminate the need for assembly and additional finishing work, simplifying inspection and repair.

This approach enhances manufacturing efficiency and the robustness of the final product. Functional integration in 3D printing (see Fig. 5) includes strategies like embedding conduits for cooling or fluid management, which optimize performance. Lattice structures also contribute to multifunctionality in applications such as heat exchangers and biomedical implants. Overall, these design strategies improve the utility of printed parts, allowing for customization and tailored geometries that traditional manufacturing cannot achieve.

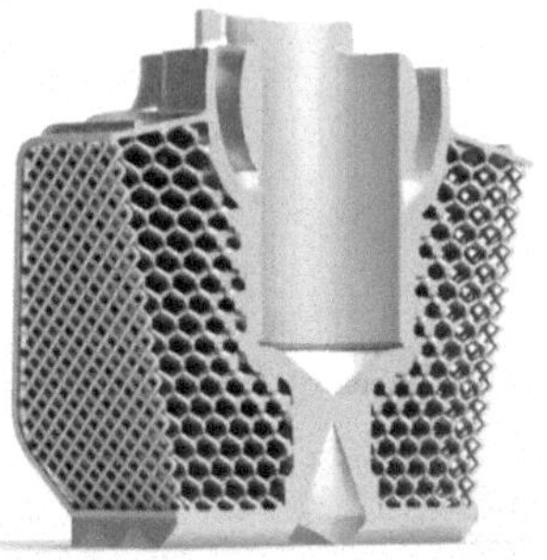

Fig. 5. Examples of Lattice Structured Heat Exchanger [16].

The integration of hardware inside 3D printed parts has ended up conceivable with progressed printing procedures. This incorporates inserting circuitry and wiring inside the printed structures, opening up conceivable outcomes for applications such as IoT sensors and keen prosthetics. Moreover, plans made with post-processing in intellect contribute to useful integration. Contemplations for afterward forms, such as machining inner strings or connection focuses, are joined into the beginning plan to include mechanical capacities consistently. In outline, accomplishing useful integration through 3D printing includes an all-encompassing approach that envelops assorted plan components to optimize the utility and execution of printed parts over different applications.

2.2 Tools and Software for Designing Additive Manufacturing (DfAM)

Generative design tools are essential in Design for Additive Manufacturing (DfAM), allowing engineers to input specific parameters and generate multiple design iterations. Tools like Autodesk Generative Design and Altair Rouse leverage advanced algorithms for optimization, enhancing the design workflow with real-time feedback. These tools exemplify the capabilities available to designers in the additive manufacturing landscape, enabling the creation of efficient and innovative structures tailored for 3D printing processes.

Topology optimization tools are essential in additive manufacturing, using advanced software to transform solid materials into efficient geometric designs based on specified load and boundary conditions. These programs, such as SIMULIA Tosca and Altair OptiStruct, analyze components to create optimized cellular structures that enhance performance and efficiency. They seamlessly integrate with finite element analysis (FEA) tools, providing comprehensive solutions for structural optimization. These technologies are widely utilized in industries like aerospace and automotive, where optimizing component performance is critical. Overall, topology optimization represents a cutting-edge solution for engineers looking to improve the functionality of components across various sectors.

Support structure tools are essential in additive manufacturing, optimizing 3D printing models effectively. Notable tools like Polydevs provide comprehensive functionalities for viewing, editing, and analyzing models, while also auto-generating support structures tailored to various geometries. Ultimaker Cura is another key tool known for its user-friendly interface, allowing users to compare factors like material usage and print time. Together, these tools enhance the 3D printing process, ensuring the successful creation of complex models with optimal support and streamlined post-processing.

2.3 Challenges and Limitations in Designing for Additive Manufacturing

There is ongoing debate about whether Design for Additive Manufacturing (DfAM) offers design flexibility or imposes limitations on engineers, as they must work within the constraints of traditional additive manufacturing systems. While challenges such as material properties, print resolution, and post-processing requirements exist, efforts are being made to address these issues through research and development of advanced materials and optimized printing parameters. Current material options for 3D printing are limited compared to conventional methods, particularly for high-strength applications. Additionally, the high costs associated with additive manufacturing equipment and materials pose challenges for cost-effective production. Companies are working to make these technologies more accessible and affordable, ultimately aiming to enhance the capabilities of additive manufacturing in various industries.

3 Design for Additive Manufacturing (DfAM) in the Automobile Industry

Additive design encompasses two main approaches: modifying existing designs for 3D printing and starting from a Design for Additive Manufacturing (DfAM) perspective. The primary goals of these designs are assembly consolidation and lightweight, which leverage the advantages of 3D printing technology (See Fig. 6). Additively manufactured components often draw inspiration from nature through biomimicry and incorporate complex geometries that optimize material use. Techniques like generative design and topology optimization enhance structural efficiency while allowing for intricate shapes. These strategies improve manufacturing processes, particularly in industries such as aerospace and automotive, where reducing weight is crucial for performance and fuel efficiency. Overall, DfAM enables innovative designs that maximize the capabilities of additive manufacturing.

The early stages of the additive manufacturing design process require careful attention to factors such as printing technology, material compatibility, and support structures. This phase also includes downstream processes like depowering and post-processing, which are vital for refining the final product. Additive design can either adapt existing

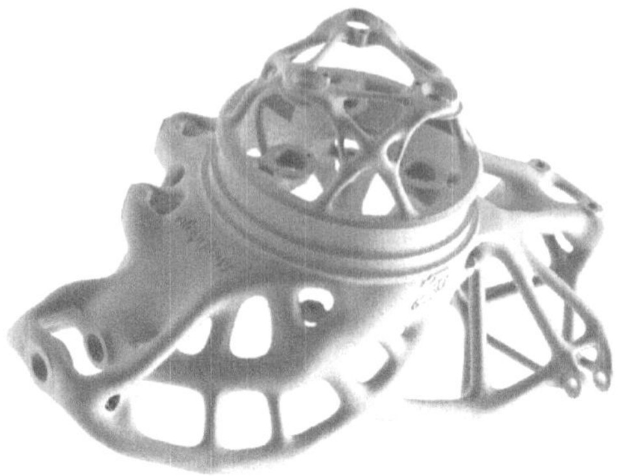

Fig. 6. Lightweight and organic-looking topology-optimized automotive upright bracket designed by Lions Racing Team eV [15].

designs for 3D printing or originate from a DfAM perspective, focusing on assembly consolidation and lightweight through advanced techniques like biomimicry and intricate geometries. As competitive market demands rise, innovative design methodologies, such as Design for X (DfX), have emerged to enhance product development and manufacturing efficiency. The X in DfX signifies a specific aspect targeted for improvement throughout the design process [18]. This strategic approach analyzes the entire product lifecycle from the early stages of design to achieve defined objectives, aiming to encapsulate essential knowledge about the process, materials, and product during the design phase [18].

Within the realm of design strategies, Design for Additive Manufacturing (DfAM) represents the application of DfX principles to the additive manufacturing (AM) process, aspiring to fully exploit the capabilities of AM technology for design purposes [19]. DfAM encompasses a suite of tools oriented toward AM design methodologies, with three prevalent methodological approaches: the opportunistic approach, the restricted approach, and the dual approach [20–22]. The opportunistic approach places no limits on the complexity of AM, while the restricted approach considers boundaries related to materials, performance, machine characteristics, and manufacturability. The dual approach combines both opportunistic and restrictive techniques. In the automotive industry, DfAM approaches are examined and reintroduced through two primary methods: Component DfAM and Assembly DfAM.

In essence, DfAM serves as a generalized design strategy or tool that enables the optimization of functional performance and critical product lifecycle aspects, such as manufacturability, cost, and reliability, within the constraints of additive manufacturing technology, particularly tailored for the automobile industry [23]. This approach persists due to the substantial design freedom offered by AM technology, and DfAM methods are essential to fully harness the specific characteristics of AM processes [24].

Typical DfAM methods or tools include topology optimization (determining the best layout of structural components in a vehicle), design for multiscale structures (incorporating lattice or cellular structures), multi-material design (enabling optimal material choices in each structural component for enhanced performance and cost-effectiveness), mass customization, part consolidation, and other design processes that benefit from AM-enabled features [25]. It is crucial to note that DfAM will applied to various 3D printing technologies in diverse ways, as different technologies require distinct optimization strategies. DfAM stands as a well-deployed additive solution viewed as a new tool to enhance manufacturing functionality and cost. For Design for Additive Manufacturing (DfAM) to be effective in the automotive sector, it requires a reciprocal relationship between management and operations. Simply establishing a budget for production and starting the printing process often lacks the necessary feedback for optimization. A successful DfAM implementation should facilitate communication and collaboration throughout the workflow, ensuring that both management insights and operational realities inform the design process. This two-way approach allows automakers to fully leverage DfAM's potential, as outlined in Table 1, to enhance production efficiency and product quality.

Table 1. Basic principles of DfAM in line with the automotive sector

Principles of DfAM	**Resource reduction**			**Quality enhancement**		
	Weight & cost	**Printing time**	**Touch time**	**Strength**	**Aesthetic**	**Functionality**
Additive thinking	■	■		■	■	
Design for orientation		■	■	■	■	
Contour design	■			■		
Segment parts	■	■				
Adding hardware			■	■	■	■
Minimizing complication	■	■	■		■	
Critical surface treatment						■

The extensive production durations, post-processing requirements, and surface quality constraints associated with Additive Manufacturing (AM) have traditionally limited its integration into mainstream automotive manufacturing, often confining it to the prototype stage. Addressing these limitations requires substantial advancements, and recent breakthroughs in Design for Additive Manufacturing (DfAM) have displayed significant potential [14]. Combining topology optimization with an offset and a build time limit emerges as a promising approach to maximize the benefits of AM while mitigating concerns related to support material wastage and extended print times.

A case study illustrates the practical application of this approach in the design of a high-performance vehicle upright, highlighting the efficacy of topology optimization [14]. Topology optimization and the development of lattice structures represent DfAM approaches aimed at reducing material usage in automotive components with high stiffness requirements. To further optimize printing time and material volume, DfAM methods such as design by direct assembly, build orientation, and support reduction are employed. These DfAM technologies, particularly well suited for extrusion-based additive manufacturing techniques, empower design engineers to create lightweight components for diverse applications, spanning automotive, aerospace, and electronics [26].

While AM techniques in traditional automobile manufacturing struggle to match the production rates achieved by conventional methods like injection molding die casting, or stamping, there is a notable potential to leverage AM for building the necessary tools for these processes. Traditional tooling involves significant investments due to the complexity and time required for machining durable materials. However, certain AM processes present a promising avenue for efficiently developing various types of molds and dies at a reduced cost and within a shorter timeframe [27].

Topology optimization has garnered considerable interest for its inherent weight savings potential during the design phase. Numerous instances, including vehicle frames/chassis, suspensions, engine support, transmission, cross-car beams, doors, engine cradles, brake calipers, etc., have been explored for their optimization using topology optimization approaches [28–35]. The Solid Isotropic Material Penalization scheme (SIMP) is one of the widely utilized topology optimization methods, demonstrated in optimizing the Body-In-White (BIW) of an electric vehicle [36]. This approach, like others, aims to reduce compliance (increase stiffness) while adhering to constraints such as mass or volume. Another study displayed the optimization of a vehicle's spring lower seat using the SIMP method, effectively reducing mass while maintaining strength through a combination of topology and topography optimization approaches [37]. The hardware in 3D printers has evolved to the point where it is no longer the bottleneck of performance; rather, the part's shape often becomes the constraining element in its overall performance. Consequently, a significant research debate on DfAM has emerged [38]. While the design requirements for various 3D printing technologies serve as a crucial starting point for DfAM, they do not encapsulate the entire picture, a fact that is not widely reported. DfAM comprises three major concepts assembly and sub-assembly part consolidation, geometric freedom, and biomimicry with considerable overlap between them. These concepts collectively contribute to unlocking the full potential enabled by 3D printing technology in the realm of additive manufacturing for the automotive sector.

4 Materials for AM in Automobile

Design for Additive Manufacturing (DfAM) enables the selection of optimal materials to meet specific performance requirements. The market continuously introduces new additive manufacturing (AM) materials with precise chemical and physical properties, which can pose challenges for traditional machining but can be effectively evaluated through rapid prototyping. Recent advancements in materials technology have led to the availability of certified high-performance polymers, carbon-fiber-reinforced thermoplastics, and metals for automotive parts. Original Equipment Manufacturers (OEMs) are increasingly using lightweight materials like carbon fiber and aluminum to enhance vehicle performance and facilitate the creation of functional prototypes. The use of adapted materials is crucial for implementing AM in demanding sectors such as automotive manufacturing, where high-performance characteristics are essential. Commonly utilized 3D printing materials include polypropylene, which is used for interior components, and Polyamide 6 (PA6), known for its heat resistance and strength, making it suitable for various automotive applications.

5 Discussion

To fully harness the potential of additive manufacturing, it is essential to incorporate Design for Additive Manufacturing (DfAM) from the very beginning of the design process. Historically, 3D printing was viewed primarily as a method for rapid prototyping or producing spare parts with minimal changes to existing designs. However, additive manufacturing has advanced significantly and is now capable of producing critical components that meet or exceed stringent engineering standards. This progress can be achieved by leveraging the unique advantages of this technology. Skilled engineers specializing in DfAM can use various strategies to enhance product ergonomics and aesthetics, particularly in consumer goods, while also improving performance and functionality in sectors like aerospace. Incorporating specific examples of successful DfAM applications across industries would further illustrate its effectiveness, along with discussing the challenges and considerations involved in adopting these practices for a more comprehensive understanding of its implementation.

Based on our research in the application of artificial intelligence in the field of mechanical engineering [39], the integration of artificial intelligence (AI) into Design for Additive Manufacturing (DfAM) presents exciting possibilities for the automotive industry. AI can enhance the design process by enabling advanced generative design algorithms [40] that analyze vast datasets to produce optimized geometries tailored for additive manufacturing. This capability [41] allows designers to explore a wider range of design alternatives that might not be immediately apparent through traditional methods. Additionally, AI can facilitate real-time decision-making during the design phase by predicting performance outcomes based on simulated conditions, thus reducing the need for extensive physical prototyping [42]. As this research field evolves, the collaboration between AI and DfAM is expected to redefine automotive manufacturing, fostering innovation in lightweight structures and complex assemblies while enhancing overall efficiency and sustainability in production practices.

6 Conclusion

In conclusion, Design for Additive Manufacturing (DfAM) provides a range of capabilities that exceed those of traditional subtractive methods, particularly in complexity, customization, and optimization. These advantages have been validated through successful applications in industries such as aerospace, medical devices, and automotive. While challenges remain, they are not fundamental barriers; rather, they can be addressed through ongoing innovation, research, and technological advancements. By continually exploring new possibilities, the potential of additive manufacturing will expand, revolutionizing product design and manufacturing processes. Furthermore, DfAM represents a transformative approach in the automotive sector, shifting additive manufacturing from merely a rapid prototyping tool to a viable production method. Specific examples of innovations achieved through DfAM could further illustrate its impact and encourage industry stakeholders to adopt this paradigm shift.

References

1. Thomas, D.: Costs, benefits, and adoption of additive manufacturing: a supply chain perspective. Int. J. Adv. Manuf. Technol. **85**(5), 1857–1876 (2016)
2. Maddikunta, P.K.R., Pham, Q.-V., Prabadevi, B., Deepa, N., Dev, K., Gadekallu, T.R., Ruby, R., Liyanage, M.: Industry 5.0: a survey on enabling technologies and potential applications. J. Ind. Inform. Integr. **26**, 100257 (2022)
3. Sarfraz, M.S., Hong, H., Kim, S.S.: Recent developments in the manufacturing technologies of composite components and their cost-effectiveness in the automotive industry: a review study. Compos. Struct. **266**, 113864 (2021)
4. Angelopoulos, A., Michailidis, E.T., Nomikos, N., Trakadas, P., Hatziefremidis, A., Voliotis, S., Zahariadis, T.: Tackling faults in the industry 4.0 era—a survey of machine-learning solutions and key aspects. Sensors. **20**(1), 109 (2020)
5. Meng, L., Zhang, W., Quan, D., Shi, G., Tang, L., Hou, Y., Breitkopf, P., Zhu, J., Gao, T.: From topology optimization design to additive manufacturing: today's success and tomorrow's roadmap. Arch. Computat. Methods Eng. **27**(3), 805–830 (2020)
6. Singh, C.D., Khamba, J.S.: Manufacturing Competency and Strategic Success in the Automobile Industry. CRC Press (2019)
7. Lin, D., Lee, C.K.M., Lau, H., Yang, Y.: Strategic response to industry 4.0: an empirical investigation on the Chinese automotive industry. Ind. Manag. Data Syst. **118**(3), 589–605 (2018)
8. Alhamad, I.M., Ahmed, W.K., Ali, H.Z., AlJassmi, H.: 3D Printing Applications in Mechanical Engineering Education. Brill (2019)
9. Stoycheva, S., Marchese, D., Paul, C., Padoan, S., Juhmani, A., Linkov, I.: Multi-criteria decision analysis framework for sustainable manufacturing in the automotive industry. J. Clean. Prod. **187**, 257–272 (2018)
10. Lecklider, T.: 3D printing drives automotive innovation. EE Evaluat. Eng. **56**(1), 16–20 (2017)
11. Singh, O.P., Ahmed, M., Sheikh, and Abhilash, M.: Modern 3D printing technologies: future trends and developments. Rec. Pat. Eng. **9**(2), 91–103 (2015)
12. Attaran, M.: The rise of 3-D printing: the advantages of additive manufacturing over traditional manufacturing. Bus. Horiz. **60**(5), 677–688 (2017)
13. Attaran, M.: Additive manufacturing: the Most promising technology to Alter the supply chain and logistics. J. Serv. Sci. Manag. **10**(03), 189–206 (2017)
14. Jankovic, D., Barari, A.: Customization of automotive structural components using additive manufacturing and topology optimization. IFAC-PapersOnLine. **52**(10), 212–217 (2019)
15. Advancing eVTOL with additive. Features. (18 November 2022) https://www.aero-mag.com/advancing-evtol-with-additive
16. Culp, E.: A closer look at the advantages of conformal cooling. Technology. (10 May 2013) https://www.etmm-online.com/a-closer-look-at-the-advantages-of-conformal-cooling-a-404359
17. Riad, K.: Photocuring epoxy with quantum dots for 3D printing, Thesis for: M.Sc Individualized Program (Engineering)Advisor: Paula M. Wood-Adams, Rolf Wuthrich and Jerome Claverie, February 2016
18. Briard, T., Segonds, F., Zamariola, N.: G-DfAM: a methodological proposal of generative design for additive manufacturing in the automotive industry. Int. J. Interact. Des. Manuf. **14**(3), 875–886 (2020)
19. Vaneker, T., Bernard, A., Moroni, G., Gibson, I., Zhang, Y.: Design for additive manufacturing: framework and methodology. CIRP Ann. **69**(2), 578–599 (2020)
20. Markus, D., Lorin, A., Thomas, N., Sven, M.: Identifying an opportunistic method in design for manufacturing: an experimental study on successful manufacturability and manufacturing effort of design concepts. Proc. CIRP. **100**, 720–725 (2021)

21. Prabhu, R., Simpson, T.W., Miller, S.R., Meisel, N.A.: Fresh in my mind! Investigating the effects of the order of presenting an opportunistic and restrictive design for additive manufacturing content on students' creativity. J. Eng. Des. **32**(4), 187–212 (2021)
22. Prabhu, R., Leguarda, R.L., Miller, S.R., Simpson, T.W., Meisel, N.A.: Favoring complexity: a mixed methods exploration of factors that influence concept selection when designing for additive manufacturing. J. Mech. Des. **143**(10) (2021)
23. Orquéra, M., Campocasso, S., Millet, D.: Design for Additive Manufacturing Method for a mechanical system downsizing. Proc. CIRP. **60**, 223–228 (2017)
24. Kumke, M., Watschke, H., Vietor, T.: A new methodological framework for design for additive manufacturing. Virtual Phys. Prototyp. **11**(1), 3–19 (2016)
25. Jung, Y., Lim, S., Kim, J., Min, S.: Lightweight design of electric bus roof structure using multi-material topology optimization. Struct. Multidisc. Optim. **61**(3), 1273–1285 (2020)
26. Sakthivel, M.R., Vinodh, S.: Application of design for additive manufacturing to an automotive component. In: Industry 4.0 and Advanced Manufacturing, pp. 169–183 (2021)
27. Mantovani, S., Barbieri, S., Giacopini, M., Croce, A., Sola, A., Bassoli, E.: Synergy between topology optimization and additive manufacturing in the automotive field. Proc. Inst. Mech. Eng. B J. Eng. Manuf. **235**(3), 555–567 (2021)
28. Chuang, C.-H., Chen, S., Yang, R.-J., Vogiatzis, P.: Topology optimization with additive manufacturing consideration for vehicle load path development. Int. J. Numer. Methods Eng. **113**(8), 1434–1445 (2018)
29. Tang, L., Wu, J., Liu, J., Jiang, C., Shangguan, W.-B.: Topology optimization and performance calculation for control arms of a suspension. Adv. Mech. Eng. **6**, 734568 (2014)
30. Marchesi, T.R., Lahuerta, R.D., Silva, E.C.N., Tsuzuki, M.S.G., Martins, T.C., Barari, A., Wood, I.: Topologically optimized diesel engine support manufactured with additive manufacturing. IFAC-PapersOnLine. **48**(3), 2333–2338 (2015)
31. Liang, M., Hu, J., Li, S., Chen, Z.: Topology optimization of transmission gearbox under multiple working loads. Adv. Mech. Eng. **10**(11) (2018)
32. Li, C., Kim, I.Y.: Topology, size, and shape optimization of an automotive cross car beam. Proc. Inst. Mech. Eng. Part D J. Automobile Eng. **229**(10), 1361–1378 (2015)
33. Li, G., Xu, F., Huang, X., Sun, G.: Topology optimization of an automotive tailor-welded blank door. J. Mech. Des. **137**(5) (2015)
34. Li, C., Kim, I.Y., Jeswiet, J.: Conceptual and detailed design of an automotive engine cradle by using topology, shape, and size optimization. Struct. Multidisc. Optim. **51**(2), 547–564 (2015)
35. Tyflopoulos, E., Lien, M., Steinert, M.: Optimization of brake calipers using topology optimization for additive manufacturing. Appl. Sci. **11**(4), 1437 (2021)
36. Yang, S., Qi, C., Hu, P., Wei, Z.Y., Wang, Y.L.: Topology optimization of electric vehicle body in white based on SIMP method. Adv. Mater. Res. **308–310**, 606–609 (2011)
37. Kong, Y., Abdullah, S., Omar, M., Haris, S.M.: Topological and topographical optimization of automotive spring lower seat. Latin Am. J. Solids Struct. **13**(07), 1388–1405 (2016)
38. Scott, C.: 3D Printing an Improved DMLS Automotive Component Using Topology Optimization and DfAM. 3DPrint.com (2018)
39. Dridi, I., Hamza, A., Ben Yahia, N.: Control of an active suspension system based on long short-term memory (LSTM) learning. Adv. Mech. Eng. **15**(2) (2023)
40. Dridi, I., Hamza, A., Ben Yahia, N.: A new approach to controlling an active suspension system based on reinforcement learning. Adv. Mech. Eng. **15**(6) (2023)
41. Bousnina, K., Hamza, A., Ben, Y.N.: An integration of PSO-ANN and ANFIS hybrid models to predict surface quality, cost, and energy (QCE) during milling of alloy 2017A. J. Eng. Res. (2023) ISSN 2307-1877
42. Bousnina, K., Hamza, A.: Reducing the energy consumed and increasing energy efficiency in the turning process. Int. J. Mod. Manuf. Technol. **12**(2), 23–28 (2020)

Investigating Compensation Techniques for Extruded Microstructured Optical Fibers Preforms Using 3D Printed Die

Manel Abda[1], Mohamed Trabelssi[2(✉)], and Noureddine Ben Yahia[1]

[1] Applied Mechanics and Systems Research Laboratory (LASMAP), ENSIT The Higher National Engineering School of Tunis, 05 Ave Taha Hussein, Tunis, Tunisia

[2] Applied Mechanics and Systems Research Laboratory, Tunisia Polytechnic School, University of Carthage, B.P. 743, La Marsa 2078, Tunisia

mohamed.trabelsi@ept.ucar.tn

Abstract. This manuscript aims to investigate the use of 3D-printed dies to improve the quality of Microstructured Optical Fibers (MOFs) extruded preforms. Therefore, two compensation procedures are introduced to adjust the geometry of the die based on the different deformation aspects of the preform. The first method, the direct method, is based on a conventional approach to adjusting the die geometry. The second method, the indirect method, compensates for each aspect of deformation separately. Using quality tools from literature (Disjoint Area Percentage (DAP), Distance Standard Deviation (DSD)), the direct method achieved a reduction of up to 13% and 5.66% in drift and distortion respectively slightly outperforming the indirect method. These outcomes highlight the potential of this compensation procedure to enhance fiber quality, minimize signal loss, and optimize manufacturing processes, with promising implications for applications in telecommunications and fiber optics technology.

Keywords: Microstructured optical fiber · extrusion · glass preform · 3D printed dies · compensation · deformation analysis · quality control

1 Introduction

Unlike conventional optical fibers that are generally made of a core and a cladding, Microstructured optical fibers are made from a single material [1, 2]. This constitutes a significant advantage compared to regular fiber. Conventional fibers typically face limitations related to thermal, chemical, and optical compatibility between the core and cladding glass types. Therefore, Microstructured Optical Fibers (MOFs) technology has emerged as a thriving area of research in the advancement of future optical fibers [1, 3]. The MOFs are processed in two steps [4]: first, a preform with the desired geometrical configuration is manufactured and then, drawn into a micro-scaled fiber. The performance of MOFs heavily relies on accurate geometric features and cross-sectional structures. However, fabricating these fibers presents significant challenges. Several MOF

M. Soula et al. (Eds.): AIAM 2024, LNME, pp. 244–259, 2025.
https://doi.org/10.1007/978-3-031-94889-3_20

manufacturing techniques, such as stacking [5], drilling [6], and casting [7], have limitations in achieving the desired final cross-sectional configuration in terms of geometry and surface quality [1, 5]. Extrusion is a promising alternative technique that's been shown to be a flexible reproducible single-step approach to polymer and soft glass preforms fabrication [1, 6, 7]. However, the preform extrusion technique does not yield precise control over the final preform geometry [1, 8]. Such limitations can generally be traced to the die-swell phenomenon. Die swell refers to the tendency of the extruded material to expand beyond the die opening diameter resulting in distortions and periodicity artifacts in the preform structure. This expansion occurs due to the complex flow pattern at the die exit and typically requires multiple adjustments to the die and extrusion parameters to achieve the desired preform shape. Minor optimizations, such as adjusting the feed holes' sizes, can alleviate these distortions.

Lately, the 3D printing technique has been used to manufacture extrusion die offering an innovative approach to overcome these aforementioned challenges [10]. This technique allows more sophisticated optimizations to be conducted. For instance, modifying the die pins to slightly non-circular shapes, which is hard to achieve with conventional manufacturing methods, can now be considered thanks to additive manufacturing. This flexibility has made the extrusion process more adaptable, motivating the development of compensation techniques to further optimize preform shapes [12].

The compensation concept originates from molding, casting, and metal forming processes. Compensation is generally an iterative procedure where the mold shape is altered to account for specious mechanical behavior that prevents the final product from perfectly matching the shape of the mold. A similar methodology is applied to the MOF extrusion process. Compensation in the context of MOF preform manufacturing addresses hole distortion during the extrusion processes. By compensating for hole distortion and drift, the shape and structural integrity of the preform can be improved which ultimately enhances the MOF structure and performance [15, 16]. To reduce hole distortion and improve structural integrity, the shapes of the pins and their respective positions are altered to compensate for preform distortion. However, in this case, scaling these deformations to match the dimensions of the die is required. This scaling operation is explained in [12] and it is necessary to eliminate the contribution/bias of the die swell.

Structural distortions in optical fibers, particularly those in hollow-core and photonic bandgap fibers, have long been a subject of research due to their significant impact on performance. Studies by Li et al. [15] and Fokoua et al. [14] have highlighted how these distortions can severely affect fiber efficiency, leading to issues like higher attenuation and poorer signal transmission. By reducing hole distortions, it's possible to better confine light within the fiber core, which not only decreases attenuation but also improves overall signal quality. These improvements have clear, real-world implications. For instance, in chemical and biochemical sensing, MOFs have been used for detecting various compounds, as discussed by Calcerrada et al. [16] in Laser Photonics Reviews. Similarly, in telecommunications, for instance, more stable fiber structures can support longer transmission distances and faster data rates, which is crucial for meeting the growing demands of modern communication networks. In other fields like medical imaging and sensing, reducing distortions can enhance the precision and sensitivity of optical systems, improving the reliability of diagnostics and making environmental monitoring more accurate.

The present manuscript aims to highlight the improvements that can be achieved through compensation of the preform shape. Compensation methods play a crucial role in addressing challenges faced by the optical fiber industry, particularly in improving the consistency and performance of fiber production. Following the introduction, the process of die compensation is presented. Two compensation methods are introduced in this manuscript. The first method, called the direct method, is based on the classical interpretation of the compensation procedure while the second one focuses on the selective aspects process based on the normalized deformation variables [17]. Normalization, which was introduced in earlier studies of the MOF extrusion [17] helps isolate specific aspects of the preform deformations, distinguishing them from the general expansion of the preform. In the first part of this paper, the two compensation approaches are presented. In the following section, the improvement achieved by both methods is presented and the results are discussed. Finally, concluding remarks and future perspectives are presented.

2 Compensation Process of MOF Die

2.1 Direct Method

The compensation procedure follows the same steps of the lens molding studied by Ananthasayanam et al. [17]. The compensation direct approach initially explored in [12], is further expanded and detailed in this study. As shown in Fig. 1, the coordinates of each point of the scaled perform holes section are compared to their corresponding points in the pin. The pins are then altered accordingly. This iterative procedure is presented in detail in this manuscript. A step coefficient can be defined to scale the difference between the hole shapes and the pins, allowing precise adjustments. To help reduce the amplitude of the alteration and avoid excessive curvatures and concave forms, as glass flow may not tolerate such defects, the iteration is performed on a die with adjusted feed holes. This ensures that the changes are controlled, and the final geometry is improved for both performance and manufacturability.

The compensation process is based on three parametric curves. These curves are shown in Fig. 1:

1. $x_i(t)$ and $y_i(t)$ be two functions that describe the parametric curve of $Hole_i$. These functions are presented with a dashed line in Fig. 1 along with their corresponding parametric curve,
2. $x_{0,i}(t)$ and $y_{0,i}(t)$ be two functions that describe the parametric curve of Pin_i. A solid line is used for these functions in Fig. 1,
3. $x_{1,i}(t)$ and $y_{1,i}(t)$ be two functions that describe the parametric curve of *Altered* Pin_i shown with a dotted line in Fig. 1.

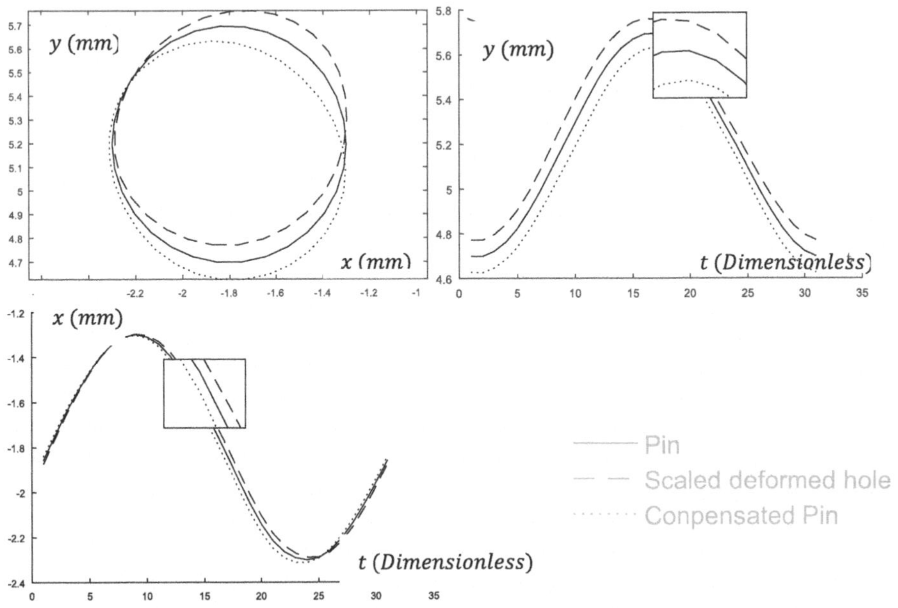

Fig. 1. First Compensation procedure performed on Hole 5: Calculated Pin shape.

The altered pin parametric functions $x_{1,i}(t)$ and $y_{1,i}(t)$ are obtained by comparing the pin with the scaled hole geometries. This process is performed based on the following equation.

$$x_{1,i}(t) = x_{0,i}(t) + \alpha\left(x_{0,i}(t) - \frac{x_i(t)}{\delta \mathrm{D}}\right)$$
$$y_{1,i}(t) = y_{0,i}(t) + \alpha\left(y_{0,i}(t) - \frac{y_i(t)}{\delta \mathrm{D}}\right) \quad (1)$$

where α is the step coefficient and δD is the expansion coefficient. Technically, α controls the intensity of the adjustment of the die geometry. The expansion coefficient δD is selected here as the ratio of the preform and die radii. The use of an expansion coefficient ensures that the compensation does not correct for the size change due to preform expansion, but only corrects the shape deformations. Therefore, the scaled perform has the same average radius as the die. Figure 1 shows the computed result for the optimized case when α is set to 1. Technically, the selection of α coefficient is based on several trials depending on the severity of the deformation. One can note that this method only allows the control of a single parameter, for instance, α.

3 Indirect Method: Adjustable Compensation Process of MOF Die

The direct method does not allow the compensation of each aspect of deformation separately. For instance, the direct procedure does not allow the holes' positions and their corresponding sizes to be compensated with different step coefficients. For this reason, a second approach is developed. It aims to separate the compensation of each aspect of the deformation allowing it to scale their corresponding alterations separately.

As in the previous approach, scaling the deformed shape to match the dimensions of the die is required. Therefore, all preform dimensions should be divided by the coefficient of expansion when used in computing the compensated dimensions. The variation of the radius hole which presents the hole shape is compensated as a first step. A reference pin is placed at the centroid of each hole to compensate for the radius as shown by the following equation.

$$r_{1,i}(\theta) = r_{0,i} + \alpha_r\left(r_{0,i} - \frac{r_i(\theta)}{\delta \mathrm{D}}\right) \tag{2}$$

where α_r is the step coefficient, δD is the expansion coefficient, $r_i(\theta)$ is radius $Hole_i$, $r_{0,\,i}$ is the radius of the reference Pin_i and $r_{1,\,i}(\theta)$ the radius of *Altered* Pin_i. Technically, α_r controls the radius of the holes without any effect on their positions. This operation is shown in Fig. 2.

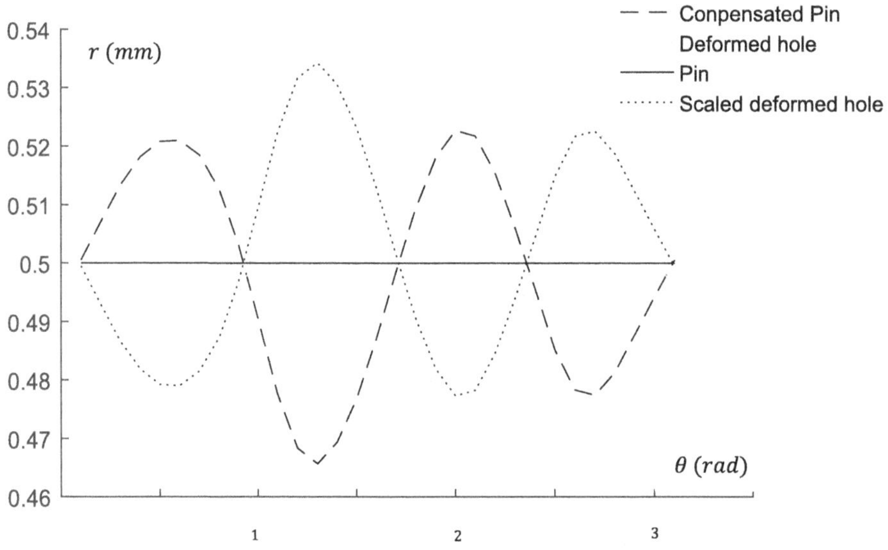

Fig. 2. Radius Compensation performed on Hole 5.

In Fig. 2 both scaled and unscaled radii of $Hole_i$, are presented to highlight the importance of the scaling operation. It can be observed that neglecting the expansion coefficient may result in an under-compensated pin. This is because the hole-to-preform radii ratio shrinks during the extrusion process.

The position of the centroids is compensated separately based on the position of the pins. Each operation can be adjusted using a separate step coefficient.

$$X_{1,i} = X_{0,i} + \alpha_O\left(X_{0,i} - \frac{X_i}{\delta D}\right)$$

$$Y_{1,i} = Y_{0,i} + \alpha_O\left(Y_{0,i} - \frac{Y_i}{\delta D}\right) \quad (3)$$

where α_O is the drift step coefficient, δD is the expansion coefficient,X_i and Y_i are the coordinates of the centroid of $Hole_i$, $X_{0,\,i}$ and $Y_{0,\,i}$ are the coordinates of the centroid of Pin_i and $X_{1,\,i}$ and $Y_{1,\,i}$ are the coordinates of the centroid of *Altered* Pin_i. Technically, α_O controls only the drift of the holes.

Figure 3 shows that if all step coefficients are set to 1, there is not a major difference between the presented compensation procedures. Unlike the previous method, the indirect method introduces two parameters (α_O and α_r) that could control each compensation process separately.

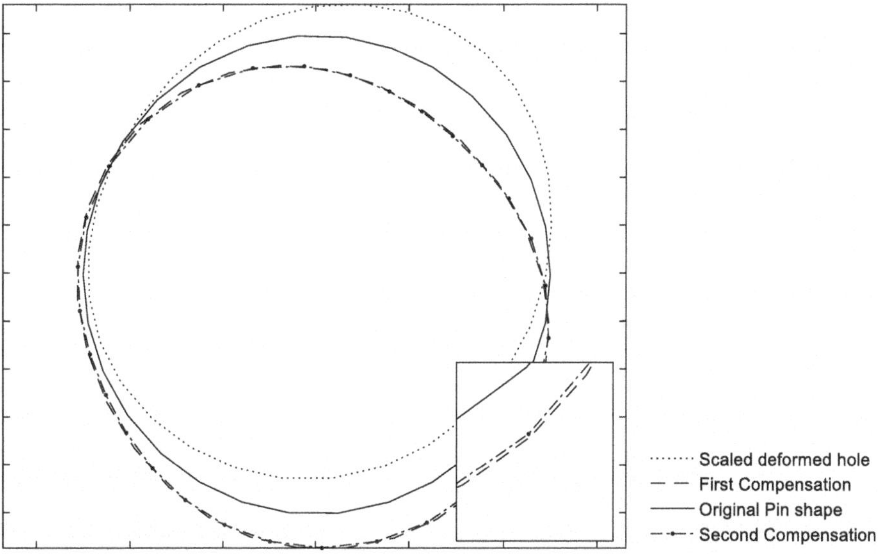

Fig. 3. Comparison between the composition procedures for hole 5: Calculated Pin shape.

In conclusion, the direct and indirect compensation methods are critical for optimizing the geometry of MOF dies. These processes ensure that defects, such as excessive curvatures or concave forms, are minimized, ultimately improving the quality and performance of the final MOF products. By enabling more precise control over die geometry, these methods not only enhance the processability of MOFs but also contribute to meeting optical performance requirements. The successful application of these compensation techniques holds significant potential for advancing MOF production, making it more efficient, reliable, and scalable for a wide range of optical applications.

4 Results and Discussion

In the present section, the results of both compensation methods are obtained based on a numerical simulation conducted on POLYFLOW. The numerical model was already validated in previous papers based on experimental data [4, 12, 16, 17].

4.1 Direct Method Results

To examine the impact of the compensation process, the results are first normalized by the expansion coefficient to suppress the impact of the die swell, i.e., they are scaled to the die dimensions. Such operation is carried out for Hole 5 in Fig. 4. Using this procedure, the hole size, position, and shape can be compared to the target geometry as well as to the initial case.

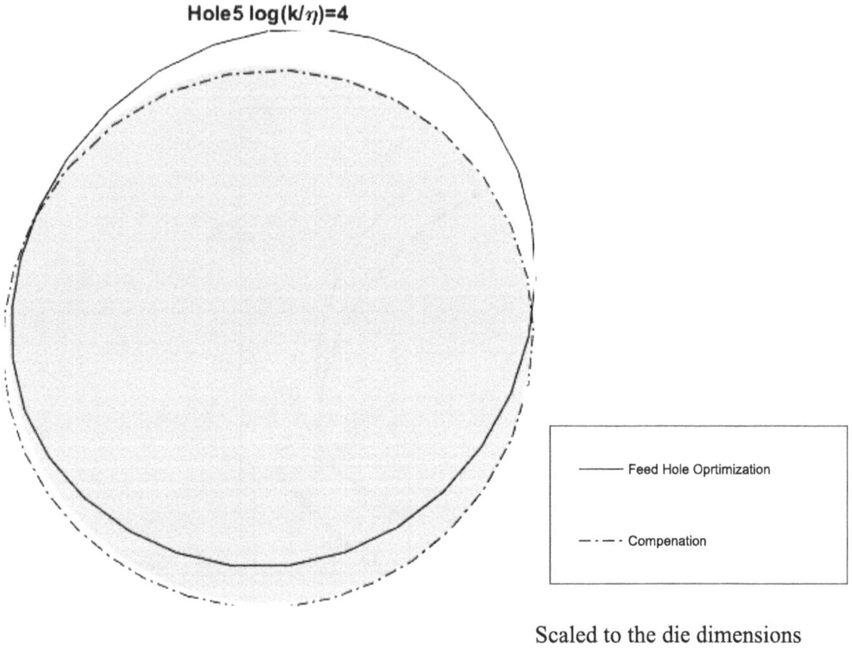

Fig. 4. Compensation procedure performed on Hole 5: Hole shape (the Target shape is presented by the gray circle)

As can be seen from Fig. 4, the compensation further improves the results and significantly reduces both the centroid drift and the distortion.

Though Fig. 4 offers visual feedback on the quality improvement of the preform shape, it does not provide an objective mathematical measure for such improvement. For this aim, quality measures need to be introduced. Among the improvement assessment measures, two of them are selected and shown in Fig. 5 [18]. The Disjoint Area Percentage (DAP) is defined as the total area outside the intersection of the two shapes referred to as the disjoint area. In this case, these shapes are the hole shape and a circle of the same area placed at its center. This area is then normalized by the hole area [18]. The DAP parameter is used to quantify the circularity of the holes. It is shown in Fig. 5 A and defined by:

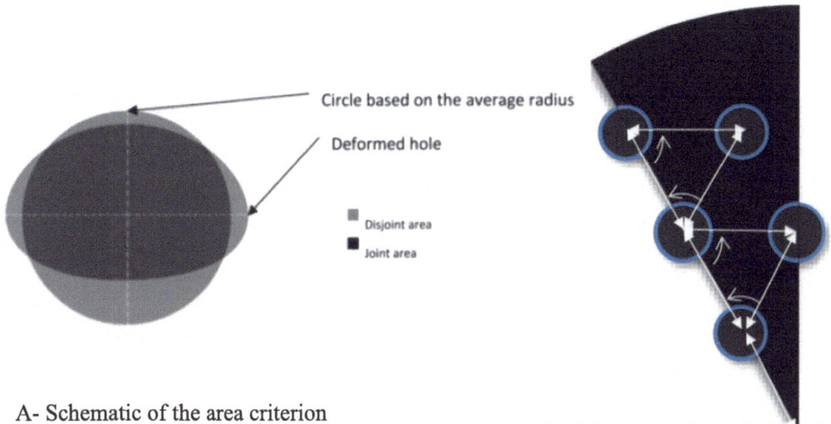

A- Schematic of the area criterion used to define the Disjoint Area Percentage

B- Schematic of standard deviation criterion

Fig. 5. Cross-section quality measures (a) Schematic of the area criterion used to define the Disjoint Area Percentage, (b) Schematic of standard deviation criterion.

$$DAP = \frac{Disjoint\ Area}{Area\ of\ circle} \times 100\% \tag{4}$$

The DAP measures for the initial base case (Feed hole optimized die) and the compensation results are presented in Table 1.

Table 1. Circularity results of the first compensation procedure

	Hole 1	Hole 2	Hole 3	Hole 4	Hole 5	Avg^{+}	S.Dev.*
Feed hole optimization	6.040%	5.137%	6.845%	6.317%	6.942%	6.256%	0.651%
Compensation results	0.508%	1.034%	1.185%	0.955%	1.799%	1.096%	0.418%

$^{+}$ DAP Average
* DAP Standard Deviation

Along with DAP, Distance Standard Deviation (DSD) is another interesting factor in assessing the cross-section quality. It is used to quantify the periodicity of holes in terms of distances. These distances are presented in Fig. 5b. To calculate the DSD, these distances are normalized by their average, and then their standard deviation is calculated [18]. Hence, the closer the DSD is to zero, the better the periodicity. To better assess the improvement brought by the compensation process, DSD percent improvement can be calculated. It measures the percentage difference between the base case and the altered die preforms. DSD criterion is defined by:

$$DSD = \frac{1}{\overline{d}}\sqrt{\frac{1}{N}\sum\nolimits_{i=1}^{N}\left(d_i - \overline{d}\right)^2} \quad (5)$$

Where: N is the total number of distances, d_i refers to each distance $i = 1..7$ and $\overline{d}$ is the distances average. The DSD percent improvements are presented in Table 2. Three DSD improvements are listed. First the improvement brought by feed hole optimization over the original case. Then the one brought by compensation from the feed hole optimization. Then Finally, the total improvement from the original case.

Table 2. Structure periodicity results of the first compensation procedure

	DSD/DSD Original case	Improvement to the Original case
Feed hole optimization	0.26	74%
Compensation	0.13	87%

Table 1 clearly shows a significant improvement with a Few DAP values dipping below 1%. Resulting in a considerable drop in the DAP average values. This improvement can be assessed by comparing the results to the feed hole optimization data. Knowing that the feed hole enhancement that constitutes the base case upon the composition procedure is applied, it is important to note that each hole responds differently to the compensation process which does not guarantee a DAP below 1%. This can be seen from values of DAP Standard Deviation which does not drop as much as the average. Hence an iterative process may be used to further improve the results. It is also crucial to note that the compensation process is limited to the number of modifications that can be applied to the pin shape. All these improvements can be visualized in Appendix A. Furthermore, in Appendix B, all hole shapes are normalized by their respective average radii. This representation hides all expansion effects and provides a better visual assessment of the improvement of the circularity of the hole.

In Table 2, the ratios of the measured DSD to the DSD of the original case as well as the improvement in the DSD results with respect to the original case are listed. Here it can be seen that the feed hole optimization yields an impressive 74% improvement in the preform periodicity. Thanks to Compensation this figure is raised to 87% compared to the original case. Hence, the compensation procedure yields better DSD over the feed hole optimization. Though this figure is far less than the improvement achieved by feed hole optimization, Appendix A shows that most holes are located very closely to their target positions.

4.2 Indirect Method Results

As stated earlier, the indirect method is a selective approach that compensates for each aspect of the deformation independently. In this pilot case, all compensation coefficients are selected to be equal to 1. This should facilitate the comparison between both methods. The presented compensation method enhanced the results by reducing the DAP similarly to the direct method. These improvements can be compared to the previous approach outcome as shown in Fig. 6 for Hole 5. Here, both methods show similar improvement for all shapes' sizes and locations of Hole 5. However, further inspection of the DAP data presented in Table 3 shows that these methods are not equivalent as the DAP for a few holes is quite different. Despite the improvement of recorded the DAP average compared to the base case, the DAP average for the indirect method is still a little higher than the direct method. Nevertheless, for a few holes, the indirect method perform better than the direct method. That said, even though the indirect method failed to bring a significant improvement over the direct method, it still brings better tools to adjust the die shape and takes advantage of 3D-printed dies. In fact, given the level of adjustability introduced by such an approach, each of the compensation coefficients can be selected based on an optimization process through a machine learning algorithm. This should improve the performance of such a method especially for dies with limited feed hole adjustability or for dies that yield large deformation.

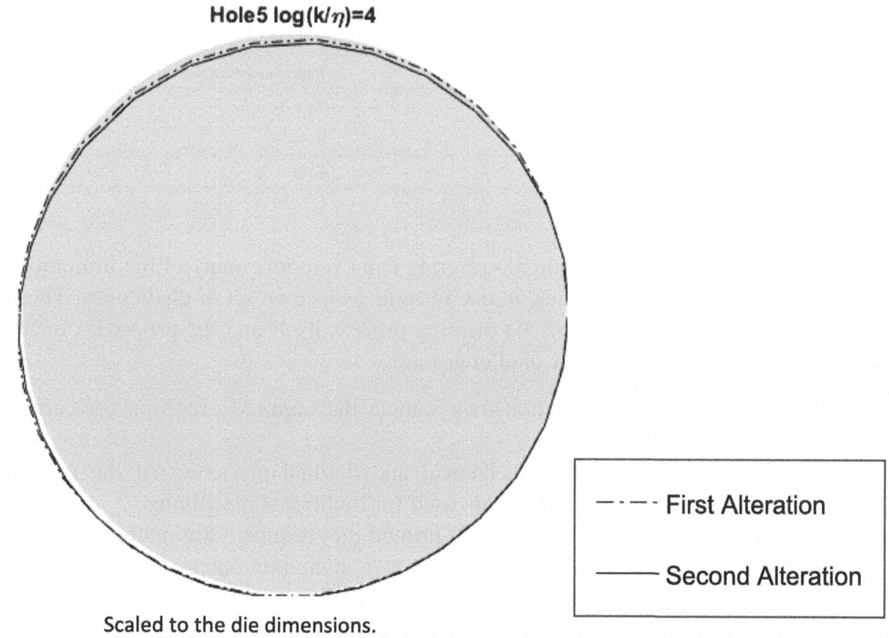

Fig. 6. Second compensation procedure performed on Hole 5: Hole shape (the Pin is presented by the gray circle).

Table 3. Circularity results of the second compensation procedure

	Hole 1	Hole 2	Hole 3	Hole 4	Hole 5	Avg+	S.Dev.*
Feed hole optimization	6.040%	5.137%	6.845%	6.317%	6.942%	6.256%	0.651%
First alteration	0.508%	1.034%	1.185%	0.955%	1.799%	1.096%	0.418%
Second alteration	2.266%	1.200%	2.690%	0.792%	1.758%	1.741%	0.688%

+ DAP Average
* DAP Standard Deviation

Inspecting Table 4 reveals a similar conclusion with respect to the DSD improvement. However, here it should be noted again that the feed hole optimization step is reprehensible for the majority of the recorded improvement. Thus, both Compensation methods seem to give similar results. Studies show that the preform's geometry strongly influences the optical fiber's ability to transmit light efficiently and with minimal loss [15, 16]. Numerical studies reveal that the tunneling grows significantly, while the bandwidth of the real fiber may reduce by a factor of two when compared to an ideal cross-section. Therefore, the resulting improvements to the preform geometry are of the most important significance. A comparison between direct and indirect compensation results is shown in Appendix C.

Table 4. Structure periodicity results of the first and second compensation procedure

	DSD/DSD Original case	Improvement to the Original case
Feed hole optimization	0.26	74%
Direct method compensation	0.13	87%
Indirect method compensation	0.19	81%

It's important to note that while 3D printing can overcome many of the limitations of traditional machining techniques, it also introduces its own set of challenges. These might include issues related to the 3D printing process itself and the properties of the printed materials. Some of the key challenges are:

- Printer resolution: which may not always meet the required precision for certain complex die geometries
- Material properties: when the mechanical and thermal properties of the printed material may not match those materials used for traditional machining.
- Post-processing requirements: where 3D printed dies require additional surface finishing steps to enhance surface quality. However, some internal areas of the die are particularly difficult or impossible to reach effectively.
- As demonstrated by Ebendorff-Heidepriem et al. [10], smoothing the internal upstream die surfaces can significantly reduce surface roughness. However, this may end up introducing more variables to the simulation process or requiring the use of softer glass.

Furthermore, 3D printed parts generally suffer from lower mechanical strength which needs to be accounted for in the die design.

While 3D printing presents opportunities for overcoming some of the limitations of traditional die manufacturing methods, it also requires careful consideration of process-related challenges such as resolution, material properties, and post-processing. Future research in die manufacturing using 3D printing should focus on addressing these challenges, particularly by developing materials with enhanced mechanical properties and refining post-processing techniques to improve the overall performance and applicability of 3D-printed dies in practical applications.

5 Conclusion

MOF performance is highly dependent on the quality of preforms. Improved die design and better control of the extrusion process produce preforms with minimum distortion. This results in fibers with reduced losses and improved optical characteristics. The study focuses on die compensation to enhance the performance of final cross-sectional geometry. This compensation is applied as a last optimization step after the adjustment of feed holes, which aims to better control the glass flux in the welding chamber. The tunings of the feedholes are meant to reduce the impact of the cross-sectional geometry distortions due to die swell which is deemed a major factor for deformations. However, this process is not able to fully negate the deformation of the holes. Compensation provides an additional procedure that allows further minor enhancements to the die shape improving the quality of the preform holes. This involves adjusting the pins' shapes to a certain achievable limit of non-circularity. Such a procedure is attainable through the use of 3D-printed dies.

This paper aims to apply compensation approaches to assess their capacity to further improve the shape of the preform. The study presents two compensation methods which both require normalized variables where the preform shape is scaled to the die shape before the compensation process begins. The first method is direct, involving a point-to-point comparison between the holes and pins' shapes. The second method is based on a series of normalized parameters that can be adjusted separately. Using either method, the circularity of the few holes is significantly improved surpassing 5% for a few cases. The resulting DAP values were mostly below 2.5% and the holes appeared mostly circular. The structure's periodicity improved by 13% and 7% for the first and second methods, respectively. Both considered procedures led to improvements in the final performance quality. While the direct method yielded slightly better results, the indirect method is more flexible and adaptable. Though the present results are promising, it should be noted that the process cannot be applied before feed hole optimization. Compensating holes with excessive deformation may lead to concave forms unless adjustments to the compensation method are made. Furthermore, the surface quality of 3D-printed dies is generally lower than machined dies which can potentially negate the benefits of the compensation process. Finally, further investigations are needed to test the robustness of the compensation process under varied conditions. To conclude, compensation provides a promising improvement to the MOF preform extrusion process. However, this technique still needs to be assessed across a broader application range to obtain more comprehensive experimental feedback and better understand the underlying challenges. Future research should focus on optimizing the compensation process for different types of deformations and exploring potential modifications to improve the die surface quality. Specific experimental setups could involve testing the compensation methods with varied extrusion parameters, glass types, and pin shapes. Addressing the limitations of 3D-printed die surfaces and evaluating the process in different extrusion setups, would provide valuable insights into how the compensation technique can be adapted for industrial-scale production.

Appendix A

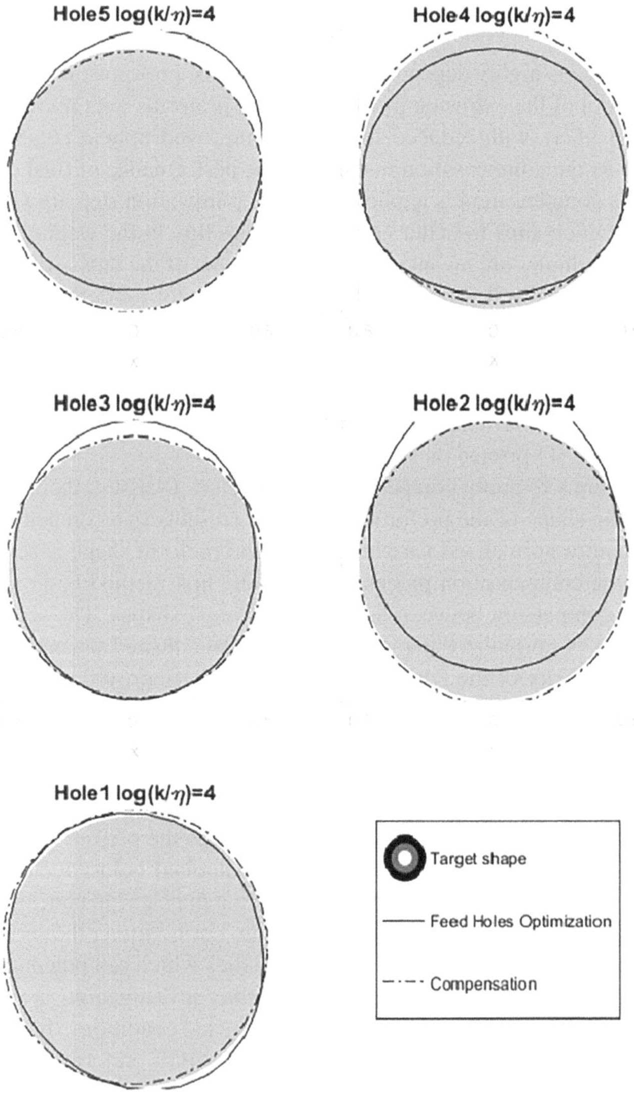

Compensation: hole shapes scaled to the die dimensions

Appendix B

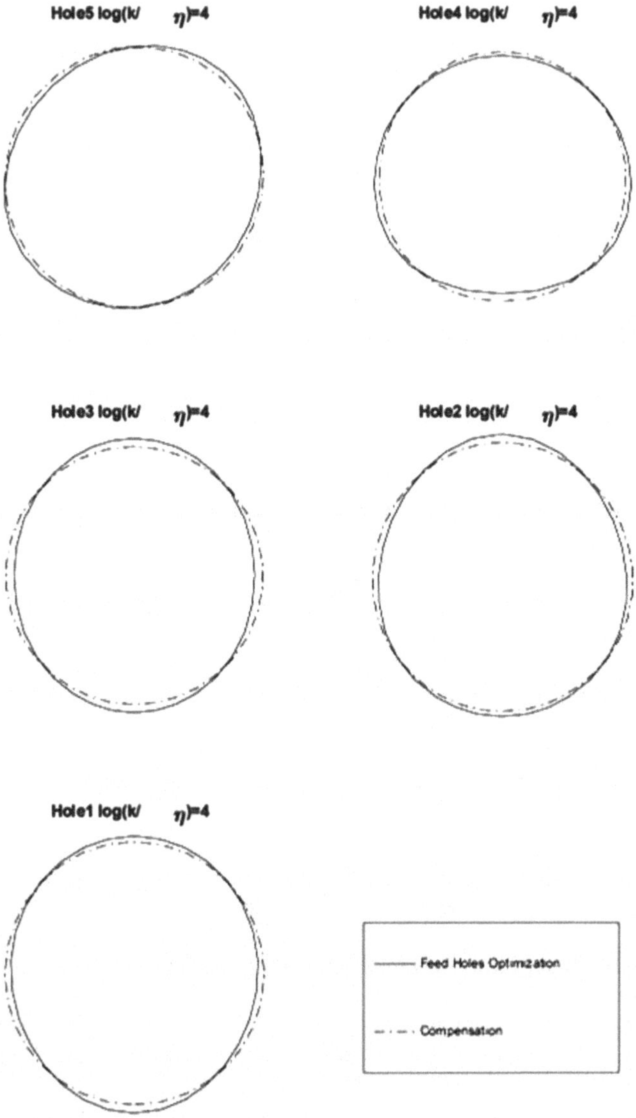

Compensation: hole shapes scaled to the pin dimensions

Appendix C

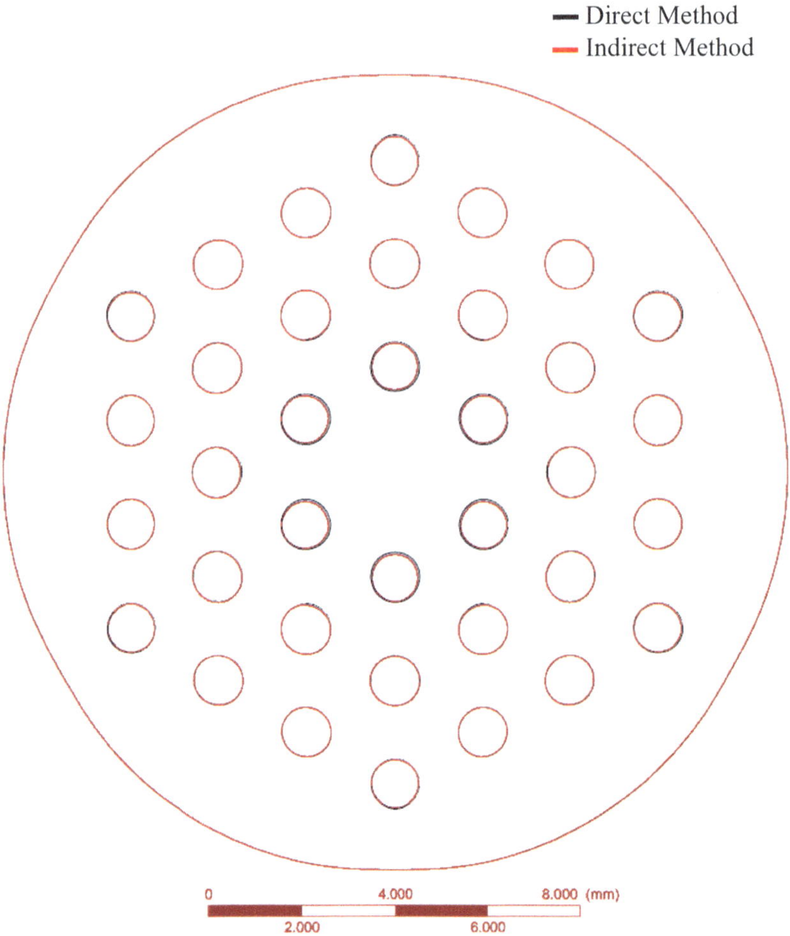

Superposition of Direct & Indirect methods resulting preform shapes

References

1. Richardson, D.J., et al.: Advances in Microstructured Fiber Technology. IEEEXplore (2005)
2. Kostecki, R., Ebendorff-Heidepriem, H., Warren-Smith, S.C., Monro, T.M.: Predicting the drawing conditions for microstructured optical fiber fabrication. *Opt. Mater. Express.* **4**(1), 29 (2014). https://doi.org/10.1364/ome.4.000029
3. Poli, F., Cucinotta, A., Selleri, S.: Basics of photonic crystal fibers. In: Photonic Crystal Fibers, pp. 7–52. Springer, Dordrecht. https://doi.org/10.1007/978-1-4020-6326-8_1
4. Trabelssi, M., Ebendorff-Heidepriem, H., Richardson, K.A., Monro, T.M., Joseph, P.E.: Computational modeling of hole distortion in extruded microstructured optical fiber glass preforms. J. Lightwave Technol. **33**(2), 424–431 (2015). https://doi.org/10.1109/JLT.2015.2388733

5. Russell, P.: Photonic crystal fibers. Science. **299**(5605), 358–362 (2003). https://doi.org/10.1126/science.1079280
6. Feng, X., Mairaj, A.K., Hewak, D.W., Monro, T.M.: Nonsilica glasses for holey fibers. J. Lightwave Technol. **23**(6), 2046–2054 (2005). https://doi.org/10.1109/JLT.2005.849945
7. Coulombier, Q., et al.: Casting method for producing low-loss chalcogenide microstructured optical fibers. Opt. Express. **18**(9), 9107–9112 (2010). https://doi.org/10.1364/oe.18.009107
8. Monro, T.M., Ebendorff-Heidepriem, H.: Progress in microstructured optical fibers. Annu. Rev. Mater. Res. **36**(1), 467–495 (2006). https://doi.org/10.1146/annurev.matsci.36.111904.135316
9. Ebendorff-Heidepriem, H., Monro, T.M.: Analysis of glass flow during extrusion of optical fiber preforms. Opt. Mater. Express. **2**(3), 304 (2012). https://doi.org/10.1364/ome.2.000304
10. Ebendorff-Heidepriem, H., Schuppich, J., Dowler, A., Lima-Marques, L., Monro, T.M.: 3D-printed extrusion dies: a versatile approach to optical material processing. Opt. Mater. Express. **4**(8), 1494 (2014). https://doi.org/10.1364/ome.4.001494
11. Ebendorff-Heidepriem, H., Monro, T.M.: Extrusion of complex preforms for microstructured optical fibers. Opt. Express. **15**(23), 15086 (2007). https://doi.org/10.1364/oe.15.015086
12. Trabelssi, M., Joseph, P.F.: Hole distortion and drift in extruded microstructured optical fiber glass preforms: part II-optimization. J. Non-Cryst. Solids. **481**, 212–223 (2018). https://doi.org/10.1016/j.jnoncrysol.2017.11.021
13. Li, M.J., West, J.A., Koch, K.W.: Modeling effects of structural distortions on air-core photonic bandgap fibers. J. Lightwave Technol. **25**(9), 2463–2468 (2007). https://doi.org/10.1109/JLT.2007.902744
14. Fokoua, E.N., Richardson, D.J., Poletti, F.: Impact of structural distortions on the performance of hollow-core photonic bandgap fibers. Opt. Express. **22**(3), 2735–2744 (2014). https://doi.org/10.1364/oe.22.002735
15. Li, M.J., Koch, K.W.: Modeling effects of structural distortions on air-core photonic bandgap fibers. *IEEE.* **25**(9), 2463–2468 (2007)
16. Calcerrada, M., García-Ruiz, C., González-Herráez, M.: Chemical and biochemical sensing applications of microstructured optical fiber-based systems. Laser Photon. Rev. **9**(6), 604–627 (2015). https://doi.org/10.1002/lpor.201500045
17. Ananthasayanam, B., et al.: Final shape of precision molded optics: part II—validation and sensitivity to material properties and process parameters. J. Therm. Stresses. **35**(7), 614–636 (2012). https://doi.org/10.1080/01495739.2012.674838
18. Trabelssi, M., Joseph, P.F.: Hole distortion and drift in extruded microstructured optical fiber glass preforms: Part I—Sensitivity analysis. J. Non-Cryst. Solids. **481**, 208–224 (2018). https://doi.org/10.1016/j.jnoncrysol.2017.10.045
19. Trabelssi, M., Ebendorff-Heidepriem, H., Richardson, K.C., Monro, T.M., Joseph, P.F.: Computational modeling of die swell of extruded glass preforms at high viscosity. J. Am. Ceram. Soc. **97**(5), 1572–1581 (2014). https://doi.org/10.1111/jace.12913

Addressing Geometrical and Dimensional Accuracy Challenges in 3D Concrete Printing

El Mesoudy Mouad[1(✉)], Foulki Rida[2], and Amegouz Driss[1]

[1] LTSI Laboratory, Sidi Mohamed Ben Abdellah University, Fes, Morocco
mouad.elmesoudy@usmba.ac.ma

[2] Department of Civil Engineering, ENSAM, Moulay Ismail University, Meknes, Morocco

Abstract. The rise of 3D concrete printing (3DCP) marks a significant shift in construction, redefining design flexibility and efficiency. By eliminating traditional formwork, this technology enables intricate, cost-effective architectural structures. However, achieving precise translation from virtual CAD models to physical structures remains a critical challenge due to the interplay of material properties, process parameters, and equipment limitations.

This study investigates the key factors influencing geometrical and dimensional precision in 3DCP, including layer deposition dynamics, concrete rheology, and machine configurations. The findings highlight gaps in understanding stratification effects, material shrinkage, and printing path accuracy. Addressing these challenges is crucial for advancing 3DCP's reliability and expanding its applications in sustainable and innovative construction. By offering insights into optimizing accuracy, this work serves as a foundation for improving 3DCP practices and achieving intricate, high-quality structures.

Keywords: Additive Manufacturing · 3D Concrete Printing · Architectural Structures · Printing Accuracy · Geometrical and Dimensional Precision · Process Optimization · Material Behavior

1 Introduction

In an era where sustainability is paramount, the construction industry, driven largely by concrete—the most widely used construction material [1, 2] faces critical environmental challenges. Buildings alone consume nearly 40% of global energy and contribute 36% of CO_2 emissions [3], the construction sector finds itself at a crossroads facing several challenges. One of these challenges is reducing the significant amount of wastage that can result from formworks used in the construction process [1] or just from the excess concrete left unused post-construction. Adopting 3D printing technology in concrete construction presents a potential solution to these issues [1, 2, 4].

3D concrete printing technology promises to revolutionize construction, challenging traditional formwork casting methods [1, 4]. 3D concrete printing integrates digital technology and insights from materials technology, facilitating freeform construction without the constraints of expensive formwork. This innovation enables unprecedented

M. Soula et al. (Eds.): AIAM 2024, LNME, pp. 260–270, 2025.
https://doi.org/10.1007/978-3-031-94889-3_21

architectural flexibility, allowing designs free from rectilinear constraints while reducing costs associated with formwork [1, 2, 4–6]. Despite its potential, 3D concrete printing faces significant technical challenges, even as the technology continues to evolve [2, 4]. The conversion of a virtual CAD model into a perfectly printed structure poses one of these obstacles, as geometrical accuracy and dimensional precision are crucial for achieving high-quality, aesthetically pleasing printed structures [7, 8]. These challenges emanate from the intricacies of the printing process [2, 9, 10], the behavior and properties of the materials used [11, 12], and the characteristics of the equipment employed in the process [4–6, 13].

This bibliographic research begins with an overview of 3D concrete printing technology, covering its history, layer-by-layer printing method, and current applications in construction. The second part of this work reviews key factors influencing the translation of a 3D CAD model into a printed concrete structure. This section examines factors related to the printing process, such as numerical parameters and material deposition; the properties and rheology of extruded concrete; and equipment-related factors.

2 3D Concrete Printing Technology

2.1 General Background

3D printing, or additive manufacturing (AM), encompasses techniques for creating 3D structures layer by layer from digital models, minimizing waste [14]. Additive manufacturing began in the 1980s and was quickly adopted in industries like aerospace, automotive, and biomedical sectors [1]. This success sparked interest in adapting AM to concrete in construction [5]. In the mid-1990s, 'Contour Crafting' emerged as the first 3D concrete printing technique, now one of the most widely used. Since then, a variety of strategies, robots, printer heads, and materials have been developed [2]. Interest in 3D concrete printing surged around 2012, accelerating developments and expanding research as its potential became more evident [2]. After 30 years of research, efforts continue to advance 3D Concrete Printing and transition it from laboratories to real construction sites [5].

2.2 3D Concrete Printing Process

Unlike traditional casting, 3D concrete printing uses digital and materials technology to enable freeform construction without costly formwork, expanding design possibilities [1, 2, 4–6]. Most 3D concrete printing systems include a conveying and extrusion system, a control unit, and a printer head [4]. Concrete is mixed with water and pumped through a hose to the printer head, which extrudes it at the specified speed, location, and angle (Fig. 1) [4–6].

The printer head includes a nozzle through which concrete is extruded, allowing precise placement and control over the material deposition [2]. Two main setups are used in 3D concrete printing: a gantry system with three axes of movement [1, 2, 4–6], and a six-axis industrial robotic arm for more dynamic control (Fig. 1) [4, 15].

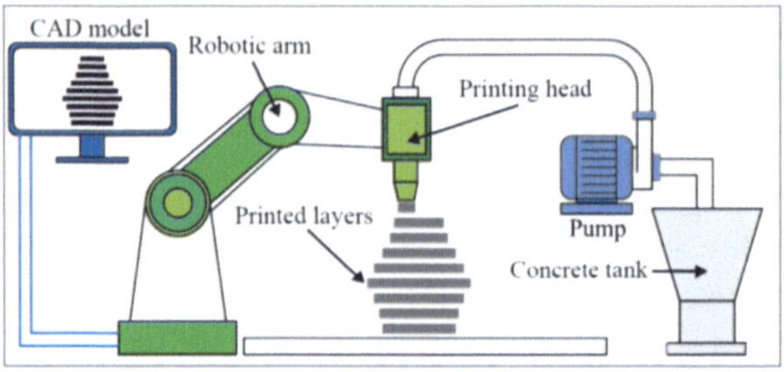

Fig. 1. 3D Concrete Printing process from a CAD model using a robotic industrial arm

2.3 Current Technologies and Advances

Recent years have seen various 3D concrete printing technologies emerge, primarily using extrusion-based and powder-based methods [1, 4–6]. We go into further detail about these techniques in the next lines. The first technique, Contour Crafting, extrudes two cementitious slurry layers to create formwork, with reinforcement connections manually added. Concrete is then poured in phases to control lateral pressure [1, 2, 5]. Concrete Printing, an extrusion-based method similar to Contour Crafting, offers 3D freedom with finer deposition resolution for detailed control over geometry. It uses high-performance fiber-reinforced concrete [1, 5].

D-shape: developed by Enrico Dini, uses a powder-based method to selectively harden sand with a binding agent, demonstrated by the rapid 3D printing of the Ferreri house [1, 2, 5].

Emerging Objects: employs a unique cement composite with selective solidification, as seen in the 2.74-meter Bloom structure made of 840 custom blocks [1]. Currently, these technologies are gaining attention in the building industry and are increasingly used to construct concrete structures (Fig. 2).

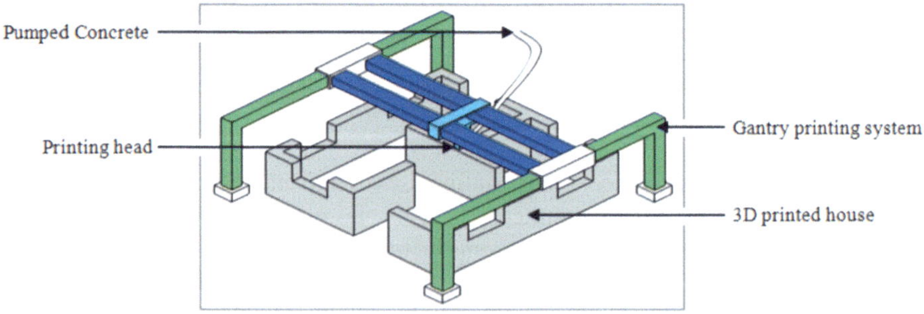

Fig. 2. 3D Concrete Printing process from a CAD model using a robotic industrial arm

Today, two leading methods of employing this emerging process in the construction field and the building industry exist. Firstly, CONPrint3D developed by TU Dresden, supports flexible, on-site printing, using standard machinery and minimizing labor requirements [1, 16]. In contrast, Large-Scale 3DCP adapts a concrete boom pump to autonomously deliver material, aiming to standardize large-scale printing for industry acceptance [1, 17].

3 Factors Influencing Geometrical and Dimensional Precision

3.1 Precision Challenges in 3D Concrete Printing

The ability to translate a virtual CAD model into a precisely printed structure is central to the success of 3D concrete printing (3DCP). However, achieving geometrical accuracy and dimensional precision is challenging due to various interdependent factors. These challenges arise from the printing process, material behavior, and equipment limitations (Fig. 3) [7, 8, 18]. A summary of the primary influencing factors is as follows:

Process-related factors: The layer-by-layer deposition method often introduces imperfections like stratification effects and imprecisions during CAD-to-print translation. Moreover, selecting appropriate printing parameters—such as nozzle speed and material flow rate—is critical to achieving optimal geometry and accuracy.
Material properties: The rheology and composition of concrete directly affect its ability to maintain form and resist deformation under load. Furthermore, shrinkage during curing can alter dimensions, necessitating advanced formulations or curing methods to mitigate this effect.
Equipment considerations: Printer precision —whether gantry systems or robotic arms—, nozzle design, and motion control systems contribute to variances in accuracy and consistency.

The upcoming sections will discuss some of the key factors that influence the final appearance of the printed object.

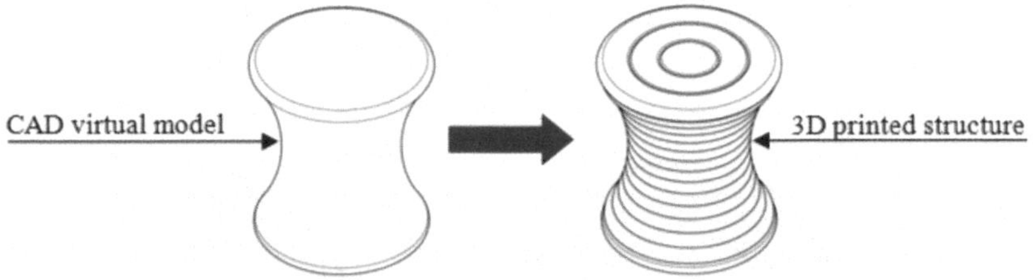

Fig. 3. Form changing after converting a 3D virtual model to the 3D printed real structure

3.2 3D Concrete Printing Process Factors

Deposition Method Layer-by-layer deposition is a hallmark of 3D printing, where the material is extruded and stacked in successive layers to form the desired structure. However, this approach can result in the *stratification effect*, also known as the "staircase effect," characterized by visible ridges on the surface of printed structures. These ridges arise due to the stepped nature of layer placement, compromising surface smoothness

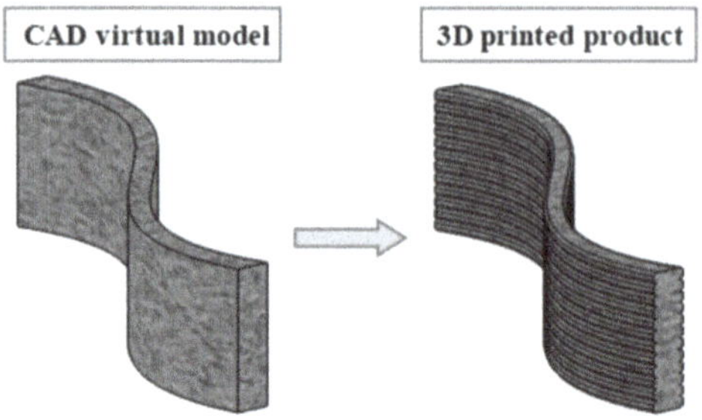

Fig. 4. Stratification effect due to the layer-by-layer deposition method

and accuracy (Fig. 4) [2, 4–6, 18]. This is distinct from 3D printing with smaller-scale filaments and poses difficulties in achieving seamless continuity in structures.

To reduce the stratification effect known also as the "staircase effect" resulting from the deposition method, many printer heads are equipped with towels or guides fixed on the extrusion nozzle to shape the material as it is deposited [5, 13]. Furthermore, inadequate bonding between layers can lead to delamination and compromise structural integrity [2, 19]. To enhance the bonding between successive layers a secondary bonding agent or an intermediate layer between layers can be utilized [19].

STL Conversion Another critical process factor that involves the numerical aspects of the technology that can be generally encountered for complex geometries [20] is the alteration of the CAD model into an STL file, which further generates a code describing the printer head's trajectory. This operation plays a crucial role in the printing process as it enables the control unit to convert the virtual model into a series of simplified geometric layers by using a slicing algorithm, which is then transformed into a printing trajectory code (Fig. 5) [9, 20]. Complex forms are often simplified during the slicing process required to generate the printing path code. This simplification can lead to imprecisions, causing them to lose their original shape after being printed [7, 18].

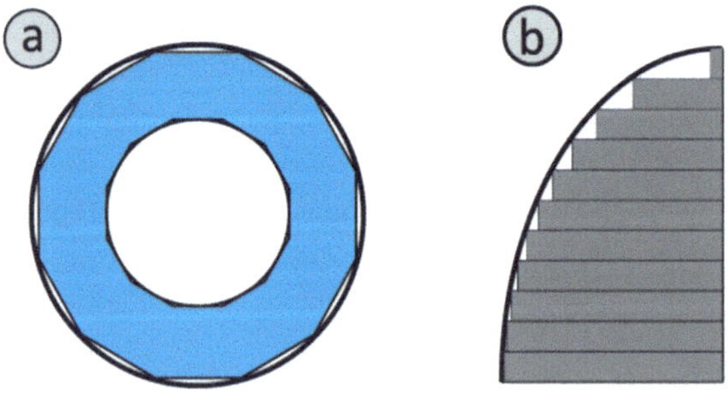

Fig. 5. (a): Conversion of a circular form to STL and (b): Conversion of a dome shape to stratified layers

To address this imprecision issue, a hybrid printing and milling solution has been proposed to address these imprecisions [7]. In this approach, the structure is initially printed and then transferred to a milling machine, which refines the shape and corrects inaccuracies. This method is particularly effective for intricate or high-tolerance structures, such as curved architectural elements or detailed decorative features. By combining additive and subtractive techniques, this workflow bridges the gap between design and execution, ensuring better fidelity to the original model.

Printing Parameters The choice of printing parameters, including the movement speed of the printer head [21, 22] and material flow defining the quantity of deposed concrete [21, 22], constitutes an additional influential factor. The velocity of the printing nozzle and the extrusion rate significantly influences the geometry of the final structure [10, 21, 22]. Balancing speed and flow rate is imperative to minimize discrepancies arising from rapid deposition, highlighting the importance of carefully calibrated printing parameters [10, 21, 22].

Balancing these parameters is essential to maintain consistent layer geometry and minimize discrepancies. For example, fine-tuning the extrusion rate ensures a uniform deposition, while adjusting nozzle velocity prevents deformation caused by rapid movements. Achieving this balance requires careful calibration and continuous monitoring during the printing process [23].

3.3 Printed Material Properties Factors

Concrete Formulation and Composition. The concrete formulation and the composition of various particles in the printed material are crucial determinants of the final form of the printed layers. Firstly, discrepancies in material composition, encompassing aggregate size, type, and proportions, affect the print's final geometry. This impact is particularly evident in the extrusion method and the choice of equipment [19, 24]. For example, the formulation of concrete can impact its setting time, which in turn affects the printing speed. It also impacts the cementitious material buildability forcing minimal layer thickness to prevent structural collapse [11, 19].

Rheological and Mechanical Properties. The rheological and mechanical properties, specifically the compression resistance of the concrete, are essential for its buildability describing the inferior printed layer's ability to resist the upper layer's weight in addition to their proper weight [11, 18, 19]. In 3D concrete printing, the compression applied to the lower layers during the printing process can influence their dimensional properties. In fact, during the printing process, the lower layers are subjected to compressive stresses applied by the upper layers [18, 19, 25]. These stresses then create deformations on the most stressed layers, increasing their thickness but decreasing their height (Fig. 6).

The rheological behavior of concrete—its flow and deformation under stress—determines its printability. Key rheological parameters include:

Yield Stress: Ensures the material maintains its shape once deposited.

Plastic Viscosity: Controls the ease of extrusion and flow during printing.

Thixotropy: Refers to the time-dependent recovery of viscosity, essential for interlayer bonding and stability [11, 18, 19].

To measure these properties, tools such as rotational rheometers or slump flow tests are used. Rotational rheometers provide precise data on yield stress and viscosity under

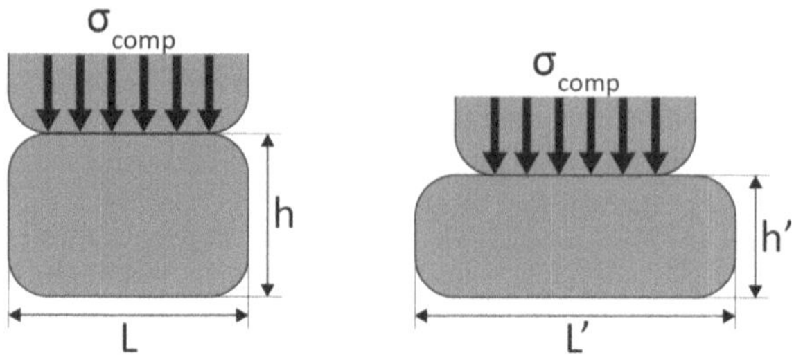

Fig. 6. Deformation of the printed concrete layer after compression

controlled shear rates, while slump tests offer a quick assessment of flowability. Standards like ASTM C143 (slump test) or ASTM C1749 (rheological measurements) are often followed in these evaluations. Proper control of rheological properties prevents issues such as layer collapse or insufficient bonding between layers [26, 27]. Mechanical performance, including compressive and tensile strengths, determines the structural integrity of the printed object. These properties are affected by the curing process, water-to-cement ratio, and the presence of additives like superplasticizers or fibers. Testing methods, such as compressive strength tests (ASTM C39) and flexural strength tests (ASTM C78), are used to evaluate these characteristics [27, 28].

Because early-age mechanical properties (such as green strength) affect the structure's capacity to sustain subsequent layers without deforming or collapsing, the timing of these tests is crucial in 3DCP. Polypropylene fibers and silica fume are examples of additives that can improve layer adhesion and green strength, which will improve precision [29].

Concrete Setting Operation. The post-printing setting operation of the concrete introduces the phenomenon of shrinkage, a natural contraction of the material during the curing process [7, 12, 19]. This phenomenon represents a significant challenge as the material shrinkage can affect both structural performance and dimensional aspects of 3D-printed concrete [19]. Recent research has focused on developing innovative approaches to address concrete shrinkage and expedite hydration and setting times. As a result, several effective curing methods have emerged, offering viable solutions that can be seamlessly integrated into a concrete 3D printer [19].

3.4 Printing Equipment Factors.

Printing Structure and Machine Precision. As mentioned earlier, there are two options for the printing structure: a gantry system [1] or robotic arms [15]. Each of these structures has its advantages and challenges, but both face issues related to machine precision. This precision can be influenced by the control system, the sensors in use, or the state of the transmission mechanical systems, such as pulley systems and ball crews. Additionally, there is a challenge related to the repeatability of the printer, signifying its ability to recreate the same form consistently from the initial layer to the subsequent upper layers [7, 18, 30].

Nozzle Form and Size (output). The second influencing factor is the nozzle's shape and size. These parameters directly impact other printing variables [7, 22]. Moreover, the design and functionality of the printing nozzle play a critical role in determining the print's accuracy [13]. Specifically, a smaller-sized nozzle can contribute to forming layers with less stratification effect, bringing the printed structure closer to the virtual model (Fig. 7).

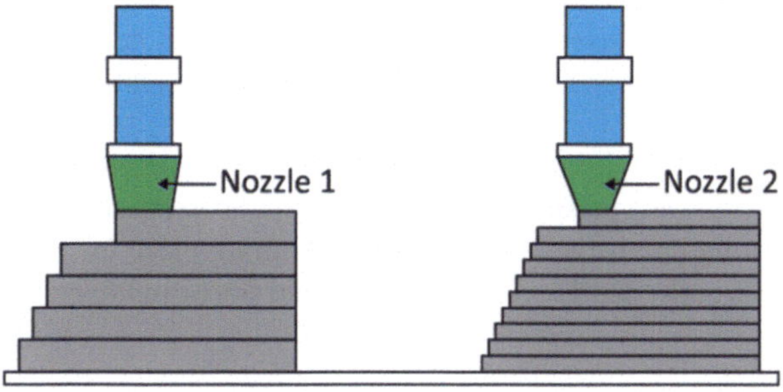

Fig. 7. Different layers forms obtained with different nozzle sizes

To prevent the imprecisions resulting from the nozzle shape and size, many printer heads are designed with a variable-size nozzle system allowing printing structures with different nozzle sizes depending on the complexity of the form [20, 22].

4 Discussion

The analysis conducted in this study highlights the multifaceted challenges that 3D concrete printing must overcome to achieve precision and reliability. Several interrelated factors significantly influence the dimensional and geometric accuracy of printed structures.

Influence of the Printing Process. The findings underscore the importance of the deposition method, trajectory code generation, and printing parameters in maintaining precision. In particular, inaccuracies during slicing and code generation can lead to discrepancies between the digital design and the physical output. Similarly, suboptimal deposition methods may result in uneven layer thicknesses, impacting both structural integrity and aesthetics. These issues are consistent with observations in the existing literature, emphasizing the critical role of technological refinement in addressing imprecision.

Material Properties and Precision. Material rheology and buildability emerged as central factors affecting dimensional accuracy. For instance, a balance between yield stress and plastic viscosity is necessary to ensure proper extrusion and layer adhesion, while controlling shrinkage minimizes post-printing deformations. The results align with previous studies on the mechanical and rheological properties of 3D-printed concrete, which emphasize the need for advanced material formulations to improve print quality.

Equipment and Structural Dynamics. The role of equipment, particularly nozzle design and printing structure, further highlights the complexity of achieving high precision. Nozzle dimensions, extrusion rates, and layer resolution must be carefully synchronized to prevent over-extrusion, under-extrusion, or misalignment. Additionally, the interplay between material and equipment dynamics demands an integrated approach to system optimization.

Hybrid techniques, such as combining printing with milling, show promise in mitigating errors associated with slicing and deposition. These approaches refine layer accuracy post-deposition, reducing surface roughness and dimensional deviations. However, further research is needed to evaluate their scalability and applicability to various structural designs.

This discussion bridges the gap between theoretical insights and practical challenges, offering a comprehensive perspective on the interconnected elements that influence 3D Concrete Printing precision. By critically examining these factors, this study contributes to a deeper understanding of the hurdles and opportunities in advancing 3DCP technologies.

5 Conclusion

This paper provides a comprehensive review of the key factors influencing precision in 3D concrete printing, shedding light on the challenges and opportunities for this transformative technology.

Key findings include:

- The precision of 3DCP is heavily dependent on the deposition method, trajectory code generation, and printing parameters, which must be meticulously controlled to ensure alignment with the original design.
- Material properties, particularly rheological and mechanical characteristics, significantly impact the dimensional and geometric accuracy of printed structures. Shrinkage control and buildability optimization are essential for achieving high-quality outcomes.
- Equipment considerations, such as nozzle design and structural dynamics, play a critical role in mitigating inaccuracies and improving layer consistency.

These findings underscore the need for an integrated approach that combines material science, equipment engineering, and advanced computational techniques to enhance the precision and reliability of 3D concrete printing.

The insights provided in this study serve as a valuable foundation for future research and practical applications, supporting the construction industry's efforts to leverage 3D Concrete Printing for precise and complex structures. While significant progress has been made, continued innovation and interdisciplinary collaboration are essential for addressing the remaining challenges and unlocking the full potential of 3D concrete printing.

Funding. I would like to thank the Research Institute for Solar Energy and New Energies (IRESEN) for supporting this research work and funding the participation in the 3rd Conference on Advances in Additive Manufacturing held in Hammamet, Tunisia.

References

1. Nematollahia, B., Xiab, M., Sanjayan, J.: Current progress of 3D concrete printing technologies. In: the 34th International Symposium on Automation and Robotics in Construction, ISARC (2017). https://doi.org/10.22260/ISARC2017/0035
2. Bos, F., Wolfs, R., Ahmed, Z., Salet, T., et al.: Additive manufacturing of concrete in construction: potentials and challenges of 3D concrete printing. Virtual Phys. Prototyp. **11**(3), 209–225 (2016). https://doi.org/10.1080/17452759.2016.1209867
3. Lucian, C.: Building materials in terms of energetic efficiency for sustainable constructions. "Ovidius" University Annals Constantza. Ser. Civ. Eng. **21**(21), 127–135 (2019). https://doi.org/10.2478/ouacsce-2019-0016
4. Chen, Y., He, S., Gan, Y., Çopuroğlu, O., Veer, F., Schlangen, E.: A review of printing strategies, sustainable cementitious materials, and characterization methods in the context of extrusion-based 3D concrete printing. J. Build. Eng. **45**, 103599 (2022). https://doi.org/10.1016/j.jobe.2021.103599
5. Khan, M.S., Sanchez, F., Zhou, H.: 3-D printing of concrete: beyond horizons. Cem. Concr. Res. **133**, 106070 (2022). https://doi.org/10.1016/j.cemconres.2020.106070
6. Lim, S., Buswell, R.A., Le, T.T., Austin, S.A., Gibb, A.G.F., Thorpe, T.: Developments in construction-scale additive manufacturing processes. Autom. Constr. **21**, 262–268 (2012). https://doi.org/10.1016/j.autcon.2011.06.010
7. Buswell, R., Jie, X., De Becker, D., Dobrzanski, J., Provis, J.: John Temitope Kolawole, Peter Kinnell: geometric quality assurance for 3D concrete printing and hybrid construction manufacturing using a standardized test part for benchmarking capability. Cem. Concr. Res. **156**, 106773 (2022). https://doi.org/10.1016/j.cemconres.2022.106773
8. Haidong, T., Wei, Z., Bahrami, A., Kahla, N.B., Ahmad, A., Özkılıç, Y.O.: Recent advancements and future trends in 3D concrete printing using waste materials. Dev. Built Environ. **16**, 100187 (2023). https://doi.org/10.1016/j.dibe.2023.100187
9. Ooms, T., Vantyghem, G., Van Coile, R., De Corte, W.: A parametric modeling strategy for the numerical simulation of 3D concrete printing with complex geometries. Addit. Manuf. **38**, 101743 (2021). https://doi.org/10.1016/j.addma.2020.101743
10. Van Der Putten, J., De Shutter, G., Van Tittelboom, K.: The effect of print parameters on the (micro) structure of 3D printed cementitious materials. In: Wangler, T., Flatt, R. (eds.) First RILEM International Conference on Concrete and Digital Fabrication—Digital Concrete 2018. DC 2018. RILEM Bookseries, 19. Springer, Cham (2019). https://doi.org/10.1007/978-3-319-99519-9_22
11. John, C., Lee, J., Bui, T.Q., Park, J., Yang, I.H.: Buildability and mechanical properties of 3D printed concrete. Materials. **13, 21**, 4919 (2020). https://doi.org/10.3390/ma13214919
12. Shahmirzadi, M.R., Gholampour, A., Kashani, A., Ngo, T.D.: Shrinkage behavior of cementitious 3D printing materials: effect of temperature and relative humidity. Cem. Concr. Compos. **124**, 104238 (2021). https://doi.org/10.1016/j.cemconcomp.2021.104238
13. Zhang, N., Sanjayan, J.: Extrusion nozzle design and print parameter selections for 3D concrete printing. Cem. Concr. Compos. **137**, 104939 (2023). https://doi.org/10.1016/j.cemconcomp.2023.104939
14. ASTM F42: Standard Terminology for Additive Manufacturing Technologies. ASTM International, West Conshohocken (2015)
15. Lublasser, E., Adams, T., Vollpracht, A., Brell-Cokcan, S.: Robotic application of foam concrete onto bare wall elements - analysis, concept and robotic experiments. Autom. Constr. **89**, 299–306 (2018). https://doi.org/10.1016/j.autcon.2018.02.005

16. Mechtcherine, V.: Venkatesh Naidu Nerella, frank will, Mathias Näther, Jens Otto, Martin Krause: large-scale digital concrete construction—CONPrint3D concept for on-site, monolithic 3D-printing. Autom. Constr. **107**, 102933 (2019). https://doi.org/10.1016/j.autcon.2019.102933
17. Gosselin, C., Duballet, R., Roux, P., Gaudillière, N., Dirrenberger, J., Morel, P.: Large-scale 3D printing of ultra-high performance concrete—a new processing route for architects and builders. Mater. Des. **100**, 102–109 (2016). https://doi.org/10.1016/j.matdes.2016.03.097
18. Jie, X., Buswell, R.A., Kinnell, P., Biro, I., Hodgson, J., Konstantinidis, N., Ding, L.: Inspecting manufacturing precision of 3D printed concrete parts based on geometric dimensioning and tolerancing. Autom. Constr. **117**, 103233 (2020). https://doi.org/10.1016/j.autcon.2020.103233
19. Baduge, S.K., Navaratnam, S., Abu-Zidan, Y., McCormack, T., Nguyen, K., Mendis, P., Zhang, G., Aye, L.: Improving performance of additive manufactured (3D printed) concrete: A review on material mix design, processing, interlayer bonding, and reinforcing methods. Structures. **29**, 1597–1609 (2021). https://doi.org/10.1016/j.istruc.2020.12.061
20. Lao, W., Li, M., Tjahjowidodo, T.: Variable-geometry nozzle for surface quality enhancement in 3D concrete printing. Addit. Manuf. **37**, 101638 (2012). https://doi.org/10.1016/j.addma.2020.101638
21. Communal, R., da Silva, W.R.L., Andersen, T.J., Stang, H., Spangenberg, J.: Influence of processing parameters on the layer geometry in 3D concrete printing: experiments and modelling. In: Bos, F., Lucas, S., Wolfs, R., Salet, T. (eds.) Second RILEM International Conference on Concrete and Digital Fabrication. DC 2020. RILEM Bookseries, 28. Springer, Cham (2019). https://doi.org/10.1007/978-3-030-49916-7_83
22. Jie, X., Ding, L., Cai, L., Zhang, L., Luo, H., Qin, W.: Volume-forming 3D concrete printing using a variable-size square nozzle. Autom. Constr. **104**, 95–106 (2019). https://doi.org/10.1016/j.autcon.2019.03.008
23. Wang, X., Jia, L., Jia, Z., Zhang, C., Chen, Y., Ma, L., Wang, Z., Deng, Z., Banthia, N., Zhang, Y.: Optimization of 3D printing concrete with coarse aggregate via proper mix design and printing process. J. Build. Eng. **56**, 104745 (2022). https://doi.org/10.1016/j.jobe.2022.104745
24. Shiwei, Y., Sanjayan, J., Hongjian, D.: Effects of cement mortar characteristics on aggregate-bed 3D concrete printing. Addit. Manuf. **58**, 103024 (2022). https://doi.org/10.1016/j.addma.2022.103024
25. Huang, X., Yang, W., Song, F., Zou, J.: Study on the mechanical properties of 3D printing concrete layers and the mechanism of influence of printing parameters. Constr. Build. Mater. **335**, 127496 (2022). https://doi.org/10.1016/j.conbuildmat.2022.127496
26. STRUCTURE Magazine: https://www.structuremag.org/article/rheological-properties-of-concrete/
27. ASTM INTERNATIONAL: Cement Standards and Concrete Standards. https://www.astm.org/products-services/standards-and-publications/standards/cement-standards-and-concrete-standards.html
28. Al Khawaja, H., et al.: Investigating the mechanical properties of 3D printed components. In: In the 2020 Advances in Science and Engineering Technology International Conferences (ASET), pp. 1–7 (2020). https://doi.org/10.1109/ASET48392.2020.9118307
29. Toutanji, H.A.: Properties of polypropylene fiber reinforced silica fume expansive-cement concrete. Constr. Build. Mater. **13**(4), 171–177 (1999). https://doi.org/10.1016/S0950-0618(99)00027-6
30. Malaeb, Z., AlSakka, F., Hamzeh, F.: Chapter 6 - 3D concrete printing: machine design, mix proportioning, and mix comparison between different machine setups. In: Sanjayan, J.G., Nazari, A., Nematollahi, B. (eds.) 3D Concrete Printing Technology, pp. 115–136. Butterworth-Heinemann (2019). https://doi.org/10.1016/B978-0-12-815481-6.00006-3

Author Index

A
Abda, Manel 244
Allegue, Lamis 92
Amdouni, Marwen 105
Aouadi, Badreddine 115
Ayadi, Mahfoudh 23, 69

B
Baili, Maher 219
Bejaoui, Ali 3
Belhadjsalah, Hédi 165
Ben Amar, Mounir 186
Ben Fathallah, Brahim 115
Ben Fraj, Boutheina 186
Ben Hariz, Nada 69
Ben Khlifa, Ated 165
Ben Rhouma, Amir 57
Ben Yahia, Noureddine 181, 231, 244
Bouaziz, Slim 203
Bouaziz, Zoubeir 219
Bougadouha, Oumayma 195
Boughanmi, Oumaima 92
Boulahem, Khaled 23
Boulila, Atef 69
Bousnina, Kamel 181, 231

C
Chatti, Sami 83
Chihi, Manel 83

D
Dakhli, Chams Eddine 115
Dalverny, Olivier 57
Dhouib, Karem 149
Dhouioui, Manel 186
Dridi, Issam 231
Driss, Amegouz 260

F
Fathallah, Raouf 195
Frija, Mounir 195

G
Gharbi, Ali 149
Guizani, Amir 203

H
Haddar, Mohamed 186, 203
Hajji, Amel 149
Hammadi, Moncef 203
Hamza, Anis 181, 231
Hamzi, Salah 57
Hentati, Hamdi 186
Hichem, Habouba Mohamed 44
Hmila, Zeinab 219

I
Imen, Souli 133

J
Jaziri, Marwa 23
Jmii, Hamza 3

K
Khaled, Boulahem 44, 133
Khaterchi, Houda 57, 165
Khodja, Malika 195

L
Larbi, Baderddine 165
Larbi, Badreddine 23, 57, 133

M
Mahfoudh, Ayadi 133
Marouani, Haykel 92
Mouad, El Mesoudy 260
Mouhmed Toumi, Nasri 165
Msaddek, El Bechir 219

M. Soula et al. (Eds.): AIAM 2024, LNME, pp. 271–272, 2025.
https://doi.org/10.1007/978-3-031-94889-3

N
Nasraoui, Helmi 105
Nasri, Mohamed Toumi 23, 57, 133
Nasri, Mondher 57, 165

Q
Qureshi, Yumna 83

R
Rezgui, Mohamed Ali 3, 105
Rida, Foulki 260

S
Seddik, Shiri 44
Soula, Mohamed 115

T
Tarfaoui, Mostapha 83
Trabelsi, Ali 3, 105
Trabelsi, Olfa 23
Trabelssi, Mohamed 244

W
Wali, Mondher 10

Y
Yousfi, Larbi 203

Z
Zghal, Ali 57
Zghal, Jihed 3
Zghal, Wissem 10
Zidi, Mondher 165

The manufacturer's authorised representative in the EU is Springer Nature Customer Service Centre GmbH, Europaplatz 3, 69115 Heidelberg, Germany. If you have any concerns regarding our products, please contact ProductSafety@springernature.com

Printed and bound by CPI Group (UK) Ltd, Croydon, CR0 4YY
01/10/2026
02225911-0006